THE STRUCTURES AND REACTIONS
OF THE
AROMATIC COMPOUNDS

THE
STRUCTURES & REACTIONS
OF THE
AROMATIC COMPOUNDS

BY

G. M. BADGER

D.Sc., Ph.D., F.R.I.C.

*Reader in Chemistry in the
University of Adelaide*

CAMBRIDGE
AT THE UNIVERSITY PRESS
1957

PUBLISHED BY
THE SYNDICS OF THE CAMBRIDGE UNIVERSITY PRESS

Bentley House, 200 Euston Road, London, N.W. 1
American Branch: 32 East 57th Street, New York 22, N.Y.

First printed 1954
Reprinted 1957

First printed in Great Britain at the University Press, Cambridge
Reprinted by offset-litho by Jarrold and Sons, Ltd, Norwich

To
MY WIFE

CONTENTS

PREFACE   page xi

CHAPTER 1. THE BENZENE PROBLEM

Aromatic character	1
The structural formula of benzene	3
The nature of unsaturation in aromatic compounds	7
The application of physical methods to the benzene problem	11
(i) X-ray and electron diffraction experiments	11
(ii) Long-wave spectroscopy	13

CHAPTER 2. THE THEORETICAL SOLUTION OF THE BENZENE PROBLEM, AND A DEFINITION OF THE TERM 'AROMATIC'

The nature of carbon-carbon bonds	17
The application of wave mechanics to the benzene problem	31
Delocalization (or resonance) energy	37

CHAPTER 3. SOME PROPERTIES OF AROMATIC COMPOUNDS

Benzenoid aromatic hydrocarbons	44
Non-benzenoid aromatic hydrocarbons	53
Tropolone	58
Heterocyclic aromatic compounds containing six-membered rings	59
Heterocyclic aromatic compounds containing five-membered rings	67
The strength of aromatic heterocyclic bases	76
Molecular complexes with aromatic compounds	79
Oxidation-reduction potentials of quinones	87

Chapter 4. Addition Reactions of Aromatic Compounds

Benzene and its heterocyclic analogues	page 97
1:2 Addition reactions	102
1:4 Addition reactions	106
Elimination reactions in dihydroaromatic systems	112

Chapter 5. The Aromatic 'Double' Bond

Bond fixation in aromatic compounds	116
Cyclization experiments	117
The diazo coupling and halogenation of phenols	123
The Claisen rearrangement	126
Chelation and bond fixation	130
Electron displacements and bond fixation	131
The Mills-Nixon effect	133
Reactions with diazoacetic ester	138
The addition of ozone	147
The addition of osmium tetroxide	157
Bond-length variations in aromatic systems	161
Double-bond character and related concepts	165

Chapter 6. The Effects of Substituents

The inductive effect	184
The mesomeric effect	194
Hammett's σ constants	204
The inductomeric and electromeric effects	208
The effects of halogen substituents	210
The effects of alkyl substituents	212
The 'ortho' effect	214
The effects of substituents in polycyclic compounds	223
The effects of hetero atoms	229

CONTENTS

Chapter 7. Aromatic Substitution Reactions

Electrophilic substitution	*page* 233
Nucleophilic substitution	243
Radical substitution	249
Nitration	252
Halogenation	257
The Friedel-Crafts and related reactions	264
Alkylations	268
Acylations	278
Intramolecular acylation: cyclization	289
Diazo coupling	294
Sulphonation	301
Hydroxylation	307
Amination: the Tschitschibabin reaction	309
Introduction of the cyano group	311
The Gomberg reaction	313
The Pschorr synthesis	316
The Elbs reaction	323
The *ortho*:*para* ratio	330

Chapter 8. The Diels-Alder Reaction

Survey of the reaction with aromatic dienes	334
Quinones as dienophiles; triptycene	354
Some stereochemical considerations	358
The mechanism of the Diels-Alder reaction	360

Chapter 9. Photo-Oxidation and Photo-Dimerization

The structure of the photo-oxides	364
Photo-oxidation of polyacenes	367
Photo-oxidation of simple heterocyclic compounds	367

CONTENTS

The influence of substituents on the ease of formation of
   photo-oxides   *page* 369
The influence of substituents on the thermal stability of
   photo-oxides   372
Reactions of photo-oxides   375
Photo-dimerization   377
The mechanism of photo-oxidation   379

CHAPTER 10. ABSORPTION AND FLUORESCENCE
SPECTRA OF AROMATIC COMPOUNDS

Benzene and substituted benzenes   383
Diphenyl and the polyphenyls   388
Spectral relationships in aromatic hydrocarbons   392
Substituted polycyclic aromatic hydrocarbons   396
Spectra of non-benzenoid aromatic hydrocarbons   403
Absorption spectra of heterocyclic aromatic systems   405
Fluorescence spectra of aromatic compounds   408

CHAPTER 11. OPTICAL ACTIVITY IN AROMATIC
COMPOUNDS

Asymmetry   411
Stereochemistry of the diphenyls   412
The influence of the size and nature of substituents   424
Other examples of restricted rotation   428
Optical activity of the 4:5-phenanthrene type   437

INDEX OF SUBJECTS   441
INDEX OF AUTHORS   448

PREFACE

For more than a hundred years chemists have been intrigued by the problems of the structures, reactions and properties of aromatic compounds. About a third of all the known organic compounds are aromatic, and it cannot be said that this fascinating branch of organic chemistry has ever been seriously neglected. At first, although very many useful synthetic methods were discovered and many reactions investigated, only slow progress was made in the solution of the fundamental problems. In the last twenty or thirty years, however, real advances have been made and our understanding of aromatic character has been vastly increased. To some extent this has been due to the increased specialization of the past few decades. New experimental techniques have been widely used, and quantum mechanics has also been applied with outstanding success.

Specialization has many dangers. It is a truism that, as the field of interest becomes increasingly narrow, it becomes more and more difficult to see the wood for the trees. For myself, the only solution seemed to be a survey of the whole field of aromatic chemistry, and that is what I have aimed at in this book.

In attempting to obtain a picture of the subject as a whole, it is inevitable that some aspects have received only scant attention. Some selection of material has had to be made. For example, some of the less important substitution reactions have been neglected entirely; and synthetic methods have not been discussed at all unless they happen also to be substitution or addition reactions directly involving an aromatic ring system.

The literature available to the middle of 1951 has been consulted, but it has been possible in a few cases to include references to papers published towards the end of that year. With such a big field many important papers must have been overlooked, and I would be grateful if readers would call my attention to any serious omissions so that corrections and additions can be made in any future edition.

PREFACE

Considerable thought has been given to the best method of arranging the material. I have endeavoured to avoid the pitfall of first stating the accepted theory and then citing the supporting facts. On the contrary, in the case of benzene, I have tried to state the problem, to give the experimental facts, and only then to give the theoretical solution to the problem. In the case of the substituted compounds, I have given the experimental facts proving the existence of inductive and mesomeric effects, and only then have discussed the explanation of these effects and the way in which they modify substitution reactions.

The major portion of the book is descriptive; but some sections are, of necessity, more exhaustive and specialized than others. It is difficult to discuss subjects such as intramolecular cyclization in a useful way without giving many detailed examples. It is hoped, therefore, that the book will prove of value not only to honours students, but also to graduate students and to research workers in organic chemistry.

It is a pleasure to express my thanks to several friends for their interest in this work. I am greatly indebted to Dr J. C. D. Brand, to Dr C. Buchanan, to Mr A. N. Hambly, to Mr A. R. Palmer, to Mr R. S. Pearce, to Dr R. I. Reed, to Dr E. F. M. Stephenson and to Dr G. Thomson for reading and criticizing one or more of the chapters in the draft manuscript. I am also grateful to Mr M. Thompson for help with some of the diagrams. Finally, I would like to express my indebtedness to Miss Shea Smith for her skilled secretarial assistance during the preparation of this book.

G. M. B.

ADELAIDE, 1952

ACKNOWLEDGEMENTS

The author would like to express his thanks to the following for permission to reproduce figures: Professor C. K. Ingold and the Royal Society for Figs. 1·2 and 1·3; Professor J. Monteath Robertson, Dr H. B. Watson, Dr E. J. Bowen, Dr W. H. Mills, and the Chemical Society for Figs. 2·16, 5·1, 5·4, 5·5, 6·2, 10·9 and 11·3; Professor A. L. Sklar and the American Institute of Physics (publishers of the *Journal of Chemical Physics*) for Fig. 2·19; Professor Louis P. Hammett and the McGraw-Hill Book Company Inc. for Fig. 6·3; Dr H. M. Powell and Messrs Macmillan and Co. Ltd. (publishers of *Nature*) for Fig. 3·2; and Dr W. H. Mills and the Society of Chemical Industry for Figs. 11·1 and 11·2.

CHAPTER 1

THE BENZENE PROBLEM

AROMATIC CHARACTER

The term 'aromatic' was originally applied to certain substances of characteristic odour isolated from various natural products. Benzoic acid was obtained from gum benzoin as early as the sixteenth century, and benzaldehyde (oil of bitter almonds), cymene (oil of carraway), toluene (balsam of Tolu) and several other aromatic compounds have also been known for a long time. Benzene was discovered in 1825 by Faraday, who obtained it from a compressed illuminating gas; and it was prepared from benzoic acid by Mitscherlich in 1834. A. W. Hofmann and C. Mansfield discovered benzene in coal tar in 1845. Coal tar, which is produced as a by-product in the manufacture of coal gas by the dry distillation of coal, soon became the most important source of aromatic compounds. Toluene, xylene and other benzene homologues have been isolated from coal tar, as well as phenol, cresol, naphthalene, anthracene, diphenyl, phenanthrene, fluorene, acenaphthene, chrysene, retene, indane, thiophen, pyridine, quinoline, *iso*quinoline, pyrrole, indole, carbazole and acridine. Well over a hundred aromatic compounds have been obtained from this source.

It was soon observed that the aromatic compounds contain a higher proportion of carbon than the fatty compounds. Benzene has the formula C_6H_6, but the saturated aliphatic compound with the same number of carbon atoms has the formula C_6H_{14}. On the other hand, most of the aromatic compounds are not highly unsaturated reactive substances, but are characterized by their remarkable stability. Benzene itself can be prepared by high-temperature decarboxylation of benzoic acid, by distilling phenol with zinc dust, and even by passing acetylene through a red-hot tube.

In 1865, in his classic papers on the constitution of aromatic compounds, August Kekulé[*] proposed the well-known cyclic

---
[*] *Bull. Soc. chim. Paris*, 1865, **3**, 98; *Liebigs Ann.* 1866, **137**, 129; see also Walker, *Ann. Sci.* 1939, **4**, 34.

structure for benzene, and he suggested that the peculiar properties of the aromatic compounds are dependent on the properties of this ring system. It is now customary to include the heterocyclic compounds of analogous structure under the term 'aromatic', for all these substances exhibit 'aromatic character' in greater or lesser degree. Aromatic character cannot be rigidly defined, but the properties of the aromatic compounds may be summarized as follows:

(1) The most characteristic property of the aromatic compounds is their stability and ease of formation by pyrolytic methods. On the other hand, a few polycyclic aromatic compounds are known which disproportionate on heating to give the corresponding dihydro derivatives.

(2) Although saturated hydrocarbons are not usually attacked in the liquid phase by reagents such as nitric acid, sulphuric acid and bromine, and although olefins usually react by addition, the aromatic compounds tend to react by substitution with greater or lesser facility:

$$C_6H_6 + HO.SO_3H = C_6H_5.SO_3H + H_2O,$$
$$C_6H_6 + HO.NO_2 = C_6H_5.NO_2 + H_2O,$$
$$C_6H_6 + Br.Br = C_6H_5.Br + HBr.$$

Substitution is not, however, an essential characteristic of aromatic compounds, for many polycyclic aromatic compounds, and some simple heterocyclic aromatic compounds, undergo *addition* reactions with very great facility.

(3) (a) The aromatic amines are less basic than the aliphatic amines, and react with nitrous acid to give diazo compounds;

(b) the aromatic hydroxy compounds (i.e. phenols) are more strongly acidic than the aliphatic hydroxy compounds (i.e. alcohols);

(c) the aromatic acids are somewhat stronger acids than the corresponding aliphatic acids;

(d) the aromatic halogen compounds are much less reactive than the aliphatic halogen compounds (provided the former are not activated by suitable groups).

(4) Parent aromatic substances are, in general, remarkably stable to oxidizing agents.

THE STRUCTURAL FORMULA OF BENZENE

The problem of providing an adequate structural formula for benzene and other aromatic compounds engaged the minds of chemists for many years. It was pointed out that if six carbon atoms are linked by alternate double and single bonds in an open chain there would be eight free combining units,

$$-\underset{|}{C}=\underset{|}{C}-\underset{|}{C}=\underset{|}{C}-\underset{|}{C}=\underset{|}{C}-$$

but that if the two end carbon atoms are linked together, a closed chain with only six free combining units results:

$$\overset{\longleftarrow}{\underset{|}{C}=\underset{|}{C}-\underset{|}{C}=\underset{|}{C}-\underset{|}{C}=\underset{|}{C}}$$

Kekulé therefore proposed that benzene has a regular hexagonal structure, and that the CH groups are linked by alternate double and single bonds.* In this way he was able to show that all disubstitution products must exist in three isomeric modifications, that there are only three isomeric modifications of the general formula $C_6H_3X_3$, and that there are six of the formula $C_6H_3X_2Y$. Kekulé was also able to explain the nature of the homologues of benzene and the essential difference between substitution in the nucleus and substitution in a side chain. As Japp wrote in 1897:[†] 'Kekulé's memoir on the benzene theory is the most brilliant piece of scientific prediction to be found in the whole range of organic chemistry.'

Although very successful in explaining the number and nature of the various substitution derivatives, Kekulé's *cyclo*hexatriene formula (I) has always been regarded as only partly satisfactory. It is very difficult to explain the stability of benzene if it is assumed to have three ethylenic double bonds.

From this point of view the structure (II) proposed by Dewar[‡] is even less satisfactory, for it requires a long *para* bond. This

---

* *Ber. dtsch. chem. Ges.* 1869, **2**, 362.
† Kekulé Memorial Lecture, *Chemical Society Memorial Lectures*, 1893–1900, p. 135.
‡ *Proc. Roy. Soc. Edinb.* 1867, **6**, 82.

formula received very little attention until it was revived by Ingold,* who pointed out that such a bond would account for the *para* transmission of the directive influence of substituents.

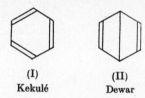

(I) Kekulé   (II) Dewar

One of the most serious objections to the Kekulé formula was pointed out by Ladenburg.† If the structure of benzene really involves alternate double and single bonds, then *ortho*-disubstituted derivatives should exist in two isomeric modifications (III) and (IV). Moreover, when the two groups are different, there should also be two isomeric *meta* derivatives (V) and (VI).

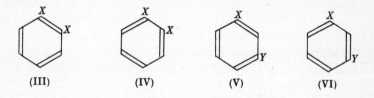

(III)   (IV)   (V)   (VI)

As such isomers were, and are, unknown, some modification of the Kekulé structure appeared to be necessary. Victor Meyer‡ suggested that the differences between such isomers might be so slight as to escape detection, but Kekulé§ further proposed that benzene has a *dynamic* structure (VII–VIII), and that the double bonds undergo a very rapid oscillation in such a way that their exact position cannot be defined.

(VII)   (VIII)

* *J. Chem. Soc.* 1922, **121**, 1133, 1143.   † *Ber. dtsch. chem. Ges.* 1869, **2**, 140.
‡ *Liebigs Ann.* 1870, **156**, 293.   § *Ibid.* 1872, **162**, 77.

THE BENZENE PROBLEM

In other attempts to solve the benzene problem, Ladenburg*
tried to interpret its peculiar stability by picturing the molecule
as a triangular prism (IX); but this suggestion is only of historical
interest, for it has now been established beyond doubt that the
molecule has a planar structure.

Claus† also abandoned the use of double bonds and proposed
a formula (X) having three *para* bonds. In this structure, each
carbon atom is supposed to be linked to three others, two in the
ortho and one in the *para* position, so that *ortho-para* direction
in substitution reactions would be the rule. In spite of the fact
that such a structure does not offer a satisfactory explanation
for the stability of benzene, it received considerable support for
a time, and in recent years it was reconsidered by Pauling‡ but
soon discarded.§

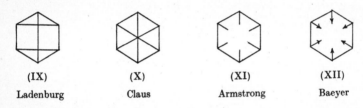

(IX)　　　(X)　　　(XI)　　　(XII)
Ladenburg　Claus　Armstrong　Baeyer

Armstrong‖ attempted to overcome the difficulty of the six
unused valencies in benzene in another way (XI). He suggested
that 'the remaining six [valencies] react upon each other—acting
towards a centre as it were, so that the "affinity" may be said
to be uniformly and symmetrically distributed'. Baeyer¶ put
forward a similar interpretation (XII), and Bamberger** extended
the hypothesis to the aromatic heterocyclic compounds, such as
pyridine (XIII), pyrrole (XIV) and other compounds.

These structures can hardly be seriously considered, for the
hypothetical bonds directed towards the centre of the ring have
no real meaning in terms of the modern electronic theory. The

---

* *Ber. dtsch. chem. Ges.* 1869, **2**, 140, 272.
† *Theoretische Betrachtungen und deren Anwendungen zur Systematik der organischen Chemie*, p. 207, Freiburg, 1867.
‡ *J. Amer. Chem. Soc.* 1926, **48**, 1132.
§ Pauling and Wheland, *J. Chem. Phys.* 1933, **1**, 362.
‖ *J. Chem. Soc.* 1887, **51**, 258. 　　¶ *Liebigs Ann.* 1892, **269**, 145.
** *Ibid.* 1893, **273**, 373.

Armstrong-Baeyer formulation was little more than a pictorial representation of the fact that the problem of the fourth valency remained unsolved.

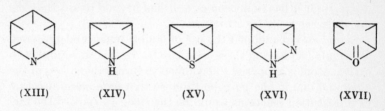

(XIII)   (XIV)   (XV)   (XVI)   (XVII)

Of all the structures proposed for benzene that of Thiele* is probably the nearest approach to the modern view. Thiele supposed that the reactivity of an olefinic linkage is due to incomplete saturation of the affinities of doubly bound carbon atoms, which may thus be assumed to have 'partial valencies', represented by dotted lines as in

$$C = C - C = C$$

In conjugated compounds the partial valencies of adjacent carbon atoms linked by a single bond are supposed to neutralize each other, the structure of butadiene therefore being represented by the formula

$$C = C - C = C$$

1:4-Addition follows naturally from such a hypothesis, and all examples of 1:2-addition in such compounds must be considered as serious exceptions. According to Thiele, benzene is a conjugated system of double bonds *par excellence*, and all the partial valencies are therefore neutralized, as in (XVIII).

As the theory of conjugated systems developed, this formula was modified, and the fully symmetrical structure (XIX) of J. J. Thomson,† involving real fractions of covalent bonds, came

---

* *Liebigs Ann.* 1899, **306**, 87.
† *Phil. Mag.* 1914, **27**, 784; Kermack and Robinson, *J. Chem. Soc.* 1922, **121**, 427; Armit and Robinson, ibid. 1925, **127**, 1604.

THE BENZENE PROBLEM

into use. However, the extension of this formula to the heterocyclic aromatic compounds, and to the polycyclic aromatic compounds, is attended by the difficulty of interpreting the very considerable differences in the apparent degree of unsaturation and 'reactivity'.

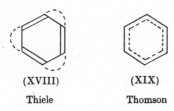

(XVIII) (XIX)
Thiele   Thomson

THE NATURE OF UNSATURATION IN AROMATIC COMPOUNDS

Evidence that benzene and other aromatic compounds are unsaturated comes mainly from a study of the addition reactions. Benzene itself can be reduced, by catalytic hydrogenation, to *cyclo*hexane, provided all traces of thiophen (which acts as a catalyst poison) are removed. The heterocyclic analogues of benzene behave similarly. Dicyclic and polycyclic compounds may give a variety of different hydrogenation products, depending on the method. Naphthalene normally gives 1:2:3:4-tetrahydronaphthalene and decalin on reduction; but conditions for the formation of 1:4-dihydronaphthalene have also been found. Similarly, anthracene (I) gives 9:10-dihydroanthracene (II) on mild reduction, but more vigorous conditions yield 1:2:3:4-tetrahydroanthracene (III), *sym*.-octahydroanthracene (IV) and perhydroanthracene (V). The ease of reduction varies very widely with the class of aromatic compound. Certain compounds exhibit very considerable reactivity and are reduced just as readily as olefinic substances.

The unsaturated character of aromatic compounds is also shown by the fact that, under certain conditions, the halogens react, not by substitution, but by addition. With benzene (VII) under the usual conditions of bromination in the presence of a catalyst such as ferric chloride, only the product of substitution, namely, bromobenzene (VI), is formed. When suitably irradiated

8 THE AROMATIC COMPOUNDS

in the absence of a catalyst, however, benzene adds six atoms of halogen to give hexabromo*cyclo*hexane (VIII); and, with chlorine, hexachloro*cyclo*hexane is formed.*

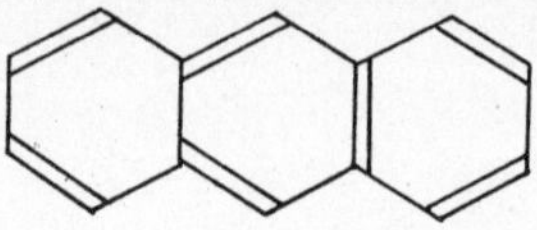

(I) 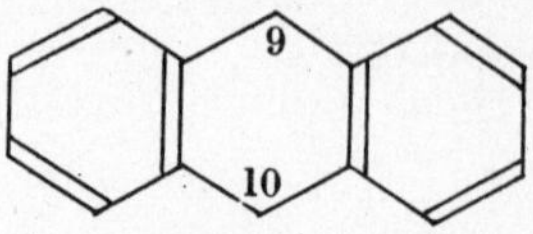　　　　　　(II)

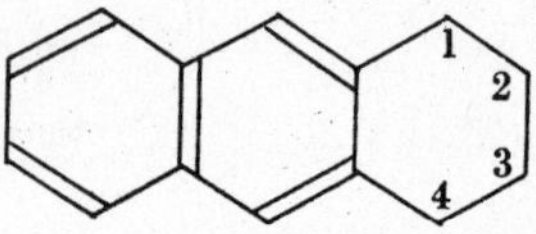

(III)

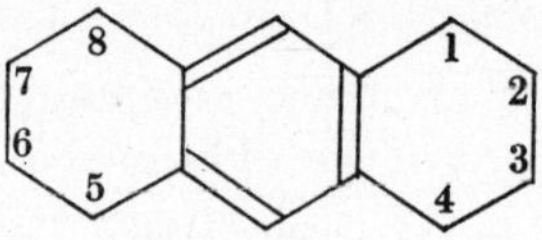

(IV) 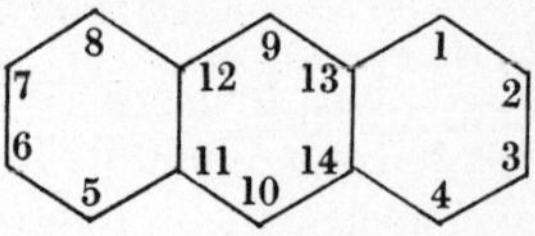　　　　　　(V)

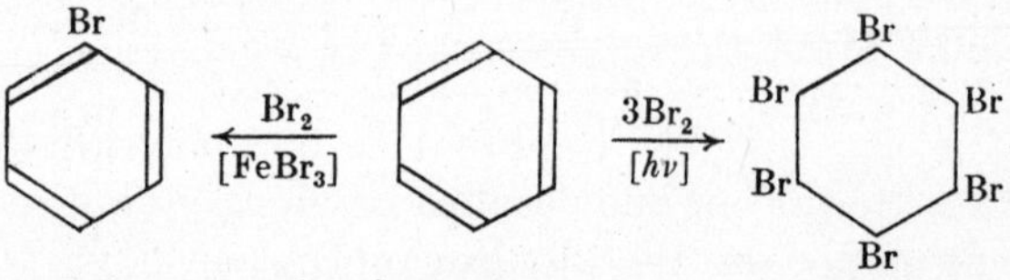

(VI) $\xleftarrow{\mathrm{Br}_2}_{[\mathrm{FeBr}_3]}$ (VII) $\xrightarrow{3\mathrm{Br}_2}_{[h\nu]}$ (VIII)

The reaction is fairly general in the whole field of aromatic hydrocarbons, and certain polycyclic compounds, such as anthracene and phenanthrene, form addition compounds with halogens even in the ordinary light of the laboratory.

Olefinic substances are normally broken down very readily by strong oxidizing agents. Indeed, the rapid reduction of alkaline permanganate in the cold was used by Baeyer† to ascertain the presence of ethylenic linkages. On the other hand, benzene is not attacked by Baeyer's reagent, nor indeed by most oxidizing

* Luther and Goldberg, *Z. phys. Chem.* 1906, **56**, 43.
† *Liebigs Ann.* 1888, **245**, 146.

agents; but a few methods of oxidative degradation are successful. Catalytic oxidation of benzene over vanadium pentoxide yields maleic acid (IX).* Furthermore, Jaffé† has shown that the metabolic oxidation of benzene in dogs and rabbits yields muconic acid (XI), and it is of some interest that, although the *cis* acid would be expected, the *trans* acid was actually isolated. *cis-cis*-Muconic acid was, however, obtained by oxidation of phenol with peracetic acid.

$$\underset{(IX)}{\overset{CH-CO_2H}{\underset{CH-CO_2H}{\|}}} \xleftarrow{[O] \text{ (catalytic)}} \underset{(X)}{\bigcirc} \xrightarrow{[O] \text{ (biological)}} \underset{(XI)}{\overset{HO_2C-CH=CH}{\underset{HO_2C-CH=CH}{}}}$$

It is also noteworthy that Cook and Schoental‡ have obtained allomucic acid, *meso*-tartaric acid and oxalic acid by the prolonged oxidation of benzene with hydrogen peroxide in *tert.*-butanol containing a little osmium tetroxide. It seems likely that the benzene double bonds become fully hydroxylated by addition, and that the hexahydroxy*cyclo*hexane (XII) formed subsequently undergoes ring fission to give allomucic acid (XIII).

$$\underset{(XII)}{\text{hexahydroxycyclohexane}} \qquad \underset{(XIII)}{\begin{array}{c} CO_2H \\ | \\ HO.C.H \\ | \\ HO.C.H \\ | \\ HO.C.H \\ | \\ HO.C.H \\ | \\ CO_2H \end{array}}$$

All these reactions appear to be consistent with the view that benzene has alternate double and single linkages.

Very convincing evidence for the presence of double bonds in aromatic compounds was obtained by a study of the reactions

* Downs, *J. Soc. Chem. Ind., Lond.*, 1926, **45**, 188T.
† *Hoppe-Seyl. Z.* 1909, **62**, 59.   ‡ *J. Chem. Soc.* 1950, p. 47.

of ozone. As is well known, ozone readily attacks ethylenic double bonds to give addition products, the exact structure of which has yet to be determined.* With benzene the reaction proceeds only slowly, but a triozonide (XV) is obtained as required by the *cyclo*hexatriene formula (XIV), and glyoxal (XVI) is formed by decomposition of the addition product with water.†

(XIV) (XV) (XVI)

The ozonization reaction has also been used in an attempt to settle the question of isomeric *ortho*-disubstituted derivatives of benzene. Levine and Cole‡ pointed out that each hypothetical isomer of *ortho*-xylene (XVIII and XIX) should give rise to its own triozonide (XVII and XX), and that while one triozonide (XVII) would be expected to give glyoxal and methylglyoxal on decomposition, the other (XX) would give diacetyl and glyoxal.

(XVII) (XVIII) (XIX) (XX)

As all three carbonyl compounds, glyoxal, methylglyoxal and diacetyl, were successfully identified among the products of decomposition of the triozonides, Levine and Cole concluded that *ortho*-xylene *does* exist in two isomeric configurations. Even so, these authors recognized the difficulties of interpretation of such experimental work, and they pointed out that the same

* Long, *Chem. Rev.* 1940, **27**, 437.
† Harries and Weiss, *Ber. dtsch. chem. Ges.* 1904, **37**, 3431; Harries, *Liebigs Ann.* 1905, **343**, 311; *Ozon und seine Einwerkung*, Springer, Berlin, 1916.
‡ *J. Amer. Chem. Soc.* 1932, **54**, 338.

THE BENZENE PROBLEM

decomposition products would also be expected if the two structures of *ortho*-xylene merely represented phases of some complex dynamic structure, and if the process of ozonization tended to stabilize these two phases. This latter explanation is nearer the modern view, for it is now believed that the interpretation given by Levine and Cole is invalid, and that *ortho*-xylene does not exist in two isomeric forms. Nevertheless, at the time, it did seem to offer a reasonable explanation of the facts (see also Chapter 5).

The different interpretations which it is possible to put on the work of Levine and Cole emphasize the difficulties which arise in attempting to determine the fine structure of benzene and other aromatic compounds by purely chemical methods, for the *reacting* molecule need not necessarily have the same arrangement of bonds as the *resting* molecule.

The essence of the difficulty in providing adequate structural formulae for benzene and other aromatic compounds is that each carbon atom is linked to only three, and not four atoms, and yet the compounds do not show the degree of instability and unsaturation which would be expected from this fact. For the reason mentioned above, the chemical evidence fails to provide an explanation; but physical methods have provided evidence of real value, and these are discussed in the following section.

THE APPLICATION OF PHYSICAL METHODS TO THE BENZENE PROBLEM

(i) *X-ray and electron diffraction experiments.* In recent years the structure of benzene has been established beyond doubt by the application of two types of physical method: X-ray and electron diffraction, and long-wave spectroscopy.

In X-ray and in electron diffraction experiments, a picture of the molecule under investigation is built up by recombining the waves which it scatters. The electron method is most easily applied to gases and easily volatile substances, and the X-ray method to crystalline compounds.*

In particular, the diffraction method lends itself to the accurate determination of interatomic distances, or bond lengths. By

---

* Robertson, *J. Chem. Soc.* 1945, p. 249.

12 THE AROMATIC COMPOUNDS

these means it has been possible to determine the normal covalent radii for a large number of atoms.* It was soon found, however, that the C—C bond length is not a constant, but varies considerably with the nature of the linkage. In diamond, and in the saturated aliphatic compounds, the C—C bond length has been found to be 1·54 Å.; in ethylene, the C—C double-bond length is about 1·34 Å.; and in acetylene, the C—C bond length is 1·20 Å. In non-conjugated compounds these lengths are found to be very constant; but in conjugated compounds considerable variations have been observed, the length of the double and triple bonds being *greater* than normal and the length of the single bonds being *shorter* than usual. These bonds of intermediate length are called *hybrid* bonds. A selection of these C—C bond lengths is given in table 1·1.

TABLE 1·1. *Carbon-carbon bond lengths*[†]

Substance	Bond length (Å.)	Remarks
Diamond	1·54	'Pure' C—C bond
Glyoxal	1·47	
Butadiene	1·46[‡]	
Furan	1·46[‡]	
Thiophen	1·44[‡]	
Graphite	1·42	
Benzene	1·39	Hybrid bonds
Pyrazine	1·39	
Resorcinol	1·39	
Hexamethylbenzene	1·39[§]	
Diacetylene	1·36[‡]	
Ethylene	1·34	'Pure' C=C bond
Acetylene	1·20	'Pure' C≡C bond

The C—C bond length in benzene and in benzene derivatives has been accurately determined both by electron diffraction and by X-ray diffraction experiments to be 1·39 Å. For example, analysis of the X-ray diffraction pattern of crystals of hexamethylbenzene gave a value of 1·39 Å. for the length of the

* Pauling, *The Nature of the Chemical Bond*, 2nd ed., Cornell University Press, 1940.
† Compiled from Wheland, *The Theory of Resonance*, John Wiley, New York, 1944.
‡ This value applies to the length of the bond conventionally represented as a 'single' bond.
§ This value applies to the bonds of the aromatic ring.

aromatic C—C bonds (fig. 1·1), and the length of the bonds linking each methyl group to the annular carbon atoms was found to be 1·53 Å.* It was also found that all the carbon atoms lie in one plane, and that the molecule possesses a centre of symmetry. Furthermore, following electron diffraction experiments with benzene vapour, Pauling and Brockway[†] were unable to confirm that all the C—C bonds are identical, but at least the results were consistent with a picture of benzene having identical sides of length 1·39 Å.

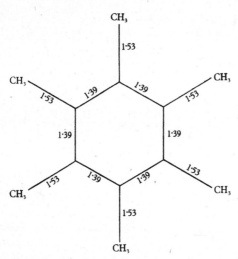

Fig. 1·1. Dimensions of the hexamethylbenzene molecule. (After Brockway and Robertson, *J. Chem. Soc.* 1939, p. 1324.)

The physical evidence of electron diffraction and of X-ray diffraction therefore indicates that benzene has a plane hexagonal structure of side 1·39 Å., and that the C—C bonds are neither single nor double, but are of intermediate or hybrid character.

(ii) *Long-wave spectroscopy.* The application of spectroscopy to the study of the structure of benzene depends on the fact that a change from a lower to a higher energy state is accompanied

---

* Brockway and Robertson, *J. Chem. Soc.* 1939, p. 1324; Lonsdale, *Trans. Faraday Soc.* 1929, **25**, 352.
† *J. Chem. Phys.* 1934, **2**, 867; see also Schomaker and Pauling, *J. Amer. Chem. Soc.* 1939, **61**, 1769; Haaijman and Wibaut, *Rec. trav. chim. Pays-Bas*, 1941, **60**, 842.

by absorption of light, and a change from a higher to a lower energy state by emission of light. The frequency of the light emitted or absorbed is determined by the magnitude of the energy change involved, and is given by the equation

$$\Delta E = E_2 - E_1 = h\nu = \frac{hc}{\lambda},$$

where ΔE is the difference in energy between the two energy levels (E_2 and E_1), ν is the frequency of light absorbed or emitted, h is Planck's constant, and c is the velocity of light. Changes in rotational energy are known to be associated with frequencies in the far infra-red region; electronic transitions are associated with light in the ultra-violet and visible regions; and the vibration frequencies of molecules are in the infra-red region.

Ingold* has pointed out that if benzene has a plane regular hexagonal structure it must have twenty fundamental vibrations, as illustrated in the accompanying diagrams (fig. 1·2). The arrows indicate motions in the plane of the ring, and the noughts and crosses motions perpendicular to the plane. By studying the spectral characteristics of benzene and of its deuterated analogues it was found possible to assign all the vibration forms of the model to particular frequency bands in the spectra.

Only those vibrations which are associated with an oscillation of dipole moment are recorded in infra-red absorption spectra.† The plane regular hexagonal model for benzene has four such vibration forms (labelled I), and four fundamental absorption bands were observed in the infra-red absorption spectrum of benzene vapour. Similarly, only those normal vibrations can appear as fundamentals in Raman spectra which involve an oscillation of the molecular polarizability. The model requires seven such vibration forms, and seven fundamental bands were observed.

The assignment of each band to a particular vibration form was achieved by comparison of the spectrum of benzene with that of hexadeuterobenzene, C_6D_6; for the sole effect of the substitution of deuterium for hydrogen is to alter the *mass* and

* *Proc. Roy. Soc.* A, 1938, **169**, 149; see also Ingold *et al. J. Chem. Soc.* 1936, pp. 912–987, 1210; 1946, pp. 222–333.
† Barnes and Bonner, *J. Chem. Educ.* 1938, **15**, 25.

THE BENZENE PROBLEM

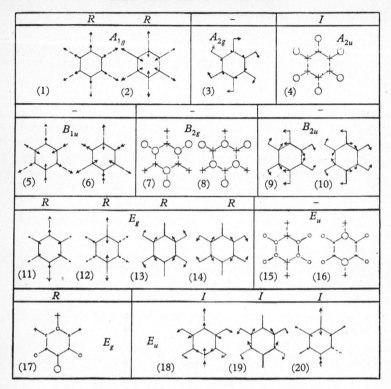

Fig. 1·2. Twenty fundamental vibration forms of benzene.
(From Ingold, *Proc. Roy. Soc.* A, 1938, **169**, 167.)

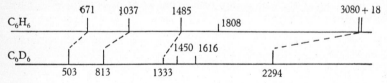

Fig. 1·3. Correlation of fundamentals in the infra-red spectra of C_6H_6 and C_6D_6.
(From Ingold, *Proc. Roy. Soc.* A, 1938, **169**, 169.)

hence the vibration frequency (fig. 1·3). Knowing the ratio of the masses, the magnitude of the frequency shifts for a given model was calculated, and the comparison of these calculated frequency shifts with those observed served the double purpose of assisting in or confirming the identification of the normal vibrations and of testing the model. Clearly the model which always gives the correct relations is proved.

The low-frequency band, 671 cm.$^{-1}$ in the infra-red absorption spectrum of C_6H_6, and the 503 cm.$^{-1}$ band in C_6D_6 have been identified with the vibration form in which the six hydrogen atoms together move perpendicularly to the plane of the ring (no. 4, fig. 1·2). Similarly, the lines at 849 cm.$^{-1}$ in the Raman spectrum of C_6H_6 and at 661 cm.$^{-1}$ in that of C_6D_6 have been assigned to the vibration form in which the plane of the hydrogen atoms rocks over the plane of the carbon ring about an axis common to both planes (no. 17, fig. 1·2). The highest frequency line at 3062 cm.$^{-1}$ in the Raman spectra of C_6H_6, and 2292 cm.$^{-1}$ in C_6D_6, is the breathing vibration of the hydrogen atoms (no. 2, fig. 1·2). The frequency shift is large because almost all the motion is in the hydrogen (or deuterium) atoms, whose mass changes in the ratio 1:2.

In this way, eleven of the twenty vibration forms of the model were investigated. The other vibration forms are not recorded in either the infra-red or the Raman spectra, and it was therefore necessary to attack the problem in other ways. One method involved a study of the fluorescence (emission) spectra, and a second method involved the investigation of the Raman and infra-red spectra of benzenes in which only some of the hydrogens are replaced by deuterium. In this way, chosen elements of symmetry can be destroyed, thus enabling additional vibration forms to appear in the spectra.

The successful conclusion of all this work has established the structure of benzene beyond all doubt. It is a plane regular hexagon of side 1·39 Å., and all the C—C bonds are entirely equivalent.

CHAPTER 2

THE THEORETICAL SOLUTION OF THE BENZENE PROBLEM, AND A DEFINITION OF THE TERM 'AROMATIC'

THE NATURE OF CARBON-CARBON BONDS*

According to the Rutherford-Bohr theory, atoms were considered to be small planetary systems consisting of a massive central nucleus having a positive charge, surrounded at a distance by a number of electrons such that the net charge was zero. The planetary electrons were each assigned to definite particular orbits, the size being regulated by Planck's quantum of action. With this model very many of the properties of atoms were satisfactorily explained for the first time. Moreover, by simple extension of the theory, Lewis was able to explain the more important types of chemical bonds. However, the Rutherford-Bohr theory took no account of the wave nature of electrons, or of the uncertainty principle. It is now recognized that the electron orbits as described by Bohr merely represent the most probable paths of the electrons.

In the modern conception, each atomic electron is assigned to an *orbital* (*atomic orbital*) described by a wave function ψ (*orbital wave function*), obtained by solution of the Schrödinger wave equation. The value of ψ for any electron varies from point to point, and the square of the wave function, ψ^2, at any point represents the probability that the electron will be found there. The square of the wave function, ψ^2, therefore represents the *probability distribution function*. In other words, $\psi^2 dV$ represents the probability that the electron will be found in the volume dV, and $4\pi r^2 \psi^2 dr$ is the probability that it will be found between the distances r and $r+dr$ from the nucleus. A planetary electron

---

* The argument in this chapter is largely indebted to the writings of Pauling (*The Nature of the Chemical Bond*, 2nd ed., Cornell University Press, 1940), Coulson (*Proc. Roy. Soc. Edinb.* A, 1941, **61**, 115; *Quart. Rev. Chem. Soc., Lond.*, 1947, **1**, 144), Penney (*The Quantum Theory of Valency*, Methuen, London, 1935), Wheland (*The Theory of Resonance*, John Wiley, New York, 1944) and of Van Vleck and Sherman (*Rev. Mod. Phys.* 1935, **7**, 167).

18 THE AROMATIC COMPOUNDS

may be approximately pictured as a charge cloud, the density of the cloud at any point being proportional to ψ^2.

Each planetary electron also has a *spin*. If the electron is conceived as a particle, this spin is similar to the rotation of the earth about its own axis; but as a wave it is much more difficult to picture. This spin can be either positive or negative with respect to a given direction. According to the Pauli exclusion principle, no two electrons can 'occupy' the same orbital *and* have the same spin. If two electrons have the same ψ, they must

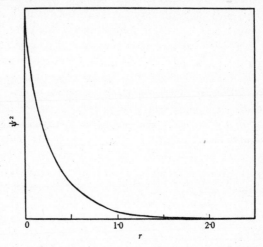

Fig. 2·1. Curve showing relationship between the probability-distribution coefficient, ψ^2, and the distance r from the nucleus, for the normal hydrogen atom. (After Pauling, *Nature of the Chemical Bond*.)

have opposed, or anti-parallel, spins. Such electrons are known as *electron pairs*. This is the case in the normal helium atom, where two electrons with opposed spins occupy the orbital of lowest energy.

In hydrogen, the nucleus is a proton having unit positive charge, and one electron occupies the orbital of lowest energy. As the periodic table is ascended, the various elements are built up by adding protons and neutrons to the nuclei, and electrons to a number of different planetary shells, which are called K, L, M, etc., shells. The K shell is that nearest the nucleus, and in all atoms other than hydrogen it is completely filled by only two

BENZENE PROBLEM, THEORETICAL SOLUTION

electrons, with anti-parallel spins. These shells are also referred to as 1, 2, 3, etc. (principal quantum numbers).

Several types of orbital are possible, depending on the energy content. They are classified as s, p, d, etc., and their shape may be determined by solution of the wave equation. For example, in the normal hydrogen atom the electron occupies the 1s-orbital. A plot of ψ^2 against r, where r is the distance from the nucleus, shows that this orbital is spherically symmetrical (fig. 2·1). Such

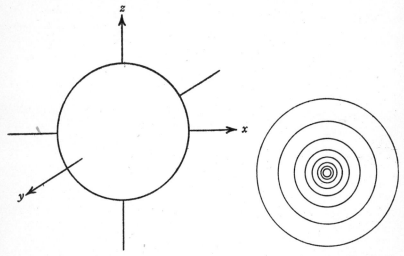

Fig. 2·2. Diagrammatic representation of the boundary surface of a 1s-orbital.

Fig. 2·3. ψ^2 contour lines for the 1s state of the hydrogen atom.

an orbital may therefore be represented diagrammatically by a spherical boundary surface such that almost all the charge cloud (conventionally 90%) lies within the contour drawn (fig. 2·2). Alternatively, the charge cloud may be represented by contour lines, as in fig. 2·3. The true contours are, of course, surfaces in three dimensions, obtained by rotating the figure shown about the vertical axis.

The values of $4\pi r^2 \psi^2$ may also be calculated for various values of r, and it is found (fig. 2·4) that for the normal hydrogen atom the maximum value of $4\pi r^2 \psi^2$ occurs when $r = 0.53$ Å. This is the radius which Bohr found for the orbit of the electron of hydrogen supposing this orbit to be a circle. It must be emphasized, however, that this radius must now be regarded only as the most probable one. The electron can be anywhere from $r = 0$ to $r = \infty$, and is most probably, or, alternatively, most frequently, at $r = 0.53$ Å.

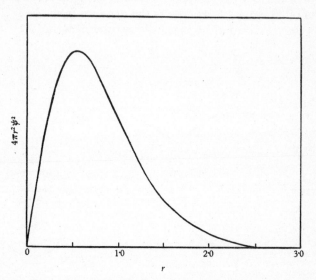

Fig. 2·4. Curve showing the relationship between $4\pi r^2 \psi^2$ and r for the normal hydrogen atom. The maximum value is at $r = 0.53$Å., which is therefore the most probable radius of the planetary electron in this atom. In contrast to fig. 2·1, which shows the probability of finding the electron in a particular direction (i.e. along a line from the nucleus), this figure shows the probability of finding it on a spherical shell at a distance r from the nucleus. (After Pauling, *Nature of the Chemical Bond*.)

Spherically symmetrical orbitals of higher energy content are referred to as $2s$, $3s$, etc.; but not all orbitals have this shape. In the L shell, for example, four orbitals are possible, and these are designated $2s$, $2p_x$, $2p_y$ and $2p_z$. A plot of the ψ^2 contour lines of a p-orbital (fig. 2·5) shows that this orbital is shaped like a dumb-bell, or hour-glass. In one half of the dumb-bell ψ is positive and in the other it is negative. There is a marked directional character in orbitals of this type, and this is recognized by means of the subscripts x, y, z. The boundary surfaces for orbitals of this type may be drawn in the same way as for the s-orbitals, and these are given in fig. 2·6. It is important to recognize that a p_x-orbital has a node in the yz-plane, and a p_z-orbital in the xy-plane.

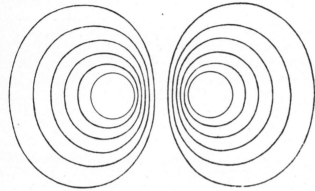

Fig. 2·5. ψ^2 contours for the 2p-orbital state of the (excited) hydrogen atom.

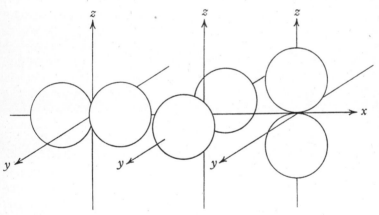

Fig. 2·6. Boundary surfaces for p_x, p_y and p_z atomic orbitals.

The boundary surfaces of the other atomic orbitals may be determined in the same way, but the s- and p-orbitals are the most important ones in organic chemistry.

The energies of the various orbitals are known to increase in the order
$$1s < 2s < 2p < 3s < 3p < 3d, \text{ etc.}$$

As one ascends the periodic table, electrons are fed, one at a time, into the various energy levels, beginning with $1s$, and allowing only two electrons, with anti-parallel spins, in each orbital. The electronic configuration of hydrogen in the ground state is $(1s)$, of helium $(1s)^2$, of lithium $(1s)^2(2s)$, of beryllium $(1s)^2(2s)^2$, and of boron $(1s)^2(2s)^2(2p_x)$. Atoms with more than one $2p$-orbital obey Hund's rules that electrons tend to avoid being in the same space orbital if possible, and that two electrons which occupy two equivalent space orbitals have parallel spins. Carbon is therefore $(1s)^2(2s)^2(2p_x)(2p_y)$; nitrogen is $(1s)^2(2s)^2(2p_x)(2p_y)(2p_z)$; oxygen $(1s)^2(2s)^2(2p_x)^2(2p_y)(2p_z)$; and so on. In the ground state carbon must be bivalent, for it has two unpaired electrons. Similarly, nitrogen is trivalent, and has, in addition, a 'lone pair' of electrons $(2s)^2$. Oxygen has two unpaired electrons, and the $(2p_x)^2$-electrons represent the 'lone pair'.

The way in which atoms combine by means of a covalent bond may be illustrated by a consideration of the nature of the bond in the hydrogen molecule.

Suppose there are two hydrogen atoms, A and B, a long way apart. The interaction between the two atoms under these conditions is zero, and the energy of the system is clearly twice the energy of a single atom. The atomic orbitals may be represented by ψ_A and ψ_B respectively. This is not the case, however, when the two atoms are separated by a distance approaching the interatomic distance in a hydrogen molecule. Each electron comes under the influence of the other nucleus as well as its own, and there is a mutual interaction, or perturbation, of the charge clouds, such that it is no longer possible to say that one electron belongs to atom A, and the other to atom B. The two atomic orbitals overlap and form two *molecular orbitals* Ψ_1 and Ψ_2.* The complete solution of Ψ for any molecule except that of hydrogen is a mathematical task of too great difficulty, but it has been shown that a molecular orbital may be approximately represented as a linear combination of atomic orbitals. In the case of a homonuclear diatomic molecule, such as hydrogen, a molecular orbital is therefore given by an expression

$$\Psi = \psi_A + \lambda \psi_B,$$

* Mulliken, *Phys. Rev.* 1932, **41**, 49.

where ψ_A is the wave function which the electron would have if it were confined to nucleus A, and ψ_B is the wave function which it would have if it were confined to nucleus B, and λ is a constant. Each electron must be equally divided between A and B, and it has been shown that λ^2 must equal unity.* Two molecular orbitals are therefore possible and take the form

$$\Psi_g = \psi_A + \psi_B$$

and

$$\Psi_u = \psi_A - \psi_B.$$

In the ground state two electrons, with anti-parallel spins, occupy the molecular orbital Ψ_g, of lower energy content. If E_A and E_B are the electronic energies of the constituent atoms a long way apart, then the total energy of the system (E) is given by

$$E = E_A + E_B + E_i,$$

where E_i is the interaction energy. It may be shown† that when the spins are anti-parallel, E_i is negative, and attraction occurs. The molecular orbital Ψ_g is known as a *bonding* orbital. The subscript g (*gerade* = even) indicates that the orbital has the same value of ψ at pairs of points diametrically opposite with respect to the centre of symmetry. The situation may be summarized as follows: a single covalent bond between atoms A and B involves two electrons, with anti-parallel spins. The atomic orbitals interact to give two molecular orbitals (which are identical except in spin). A plot of Ψ^2, the probability distribution function, along the axis AB shows that the bonding electrons are largely concentrated in the region between A and B (fig. 2·7). Diagrammatically, the orbital may be represented by the boundary surface method as a 'sausage type' of orbital (fig. 2·8), which is symmetrical about the line AB.

The molecular orbital Ψ_u is unoccupied in the ground state, but represents an excited state. Similar construction of the boundary surface and a plot of Ψ^2 shows that there is a nodal plane at right angles to the axis AB, so that Ψ changes sign. The subscript u (*ungerade* = odd) is therefore used, and this type

* Pauling, *Chem. Rev.* 1928, **5**, 173; Lennard-Jones, *Trans. Faraday Soc.* 1929, **25**, 668.
† Heitler, *Elementary Wave Mechanics*, 2nd ed., p. 100, Oxford University Press.

of orbital, in which only a small fraction of the charge cloud lies between the two atoms, is called an *anti-bonding* orbital. The boundary surface may be represented as in fig. 2·9.

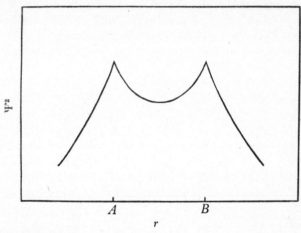

Fig. 2·7. Plot of Ψ^2, the probability-distribution function, along the axis (AB) of a homonuclear diatomic molecule such as H_2. (After Coulson, *Proc. Roy. Soc. Edinb.* A, 1941, **61**, 120.)

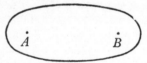 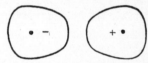

Fig. 2·8. Boundary surface of a 'sausage type' of molecular orbital for a homonuclear diatomic molecule such as H_2.

Fig. 2·9. Boundary surface of the antibonding molecular orbital for a homonuclear diatomic molecule such as H_2.

Similar molecular orbitals may be calculated for other types of covalent bonds. The formation of a hydrogen molecule involves the interaction between two $1s$ atomic orbitals; but in other cases different orbitals such as $1s$ and $2p_x$, or $2p_x$ and $2p_x$, etc., may be involved. The directional character of bonds follows from the principle that the strongest bonds are formed by maximum overlapping of atomic orbitals. Thus the H_2O molecule involves the interaction of a $2p_x$-orbital with a $1s$-orbital, and the interaction of a $2p_y$-orbital with another $1s$-orbital, so that the HOH angle should be around 90°. In fact, it is somewhat greater than this.

As already indicated, the electronic configuration of carbon in the ground state is $(1s)^2(2s)^2(2p_x)(2p_y)$. This configuration, with two unpaired electrons, corresponds to bivalency, and it is, of course, well known that carbon rarely forms compounds in this way.* The characteristic quadrivalent state can only be achieved if there are four unpaired electrons, and the various orbitals of carbon must therefore undergo some sort of rearrangement before compounds are formed. Pauling† has described how this may be achieved by a process of exciting one of the $2s$-electrons into the empty $2p_z$-orbital, followed by suitable overlapping, or *hybridization*, of the orbitals to give maximum overlapping in phase. Such a process of hybridization is necessary to account for the fact that in saturated organic compounds, such as CH_4, the four carbon valencies are equivalent, whereas, even if carbon is excited to a $(1s)^2(2s)(2p_x)(2p_y)(2p_z)$ state, the $2p$-orbitals would be directed 90° apart, and the fourth, the $2s$-orbital, would have no directional character. Moreover, the bonds which are formed from the hybrid orbitals are stronger than those formed by electrons in the constituent $2s$- and $2p$-orbitals.

It is found that when the four orbitals $2s$, $2p_x$, $2p_y$ and $2p_z$ are 'mixed', the overlapping in phase occurs in such a way that four new orbitals, t_1, t_2, t_3 and t_4, are formed, all at angles of 109° 28′ to one another. This is *tetrahedral hybridization*, and the boundary surface of each orbital of this new type may be represented diagrammatically as in fig. 2·10.

In *digonal hybridization* only the $2s$- and $2p_x$-orbitals are compounded in phase, or hybridized. Two new orbitals, which point in opposite directions along a straight line, are obtained, as in fig. 2·11.

In this type of hybridization (which is evidently present in acetylenes) the two remaining orbitals $2p_y$ and $2p_z$ are unchanged and preserve their directional character along the y- and z-axes.

It is the *trigonal* state of hybridization which is of greatest importance in aromatic chemistry. In this state the $2s$-, $2p_x$- and $2p_y$-orbitals are mixed in phase. Three entirely equivalent orbitals in the xy-plane are obtained, as illustrated in fig. 2·12.

---

* For carbon monoxide see Coulson, *Quart. Rev. Chem. Soc.*, Lond., 1947, **1**, 159.
† *J. Amer. Chem. Soc.* 1931, **53**, 1367.

26 THE AROMATIC COMPOUNDS

The three resulting coplanar orbitals are entirely equivalent, and are directed at angles of 120° to one another. The remaining $2p_z$-orbital retains its directional character along the z-axis, and is, of course, perpendicular to the plane of the other three orbitals. This type of hybridization evidently occurs in ethylene and related compounds, and also in the aromatic compounds.

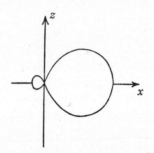

Fig. 2·10. Boundary surface for a single tetrahedral hybrid atomic orbital.

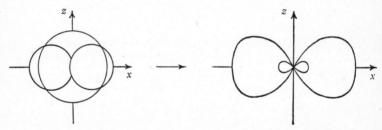

Fig. 2·11. Diagrammatic representation of the process of hybridization of a $2s$- and $2p_x$-orbital into two digonal hybrid atomic orbitals.

Using these orbitals for the carbon atom, the nature of the bonds in carbon compounds becomes clear. In methane, each tetrahedrally hybridized orbital interacts with the $1s$-orbital of a hydrogen atom in such a way that a molecular orbital is formed. The principle of maximum overlapping indicates that the bonds are directed to the corners of a regular tetrahedron.* Each of the resulting molecular orbitals may be represented diagrammatically as in fig. 2·13.

* Pauling, *The Nature of the Chemical Bond*, see p. 76; Maccoll, *Trans. Faraday Soc.* 1950, **46**, 369.

BENZENE PROBLEM, THEORETICAL SOLUTION

In ethylene, each carbon atom contributes three orbitals disposed at 120° to one another. Two of these are used for bonding to hydrogen atoms, and the third is used for bonding to the other carbon atom. Such C—H and C—C bonds are called σ-bonds.

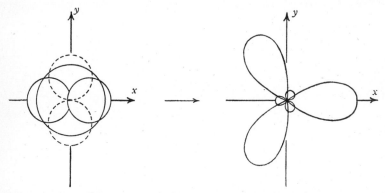

Fig. 2·12. Diagrammatic representation of the process of hybridization of $2s$-, $2p_x$- and $2p_y$-orbitals (trigonal hybridization).

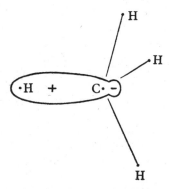

Fig. 2·13. Diagrammatic representation of a C—H molecular orbital in methane.

Each carbon atom has a remaining $2p_z$-orbital, disposed in two lobes above and below the xy-plane, that is, above and below the plane of the σ-bonds. These $2p_z$-orbitals overlap to some extent (although not to the same extent as occurs in σ-bonds) and form 'π-orbitals'. This lateral overlap results in weak bond formation, and such a bond is the classical 'double bond' of organic chemistry. The electrons forming it are π-electrons, 'mobile' electrons or 'unsaturation' electrons.* It is possible to draw the ψ^2 contour lines for the π-electrons in a molecule such as ethylene, and the π-orbitals are seen to resemble an atomic p-orbital (fig. 2·14).

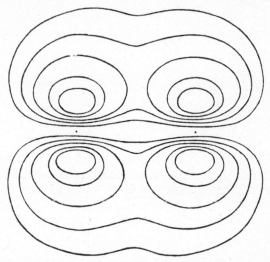

Fig. 2·14. ψ^2 contour lines for the unsaturation electrons, or π-electrons, in ethylene. (After Hückel, Z. Elektrochem. 1937, **43**, 752, 827.)

The formation of a π-bond may therefore be illustrated diagrammatically as in fig. 2·15, which shows the interaction of two $2p_z$-orbitals to form a molecular orbital, the boundary surfaces of which resemble two 'streamers' or 'buns' above and below the plane of the bonds, and having a node in the plane of the molecule.

The two bonds in a C—C double bond are therefore different. One is a σ-bond and the other a π-bond. The rigidity of such a system, and the resulting *cis-trans* isomerism in suitable derivatives, are explained by the fact that maximum overlapping of the two $2p_z$-orbitals can only occur if the lobes are parallel, that is, if they are both perpendicular to the plane of the σ-bonds. Rotation about the C_1-C_2 axis is therefore restricted. Moreover,

* Hückel, Z. Phys. 1931, **70**, 204; 1931, **72**, 310; 1932, **76**, 628; Lennard-Jones, Proc. Roy. Soc. A, 1937, **158**, 280; Mulliken, J. Chem. Phys. 1939, **7**, 339.

as the degree of lateral overlap of two $2p_z$-orbitals is not as great as occurs in the formation of a σ-bond, a π-bond is more readily broken. This occurs when a *cis* compound is converted into the corresponding *trans* derivative by rotation about the C_1-C_2 axis, the bond being reformed in the opposite configuration. It also occurs during *addition* reactions. The ethylenic bond is therefore characterized by high reactivity and by its tendency to undergo addition.

In acetylene, each carbon atom contributes two σ-electrons, which are used to form the C—H and C—C σ-bonds. The remaining $2p_y$- and $2p_z$-orbitals are directed at 90° to one another and to the axis of the C—C bond, and lateral interaction occurs in the same way.

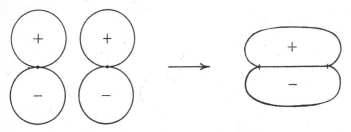

Fig. 2·15. Diagrammatic representation of the formation of a π-bond from two $2p_z$-orbitals.

It might be expected that the formation of lateral π-bonds of this type would result in some shortening of the C—C bond distance, and this has been found to be the case.

Experimentally determined electron contour maps for several typical C—C bonds have been published by Robertson.* These maps, of course, show the *total* electron distribution, both σ- and π-electrons. Each contour line represents a density increment of one electron per Å.², the one-electron line being dotted, and in each case the bond is tilted at a small angle (12–15°) to the projection plane, which gives an apparent additional shortening of about 3 % in the bond length.

Fig. 2·16a shows a single bond (CH_3—CH=, length 1·54 Å.) in sorbic acid; fig. 2·16b a shortened single bond (=CH—CH=, length 1·44 Å.) in the same structure; fig. 2·16c a double bond

* *J. Chem. Soc.* 1945, p. 256.

(—CH=CH—, length 1·34 Å.) in the same structure; and fig. 2·16d a triple bond (—C≡C—, length 1·19 Å.) in diphenylacetylene (tolan). The maps show that the electron density between the atoms rises fairly steadily from less than one electron in a single bond to just over four electrons per Å.² in a triple bond.

The treatment which has been applied to ethylene can also be extended to polyatomic molecules such as the allyl radical, butadiene, and so on.

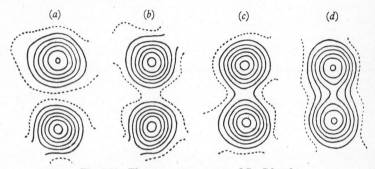

Fig. 2·16. Electron contour maps of C—C bonds.
(From Robertson, *J. Chem. Soc.* 1945, p. 256.)

In butadiene the four carbon atoms each contribute three σ-type orbitals, which by interaction, or overlapping with the $1s$ hydrogen orbitals, form the σ-bonds. Each carbon atom also contributes one π-electron. The orbitals ($2p_z$) of these electrons interact in just the same way as in ethylene; but the important point is that (if we call the four carbon atoms A, B, C and D) the $2p_z$-orbital of the B carbon atom will interact both with the similar orbital of A and with the similar orbital of C, and so on. The molecular orbitals cannot therefore be considered as 'localized' but must be 'non-localized' and embrace all four carbon atoms. Four molecular orbitals, Ψ_1, Ψ_2, Ψ_3 and Ψ_4, are possible. Each may be built up as a linear combination of atomic orbitals. That is,

$$\Psi_1 = c_{1a}\psi_a + c_{1b}\psi_b + c_{1c}\psi_c + c_{1d}\psi_d,$$

where ψ_a represents the wave function that the π-electron would have if it were confined to nucleus A, and ψ_b is the wave function

BENZENE PROBLEM, THEORETICAL SOLUTION 31

which the electron would have if confined to nucleus B, and so on; and c_{1a}, c_{1b}, etc., are constants. The four molecular orbitals have different energy contents. In the ground state two π-electrons (with anti-parallel spins) occupy the orbital of lowest energy. This orbital corresponds to the overlapping in phase of all four $2p$-orbitals and therefore has a node in the plane of the σ-bonds. The boundary surface of this orbital may therefore be represented as in fig. 2·17. The orbital of next highest energy content has an additional nodal plane at right angles to the central C—C link, and may therefore be represented as in fig. 2·18.

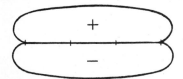
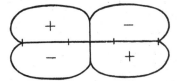

Fig. 2·17. Boundary surface for Ψ_1, the molecular orbital of lowest energy in butadiene.

Fig. 2·18. Boundary surface for Ψ_2, the molecular orbital of next highest energy in butadiene.

In the ground state, therefore, the actual state of the molecule is obtained when two electrons, with anti-parallel spins, occupy each of these molecular orbitals. Two other molecular orbitals are possible, which have additional nodal planes at right angles to the C—C bonds, and electrons pass into these excited states when the molecule absorbs light.

THE APPLICATION OF WAVE MECHANICS TO THE BENZENE PROBLEM

In benzene there are thirty valency electrons. Each carbon atom contributes four, and each hydrogen one. It is known that the molecule is a plane regular hexagon with angles of 120°, and it follows from the discussion in the previous section that the carbon atoms must be in the trigonal state of hybridization (sp^2) formed by the overlapping, or interaction, of $2s$-, $2p_x$- and $2p_y$-orbitals.* These hybrid orbitals, by overlapping with the hybrid orbitals of the neighbouring carbon atoms, form the C—C σ-bonds. The situation may be represented diagrammatically as in fig. 2·19. The six remaining hybrid orbitals

* Compare Penney, *Proc. Roy. Soc.* A, 1934, **146**, 223.

overlap with the $1s$ orbitals of the six hydrogen atoms to form the C—H bonds. Twelve electrons are thereby allocated to the six single C—C bonds (σ-bonds) and another twelve to the six C—H bonds (σ-bonds).

Simple counting shows that this leaves six electrons which are not involved in the formation of the σ-bonds. These six electrons constitute the 'aromatic sextet' of earlier workers.

By virtue of the trigonal state of hybridization, each carbon atom has a $2p_z$-orbital, in the shape of a dumb-bell or hour-glass oriented perpendicular to the plane of the ring. In a hypothetical state in which no overlapping occurs, these orbitals may be represented as in fig. 2·20.

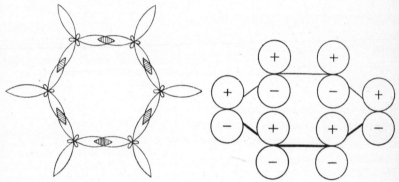

Fig. 2·19. Diagrammatic sketch showing the overlap of the trigonal type of hybrid orbitals of carbon. (From Sklar, *J. Chem. Phys.* 1937, **5**, 669.)

Fig. 2·20. Hypothetical state of the benzene molecule showing the six separate $2p_z$-orbitals, perpendicular to the plane of the ring.

In point of fact, however, these $2p_z$-orbitals must interact laterally with one another in just the same way as occurs in ethylene and in butadiene, and in other conjugated molecules. Indeed, because of its cyclic structure, the interaction must be the most perfect possible, resulting in the formation of very strong π-bonds.

From the six atomic orbitals of $2p_z$ character, six distinct molecular orbitals can be formed by linear combination of the atomic orbitals. These molecular orbitals, $\Psi_1, \Psi_2, \ldots, \Psi_6$, can each be represented by equations of the form

$$\Psi_1 = c_{1,1}\psi_1 + c_{1,2}\psi_2 + c_{1,3}\psi_3 + c_{1,4}\psi_4 + c_{1,5}\psi_5 + c_{1,6}\psi_6,$$

where ψ_1 represents the wave function which the electron would have if confined to nucleus 1, and ψ_2 the wave function which it would have if confined to nucleus 2, and so on. The constants, $c_{1,1}, c_{1,2}, \ldots; c_{2,1}, c_{2,2}, \ldots;$ etc., have various values, depending on the energy of the orbital. As before, two electrons (with antiparallel spins) occupy the molecular orbitals of lowest energy in the ground state, and the three molecular orbitals of higher energy content are only occupied in excited states.

In the lowest energy state the atomic orbitals interact in phase, and the resulting molecular orbital (Ψ_1) has a node in the plane of the ring. The resulting boundary surface therefore resembles two 'streamers' which extend right round the ring, part of the orbital being above the plane of the ring and part below it, as in fig. 2·21.

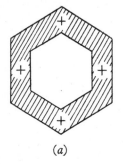 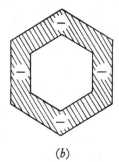

(a)   (b)

Fig. 2·21. Diagrammatic sketch of molecular orbital of lowest energy content as seen from above (a) and below (b) the plane of the ring.

In the other molecular orbitals there are additional nodal planes. The next two orbitals of higher energy have an additional nodal plane perpendicular to the plane of the ring. They are identical except that they are out of phase. The ground state of benzene may therefore be pictured as a superposition of the three molecular orbitals just described: two electrons in Ψ_1, two in Ψ_2 and two in Ψ_3. In the excited states, one or more electrons is promoted to a molecular orbital of higher energy content. The first three molecular orbitals are *bonding* orbitals, as their energies are less than that of an isolated $2p_z$ atomic orbital. The other three are of higher energy, and are *anti-bonding*.

Evaluation of the constants* gives the following wave functions for the three occupied molecular orbitals:

$$\Psi_1 = \frac{1}{6^{\frac{1}{2}}}(\psi_1+\psi_2+\psi_3+\psi_4+\psi_5+\psi_6),$$

$$\Psi_2 = \frac{1}{12^{\frac{1}{2}}}(2\psi_1+\psi_2-\psi_3-2\psi_4-\psi_5+\psi_6),$$

$$\Psi_3 = \frac{1}{4^{\frac{1}{2}}}(\psi_2+\psi_3-\psi_5-\psi_6).$$

Using these values it is possible to derive a number of useful quantities. For example, the π-electron density at an atom 1 is defined to be c_1^2, summed over all the occupied orbitals. At carbon atom 1, therefore, we have:

Orbital	Population (n)	c_1	nc_1^2
1	2	$1/6^{\frac{1}{2}}$	$\frac{1}{3}$
2	2	$2/12^{\frac{1}{2}}$	$\frac{2}{3}$
3	2	0	0

π-electron density at $C_1 = 1\cdot0$

In the same way, the π-electron density may be evaluated at all the other remaining carbon atoms. It is unity in each case. (It is unity at all carbon atoms in all unsubstituted benzenoid aromatic compounds.)

This, then, is the theoretical solution of the benzene problem. There are no pure single bonds, and no pure double bonds. It is not necessary to suppose that there is a sort of dynamic oscillation, as postulated by Kekulé. All the C—C bonds are exactly equivalent, and intermediate in character between single and double. All the properties of benzene could be expressed quantitatively by a rigorous solution of the Schrödinger wave equation, but, as already indicated, such a rigorous solution is not possible for any but the simplest molecules. The approximate solution outlined above is only possible by assuming that the wave function can be expressed as a linear combination of another set of functions, in this case the atomic orbitals.

Another approximate approach has also been devised. In this method, which is especially due to Heitler, London, Slater,

* Coulson, *Proc. Roy. Soc.* A, 1939, **169**, 413.

BENZENE PROBLEM, THEORETICAL SOLUTION

Pauling and Daudel, it is assumed that the wave function can be expanded, or approximately represented, as a linear combination of the wave functions corresponding to all the possible valence-bond structures for a conjugated molecule. In other words, the actual normal state of the molecule is considered to be represented not by any one of the alternative reasonable structures of the classical valence-bond type, but by a combination, a *hybrid*, the individual contributions of each structure being determined by their nature and stability.*

The actual normal state of the butadiene molecule may be considered as a *hybrid* of the structures (I) and (II):

$$\begin{array}{cc} \text{CH}\!\!\diagup\!\!\text{CH}_2 & \text{CH}\!\!\diagup\!\!\text{CH}_2 \\ | & \| \\ \text{CH}\!\!\diagdown\!\!\text{CH}_2 & \text{CH}\!\cdots\!\text{CH}_2 \\ (\text{I}) & (\text{II}) \end{array}$$

The dotted line in (II) represents a linkage which must be considerably longer than normal, and this is sometimes called a *formal* bond. Such a bond, although not representing an appreciable actual binding force directed between the two atoms concerned, corresponds to a pairing of spins of the electrons located on the two atoms. It is not true to say that the butadiene molecule is a 'mixture' of both structures. Neither valence-bond structure correctly represents the molecule, which is, in fact, something intermediate between the two: a hybrid. The molecule is said to be a *resonance* hybrid, or that it 'resonates'.†

Brief consideration of the two structures (I) and (II), which contribute to the butadiene molecule, makes it plain why the two double bonds in the classical structure do not behave in the same way as pure double bonds. They are represented as double bonds in only one of the contributing structures, and as single bonds in the other. Similarly, the central C—C bond is a double bond

---

* Pauling, *The Nature of the Chemical Bond*; Wheland, *The Theory of Resonance*.
† The difficulty of this approach is that, as Benfey has said (*J. Amer. Chem. Soc.* 1950, **72**, 1429), 'no matter how much explanation is given, to say that a molecule *resonates*, leaves an impression...that the molecule *does* something, that there are electronic movements between extreme positions'. Nevertheless, in spite of this difficulty, the theory of resonance has been widely applied in organic chemistry.

in structure (II) and a single bond in structure (I). This explains why this bond in butadiene has some 'double-bond character' and is shorter than a pure single C—C bond. Indeed, by determining the extent to which each structure contributes to the hybrid (for they need not, and frequently do not, contribute equally) a quantitative measure of the double-bond character can be derived.

A complete set of all the possible valence-bond structures which contribute to form a hybrid molecule is known as a *canonical set*, and the hybrid molecule has a smaller energy content than any of the contributing structures. The molecule may therefore be said to be stabilized by resonance.

The case of benzene is similar. No single satisfactory valence-bond structure can be drawn for this molecule, nor for any other conjugated compound, whether of aromatic or non-aromatic type. Pauling and Wheland* have shown that the five structures (III)–(VII) represent a complete set of valency theory electron distributions for benzene, and therefore constitute the canonical set.

The two structures of the Kekulé type are the most stable ones, and the three of the Dewar type represent other possible, but less stable, electron distributions. In these structures (V)–(VII), the formal bond does not represent any appreciable binding force, but corresponds to a pairing of spins of the electrons located on the atoms concerned. Benzene may be considered as a *hybrid* of these five structures, or, alternatively, it may be said that the molecule resonates between the five structures, the symbol ↔ implying such a resonance (cf. ⇌ for tautomerism, equilibrium, etc.).

As the Kekulé structures are more stable than the Dewar structures, they contribute to the hybrid to a greater degree,

* *J. Chem. Phys.* 1933, 1, 362.

BENZENE PROBLEM, THEORETICAL SOLUTION

and the coefficients of the wave functions represent the proportion in which the various structures are mixed, or compounded together. Thus:

$$\Psi_{\text{benzene}} = k_1(\psi_{\text{III}} + \psi_{\text{IV}}) + k_2(\psi_{\text{V}} + \psi_{\text{VI}} + \psi_{\text{VII}}).$$

In this way it can be seen that in the hybrid benzene molecule all the C—C bonds are equivalent, and that they are neither single nor double, but of intermediate character.

(VIII)   (IX)   (X)

Other aromatic molecules may be treated in the same way. The canonical set for naphthalene includes forty-two valence-bond structures. Three (VIII–X) are of the Kekulé type, and the remainder all have one or more 'formal' or 'para' bonds. The actual normal state of the molecule must be considered to be a hybrid of all forty-two contributing structures. Similarly, there are 429 contributing structures to the anthracene hybrid, and the higher polycyclic aromatic hydrocarbons have many thousand contributing structures. The actual number of contributing structures in each case is

$$\frac{n!}{(\tfrac{1}{2}n)!\,(\tfrac{1}{2}n+1)!},$$

where n is the number of π-electrons.

DELOCALIZATION (OR RESONANCE) ENERGY

It has been shown in the previous sections that the π-electrons of conjugated molecules tend to occupy molecular orbitals which embrace all the carbon atoms in the conjugated system. As a consequence of this, these π-electrons have a greater binding energy (i.e. a lower total energy) than when paired in localized bonds. This increase in bonding energy is called the delocalization or resonance energy; it is a measure of the stabilization due to resonance.

Delocalization or resonance energy may be readily calculated by the molecular orbital method, for it is simply the difference between (i) the energy which the π-electrons would have if paired in the simplest way to form pure double bonds, and (ii) the total energy of the electrons in their completely delocalized molecular orbitals.

Consider the allyl radical, $CH_2=CH-CH_2-$. If it is supposed that two π-electrons form the π-bond component of a pure double bond between carbon atoms 1 and 2, and that the third electron remains isolated on carbon atom 3, then the total energy (E) of the π-electrons would be $E = 3E_0 + 2\beta$, where E_0 is the energy of an isolated $2p_z$-orbital, and β is the 'resonance' or 'exchange integral'. This constant β, *which has a negative value*, is indicative of the difference in binding energy between ethane and ethylene; this difference is 2β. That is, 2β is the difference in energy between a pure double bond and a pure single bond.*

In point of fact, however, the two C—C bonds in the allyl radical are equivalent, and equal in length,[†] and the electrons occupy non-localized molecular orbitals. The energies of the three possible molecular orbitals may be calculated to be $E_0 + \sqrt{2}\beta$, E_0 and $E_0 - \sqrt{2}\beta$. The latter must be an anti-bonding orbital, for the energy is higher than that of the atomic orbitals. In the ground state two electrons, with anti-parallel spins, occupy the orbital with energy of $E_0 + \sqrt{2}\beta$, and one electron (unpaired) occupies that with energy E_0. The total energy of the system is therefore $3E_0 + 2\sqrt{2}\beta$. The difference between the energy of the allyl radical assuming it has the classical structure, and its energy assuming the electrons occupy non-localized molecular orbitals, is therefore $(3E_0 + 2\beta) - (3E_0 + 2\sqrt{2}\beta)$, or -0.828β. This represents the magnitude of the delocalization or resonance energy. It is the increase in bonding energy due to the delocalization, or, alternatively, the decrease (with resultant stabilization) in the total energy content. As β has a negative value the resonance energy is *positive*.

* Lennard-Jones, *Proc. Roy. Soc.* A, 1937, **158**, 280
† Coulson, *Proc. Roy. Soc.* A, 1938, **164**, 383.

The situation in benzene is very similar. If benzene is considered to be represented by a Kekulé structure, then the total energy of the system would be $6E_0 + 6\beta$, a contribution of $1E_0$ from each $2p_z$-orbital and a contribution of 2β from each double bond. Solution of the various equations for benzene shows that the six molecular orbitals ($\Psi_1, \ldots, \Psi_6$) have the following energies:

$$E_1 = E_0 + 2\beta,$$
$$E_2 = E_0 + \beta,$$
$$E_3 = E_0 + \beta,$$
$$E_4 = E_0 - \beta,$$
$$E_5 = E_0 - \beta,$$
$$E_6 = E_0 - 2\beta.$$

In the ground state the three molecular orbitals (Ψ_1, Ψ_2 and Ψ_3) of lowest energy (E_1, E_2 and E_3) are all doubly filled. These are called bonding orbitals, as their energy is less than that of an isolated $2p_z$ atomic orbital (as β is negative). The other molecular orbitals are anti-bonding. In the ground state, therefore, the total energy of the system must be

$$2(E_0 + 2\beta) + 2(E_0 + \beta) + 2(E_0 + \beta),$$

i.e. $6E_0 + 8\beta$. The resonance energy is therefore

$$(6E_0 + 6\beta) - (6E_0 + 8\beta),$$

or -2β,[*] and, as β has a negative value, the resonance energy is positive. The value of the constant β can be determined approximately by comparison between ethane and ethylene, and the resonance energy of benzene thereby evaluated. It is of the order of 36 kcal./mole.

Delocalization or resonance energy can also be evaluated experimentally,[†] and the principles of the methods involved are as follows.

Using thermochemical and spectrographic data obtained from compounds for which an unambiguous valence-bond structure

---

[*] Coulson, *Proc. Roy. Soc. Edinb.* A, 1941, 61, 115.
[†] Pauling and Sherman, *J. Chem. Phys.* 1933, 1, 606.

can be written, a table of bond energies for many different types of bond can be drawn up.* A selection of these bond energies is given in table 2·1.

TABLE 2·1. *Energy values for various bonds*†

Bond	Bond energy (kcal./mole)	Bond	Bond energy (kcal./mole)
H—H	103·4	O—O	34·9
C—C	58·6	O=O	96
C=C	100	O—H	110·2
C≡C	123	C—S	54·5
C—H	87·3	C—F	107·0
C—O	70·0	C—Cl	66·5
C—N	48·6	C—Br	54·0
C=N	94	C—I	45·5
C≡N	150		

In non-resonating molecules, the heat of formation is an additive function of the heats of formation (or bond energies) of the individual bonds, and the heat of formation of any molecule (as a gas) can be calculated from its elements.‡ Using Pauling's values for the heats of formation of a C≡C bond and C—H bonds, for example, it is possible to calculate the heat of formation of acetylene as follows:

	kcal./mole
Heat of formation C≡C bond	123
Heat of formation C—H bond	87·3
Heat of formation C—H bond	87·3
	297·6

The experimental value for the heat of formation of acetylene from its constituent atoms is 298·1 kcal./mole, so that the agreement between the calculated and experimental figures is good.

This agreement between calculated and experimental values of the heats of formation applies only to non-resonating molecules, to compounds for which an unambiguous valence-bond

---

* Pauling, *The Nature of the Chemical Bond*; Wheland, *The Theory of Resonance*; Rice, *Electronic Structure and Chemical Binding*, 1940; Dewar, *The Electronic Theory of Organic Chemistry*, Oxford University Press, 1949.
† After Pauling, *The Nature of the Chemical Bond*, pp. 53 and 131.
‡ Indeed, bond energies may be defined as quantities which may be assigned to bonds such that their sum in a molecule gives its heat of formation from its separated atoms. Compare Cottrell and Sutton, *Quart. Rev. Chem. Soc., Lond.*, 1948, **2**, 260.

structure can be written. A conjugated compound, a compound which resonates between two or more structures, is actually more stable than any one of the possible valence-bond structures, and the actual heat of formation of the molecule is always found to be *greater* than the calculated value. This means that the heat of formation of a conjugated compound is always greater than that of its unconjugated isomers.

The application of this line of argument to benzene is of considerable interest. Supposing benzene to have the Kekulé formula, its heat of formation would be the sum of the heats of formation of three C—C single bonds, three C—C double bonds, and six C—H bonds, namely, 1000 kcal./mole. Experimentally, the heat of formation of benzene is found to be 1039 kcal./mole. The difference, 39 kcal./mole, is a measure of the increased stability of the molecule due to resonance, and is the resonance energy of the molecule.

This experimental method for the determination of resonance energies is not, however, as accurate as could be desired. The bond energies of Pauling are only approximations (as they represent the *average* amount of energy required to break the bonds in a molecule, and not the actual energy required to break a specific link). Furthermore, the resonance energies are always small compared with the heats of formation, and the probable error is therefore considerable.

A more satisfactory method for the determination of resonance energies was introduced by Kistiakowsky and his collaborators.[*] This method is based on experimentally determined heats of hydrogenation. The heat of hydrogenation of ethylene is 32·8 kcal./mole; that of compounds of type $CH_2={=}CHCH_2R$ averages 30·2 kcal./mole; and that of compounds of type $RCH_2CH{=}CHCH_2R'$ averages 28·2 kcal./mole, R being either hydrogen or alkyl. These variations are ascribed to the effects of alkyl groups, and in related compounds these values are followed very closely.

It is to be expected that in dienes and polyenes the heats of hydrogenation will be simple multiples of that for the analogous compound containing only one unsaturated linkage, and in

[*] *Chem. Rev.* 1937, **20**, 181; *J. Amer. Chem. Soc.* 1936, **58**, 137, 146; 1937, **59**, 831; Williams, ibid. 1942, **64**, 1395.

unconjugated compounds this is found to be the case. Table 2·2 lists the experimentally determined heats of hydrogenation of a number of compounds, together with the values calculated assuming that the heats of hydrogenation for each double bond are additive. With unconjugated compounds the agreement is reasonably good. With *conjugated* compounds, however, the observed heats of hydrogenation are found to be less than the calculated values, and less than that of their unconjugated isomers. This is because the resonance energy of the conjugated compounds is lost in the reduction. The difference between the observed and calculated values is a measure of the resonance energy (E_R).

TABLE 2·2. *Heats of hydrogenation and resonance energies**

Compound		$-\Delta H$ obs. (kcal./mole)	$-\Delta H$ calc. (kcal./mole)	E_R (kcal./mole)
Unconjugated:	Penta-1:4-diene	60·8	60·4	—
	Hexa-1:5-diene	60·5	60·4	—
	Limonene	54·1	55·1	—
Conjugated:	Buta-1:3-diene	57·1	60·4	3·3
	Penta-1:3-diene	54·1	58·4	4·3
	α-Terpinene	50·7	53·8	3·1
	*Cyclo*hexa-1:3-diene	55·4	56·4	1·0
Aromatic:	Benzene	49·8	84·6	34·8
	Styrene	77·5	113·6	36·1
	Indene	69·9	111·6	41·7

The resonance energies determined by the two methods do not always agree exactly. This is mostly due to the approximate character of the table of bond energies.[†]

For most purposes, however, the agreement between the two methods is satisfactory; both methods indicate that the resonance energies of aromatic compounds are always considerable, and always much greater than those of closely related, but non-aromatic, systems. In agreement with Dewar,[‡] it is

---

[*] After Dewar, *The Electronic Theory of Organic Chemistry*, p. 35, Oxford University Press, 1949.

[†] For a discussion of this question see Dewar, *Trans. Faraday Soc.* 1946, **42**, 767; Bremner and Thomas, ibid. 1948, **44**, 338.

[‡] *The Electronic Theory of Organic Chemistry*, see p. 160.

therefore possible to define an aromatic compound as *a cyclic compound with a large resonance energy where all the annular atoms take part in a single conjugated system.**

TABLE 2·3. *Resonance energies of aromatic compounds*[†]

	kcal./mole		kcal./mole
Benzene	39	Pyrrole	31
Naphthalene	75	Indole	54
Anthracene	105	Carbazole	91
Phenanthrene	110	Furan	23
Pyridine	43	Thiophen	31
Quinoline	69	Azulene	46

This definition includes not only the polycyclic hydrocarbons such as anthracene, naphthacene, pentacene, etc., which undergo addition reactions very readily, but also the heterocyclic analogues, thiophen, furan, etc., and the non-benzenoid aromatic compounds such as azulene. Table 2·3 gives the resonance energies of some aromatic compounds.

* 'A molecule may be said to be of aromatic type if it is a cyclic unsaturated compound containing at least two conjugated double bonds in the ring when represented by the conventional symbols, and in which these bonds interact to a greater or lesser extent, thus bringing about a certain stabilisation of the molecule by resonance, which will in consequence be more saturated than if the double bonds were fixed and purely olefinic in character.' Wilson Baker, *J. Chem. Soc.* 1945, p. 258.

† After Pauling, *The Nature of the Chemical Bond*, p. 136; see also Heilbronner and Wieland, *Helv. chim. acta*, 1947, **30**, 947.

CHAPTER 3

SOME PROPERTIES OF AROMATIC COMPOUNDS

BENZENOID AROMATIC HYDROCARBONS

In many ways benzene is the simplest and most important aromatic compound. Its reactions have been very extensively investigated, and the concept of 'aromatic character' is largely based on the properties of this substance. Nevertheless, more complex benzenoid hydrocarbons such as naphthalene (I), anthracene (II), phenanthrene (III) and pyrene (IV) are also of considerable interest, especially as they are found to exhibit 'aromatic character' in varying degree.

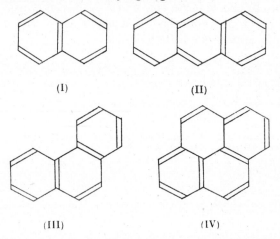

As with benzene, the reactions of naphthalene are predominantly those of substitution rather than of addition. Nitration, halogenation, sulphonation and other substitution reactions take place more readily than with benzene. With some of the polycyclic compounds, however, the reactions are predominantly those of addition and not substitution. Such compounds are usually extremely reactive, and may be said to exhibit some 'aliphatic character'.

SOME PROPERTIES OF AROMATIC COMPOUNDS

The series of linear benzologues of benzene is of special interest in this connexion. This series includes the compounds benzene, naphthalene, anthracene, naphthacene, pentacene and hexacene. Each member is derived from the one immediately preceding it by the linear 'fusion' of one additional benzene ring. The study of this series is largely due to Clar,[*] who has pointed out that the series is characterized not only by a deepening in the colour of the compounds from colourless to orange, deep blue and deep green, but also by a very marked and progressive increase in reactivity.

TABLE 3·1. *Linear benzologues of benzene*

Compound	No. of linear rings	Colour of compound
Benzene	1	Colourless
Naphthalene	2	Colourless
Anthracene	3	Colourless
Naphthacene	4	Orange
Pentacene	5	Deep blue
Hexacene	6	Deep green

The series is also characterized by a marked increase in the tendency to give addition rather than substitution products as the number of rings increases, and this increase is paralleled by a marked and progressive increase in the stability of the dihydro derivatives. Thus although naphthalene is readily substituted by bromine and other reagents, anthracene often reacts by addition, under quite mild conditions, to give 9:10 addition-products. With bromine, 9:10-dibromo-9:10-dihydroanthracene (V) is formed. With chlorine, 'anthracene dichloride' is obtained. Mild reduction gives 9:10-dihydroanthracene (VI), and other 9:10 addition-products can also be obtained with various reagents. Moreover, although dihydrobenzene is unstable, and dihydronaphthalene moderately stable, dihydroanthracene is quite stable. The stability of the dihydrides of some of the higher members of the series is very marked indeed, and the tendency to form dihydro derivatives is so marked in some of the higher members of the series that when the aromatic compound is

[*] *Aromatische Kohlenwasserstoffe*, 2nd ed., Springer-Verlag, Berlin, 1952.

heated part disproportionates to give the dihydride. Hexacene (VII), for example, is partly converted into dihydrohexacene (VIII) during vacuum sublimation.

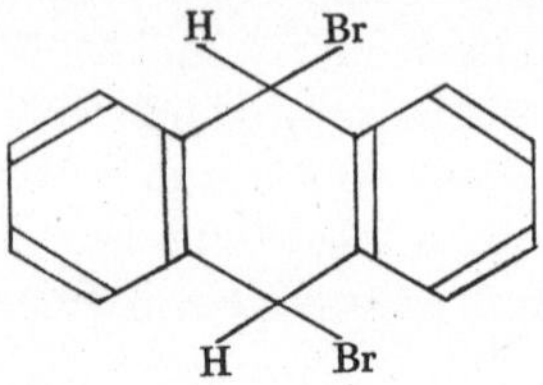

(V)

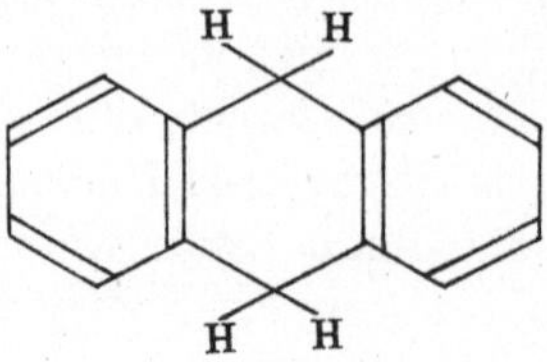

(VI)

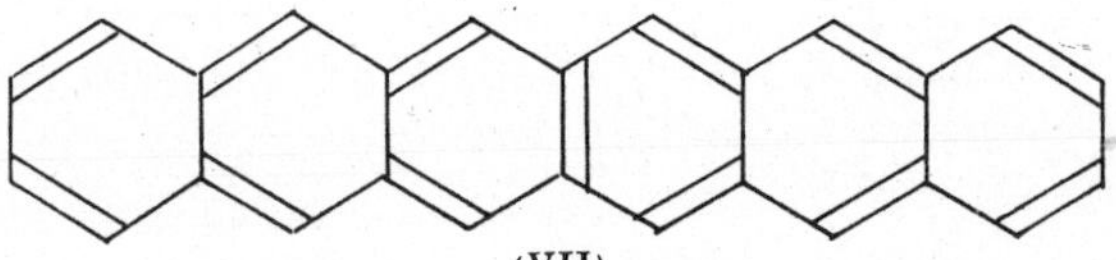

(VII)

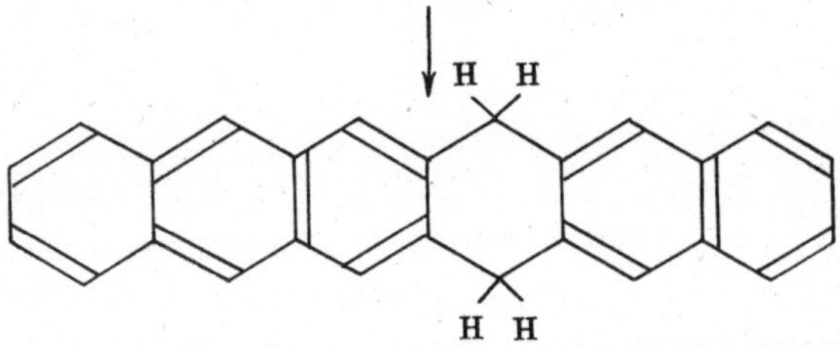

↓

(VIII)

The increasing stability of the dihydro derivatives as the series is ascended has been observed in other ways. For example, Clar has pointed out that the hydroxy derivatives of the hydrocarbons become progressively less stable as the series is ascended, and that this decrease in stability is paralleled by an increase in the stability of the keto forms. Phenol (IX) reacts entirely in the enol form and no derivatives of the keto form (X) can be prepared. Nevertheless, it is worth noting that phloroglucinol shows many of the reactions of a triketone, and even resorcinol shows some of the reactions of a diketone.

In the case of 1-naphthol (XI), the keto-enol equilibrium is again in favour of the enol form, for 1-naphthol reacts entirely as a phenol. With 9-hydroxyanthracene (9-anthrol (XIII)), however, the keto form (anthrone (XIV)) is the more stable, although both forms can be isolated, and the substance reacts in either form depending on the conditions.*

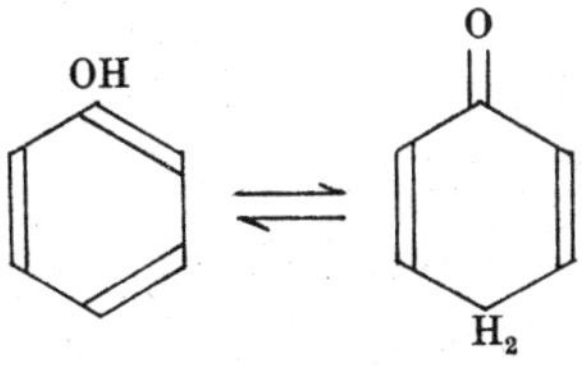

(IX) (X)

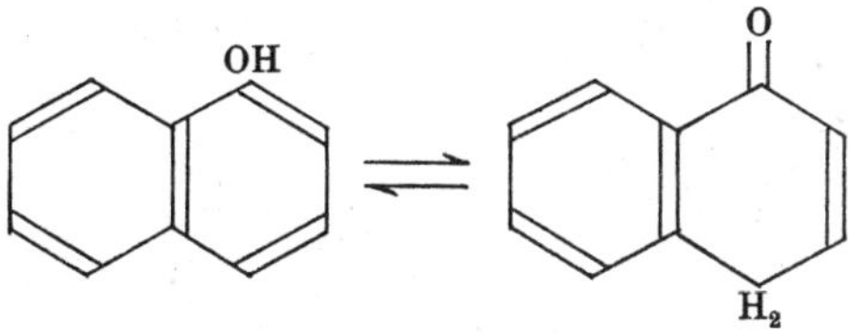

(XI) (XII)

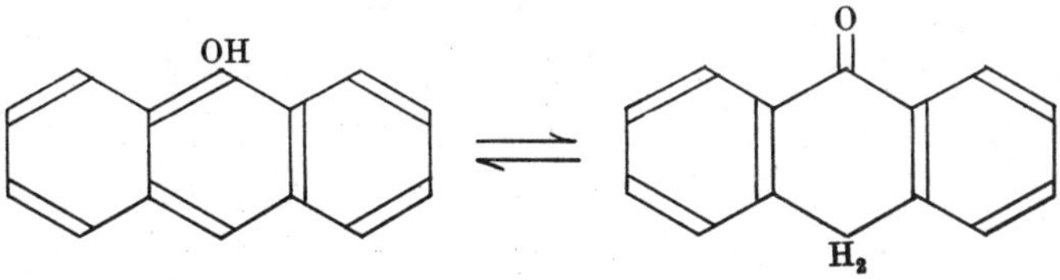

(XIII) (XIV)

In the naphthacene series, the keto-enol equilibrium is apparently displaced far to the keto side, for the substance reacts entirely as a ketone (XVI); and the same applies to the corresponding derivatives of the higher benz- homologues.

Similar *trans*-annular tautomerism has often been postulated for the methyl derivatives of this series of hydrocarbons, e.g. 9-methylanthracene (XVII). In the lower members of this series there is little doubt that the methyl derivatives (e.g. (XVII)) are very much more stable than the methylenedihydro forms (e.g. (XVIII)). Recently, however, Clar and Wright[†] have

* Meyer, *Liebigs Ann.* 1911, **379**, 37.   † *Nature, Lond.*, 1949, **163**, 921.

provided spectrographic evidence that the methyl derivative of pentacene (XIX) exists almost entirely in the methylenedihydro form (XX) at ordinary temperatures.

(XV) ⇌ (XVI)

(XVII) ⇌ (XVIII)

(XIX) ⇌ (XX)

The reason for the increasing stability of the dihydro derivatives, and the increasing prevalence of addition reactions as the series of benz- homologues is ascended, is very simple. Before benzene can be converted into 1:4-dihydrobenzene, two π-electrons have to be localized (as in (XXI)), and the energy required, the *para*-localization energy, is easily computed from the expression

$$P = E_r - E + 2\alpha,$$

where E is the π-electron energy of the original conjugated system, 2α is the energy of two isolated π-electrons, and E_r is the total π-electron energy of the 'residual' molecule. The *para*-localization energy may also be expressed in terms of resonance energies; i.e.

$$P = R - R_r - 2\beta$$

or

$$P' = R' - R'_r - 1 \cdot 6\gamma,$$

where R and R_r are the resonance energies and β and γ are the exchange integrals.*

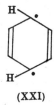

(XXI)

The *para*-localization energy for benzene is 91 kcal./mole, that for naphthalene is 82 kcal./mole, and that for the 9:10 positions of anthracene, 70 kcal./mole. It is apparent that as the series is ascended the *para*-localization energy becomes progressively smaller (fig. 3·1).

This means that when anthracene forms a 9:10 addition-compound the amount of resonance energy lost is much less than with benzene. In the latter case the tendency is towards substitution. With anthracene and the higher benz- homologues, however, the energy difference is smaller and the tendency is more towards addition.

The variation in the position of the keto-enol isomerization as the series is ascended may be explained in a similar fashion. It is well known that in the simple non-conjugated aliphatic

* Brown, *J. Chem. Soc.* 1950, p. 691.

compounds, such as acetaldehyde, the keto forms are always the more stable. The enol forms, if they exist at all, are present only in traces. In phenol, both the keto and enol forms are stabilized by resonance, but the stabilization is much greater for an intact benzene ring (IX) than for the system (X), and phenol therefore exists entirely in the enol form. With anthrol (XIII) and anthrone (XIV), however, the two resonance energies are more nearly equal. The resonance energy of anthrol is somewhat less than three times that of benzene, and the resonance energy of anthrone

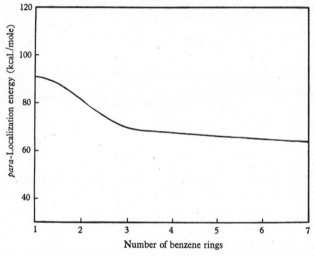

Fig. 3·1. Relationship between the number of linear benzene rings and the *para*-localization energy required to form the dihydro derivatives.

is about the same as that of benzophenone, which is somewhat more than twice that of benzene. The gain in resonance energy in passing from the keto to the enol form is much less than in the corresponding tautomerism in phenol, and the keto form is therefore more stable, as is the case with aliphatic compounds which are unaffected by resonance stabilization.

The angular 'fusion' of benzene rings also gives compounds having varying degrees of 'aromatic character'. Phenanthrene (III) is the parent compound of this type, and this compound is characterized by the pronounced reactivity and 'aliphatic character' of its 9:10 bond. Phenanthrene is hydrogenated to the 9:10-dihydride (XXII) over copper chromite catalyst at a moderate temperature.* Bromine also adds to the 9:10 bond to give 'phenanthrene dibromide' (XXIII).† This addition product is of moderate stability; it loses hydrogen bromide on prolonged standing at room temperature, or more rapidly on warming, to give 9-bromophenanthrene. Chlorine adds to phenanthrene in the same way, and this addition product is somewhat more stable.

(XXII)   (XXIII)

All the polycyclic compounds can be considered as derivatives of either anthracene, phenanthrene, or of the highly condensed hydrocarbon, pyrene (IV). Pyrene reacts predominantly by substitution. Mono-substitution occurs at the 3-position (XXIV), and the second substituent enters at either the 8- or the 10-position, to give mixtures of 3:8- and 3:10-disubstitution derivatives (XXV) and (XXVI).‡ Similarly, oxidation gives the 3:8- and 3:10-quinones. Further substitution of pyrene gives 3:5:8:10-tetrasubstituted derivatives.

The problem of assigning bond structures to naphthalene, phenanthrene, anthracene and the other polycyclic compounds proved even more difficult than with benzene. For example, it is possible to assign a bond structure to phenanthrene in which each ring has three double bonds analogous to the Kekulé benzene formula. On the other hand, with anthracene, pyrene, and with

* Burger and Mosettig, *J. Amer. Chem. Soc.* 1935, **57**, 2731; 1936, **58**, 1857; Durland and Adkins, ibid. 1937, **59**, 135; 1938, **60**, 1501.
† Fittig and Ostermayer, *Ber. dtsch. chem. Ges.* 1872, **5**, 933; *Liebigs Ann.* 1873, **166**, 361; Graebe, *Ber. dtsch. chem. Ges.* 1872, **5**, 861, 968; Price, *J. Amer. Chem. Soc.* 1936, **58**, 1834.
‡ Vollmann, Becker, Corell and Streeck, *Liebigs Ann.* 1937, **531**, 1.

THE AROMATIC COMPOUNDS

the higher benz- homologues of these hydrocarbons, it is impossible for *all* the rings to be completely benzenoid in the distribution of double bonds. One or more rings must have a 'quinonoid' distribution.

(XXIV)   (XXV)   (XXVI)

This fact led to the enunciation of the Fries rule for assigning conventional structural formulae to polycyclic compounds.* According to this rule, each aromatic ring strives to assume a benzenoid condition with three double bonds, and the structure of a polycyclic compound which has the greatest number of rings containing three double bonds has the lowest energy content and is, therefore, to be preferred.

Neglecting structures with 'formal' bonds, there are two fundamental valence-bond structures (XXVII) and (XXVIII), and their mirror images, for anthracene. In the first of these (XXVII) only one ring is of quinonoid character, but in the second structure (XXVIII) there are two rings having only two double bonds each. The former is therefore preferred.

In an interesting discussion of this question Fieser[†] associates the reactivity of a polycyclic compound with the presence of quinonoid rings. For example, Fieser attributes the reactivity of the 9:10-positions in anthracene to the fact that they correspond to the terminal positions of an *ortho*-quinonoid system (XXIX), and, as such, may be compared with *ortho*-benzoquinone itself (XXX). This simple explanation has much to commend it for most qualitative purposes.

On the other hand, the essential nature of aromatic compounds as revealed by the theoretical solution of the benzene problem

* Fries, Walter and Schilling, *Liebigs Ann.* 1935, 516, 248.
† See Gilman's *Organic Chemistry*, 2nd ed. 1, 117–213, John Wiley, New York.

SOME PROPERTIES OF AROMATIC COMPOUNDS

(Chapter 2) must not be overlooked. There can be no true 'double bonds' in any aromatic compound, and yet there are vast differences in reactivity and in the tendency to give addition products. These and other problems are discussed in succeeding chapters.

(XXVII)   (XXVIII)

(XXIX)   (XXX)

NON-BENZENOID AROMATIC HYDROCARBONS

The simplest hydrocarbon which might exhibit aromatic stability and properties is *cyclo*butadiene (I),* and although no authentic preparation of this substance has yet been described, there appears to be no reason why it should not be capable of existence. A number of derivatives of this ring system have been prepared. Dibenzo*cyclo*butadiene (II), or biphenylene, was prepared by Lothrop[†] by distilling 2:2′-dibromodiphenyl, or 2:2′-diiododiphenyl, with cuprous oxide. Some derivatives of biphenylene have also been described.[‡]

The structure of biphenylene has been completely established by electron diffraction[§] and by X-ray analysis of the crystal structure,[‖] and the ultra-violet absorption spectrum is also consistent with the tricyclic aromatic structure.[¶]

---

* Baker, *J. Chem. Soc.* 1945, p. 258; see also *Endeavour*, 1950, **9**, 35.
† *J. Amer. Chem. Soc.* 1941, **63**, 1187.
‡ Lothrop, ibid. 1942, **64**, 1698.
§ Waser and Schomaker, ibid. 1943, **65**, 1451.
‖ Waser and Chia-Si Lu, ibid. 1944, **66**, 2035.
¶ Carr, Pickett and Voris, ibid. 1941, **63**, 3231.

*cyclo*Pentadiene (III), with an odd number of carbon atoms, is not aromatic. Its resonance energy, calculated from the heat of hydrogenation, is only 3 kcal./mole, which is no greater than that for comparable open-chain compounds. *cyclo*Pentadiene is therefore predominantly aliphatic in character. The CH_2 group is, of course, completely saturated and cannot contribute electrons to the common pool. On the other hand, the removal of a proton from *cyclo*pentadiene gives a cyclic compound having a sextet of electrons. The *cyclo*pentadiene *anion* (IV) is aromatic, and in view of the increased stability and resonance energy which results, *cyclo*pentadiene forms salts with alkali metals.

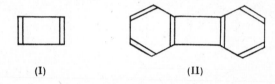

(I)   (II)

*cyclo*Heptatriene (V) is also aliphatic. It has an odd number of carbon atoms, and its resonance energy as determined from the heat of hydrogenation is 6·7 kcal./mole.

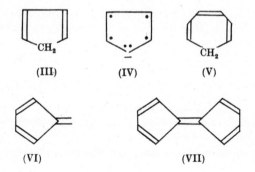

(III)   (IV)   (V)

(VI)   (VII)

Fulvene (VI) is very closely related to *cyclo*pentadiene. The conjugation, although not completely cyclic, is considerable, and the fulvenes are fairly stable. In particular, the unknown hydrocarbon fulvalene (VII) would be expected, on theoretical grounds, to exhibit aromatic stability.*

* Brown, *Trans. Faraday Soc.* 1949, **45**, 296.

SOME PROPERTIES OF AROMATIC COMPOUNDS

The higher benzologues of *cyclo*pentadiene are indene (VIII) and fluorene (IX). In both of these compounds the methylene group is reactive. Like *cyclo*pentadiene, both compounds form sodium salts, and advantage is taken of this fact in the separation of these substances from coal tar.

(VIII) (IX) (X)

These compounds are therefore not fully aromatic. Indene itself is rather unstable. It is autoxidizable, and readily polymerizes. Fluorene, on the other hand, is stable, and under selected conditions can be substituted either in the methylene group or in one of the benz-rings.

*cyclo*Octatetraene (X) was first prepared by Willstätter and his co-workers by exhaustive methylation of the pomegranate alkaloid, pseudopelletierine.* Willstätter showed that it is an extremely reactive hydrocarbon. Indeed, Vincent, Thompson and Smith[†] called attention to the striking similarity between Willstätter's hydrocarbon and styrene. Recently, however, *cyclo*octatetraene has become available in large quantities, a method for its preparation by the polymerization of acetylene in the presence of nickel catalysts having been devised. The chemistry of this substance has therefore been investigated in some detail, and there is no doubt that Willstätter's hydrocarbon was *cyclo*octatetraene.[‡]

Although *cyclo*octatetraene is a cyclic compound with alternate double and single bonds, its properties are those of an aliphatic polyene, and not of an aromatic compound. According to Prosen, Johnson and Rossini,[§] *cyclo*octatetraene is less stable than styrene by 34 kcal./mole.[‖]

* Willstätter and Waser, *Ber. dtsch. chem. Ges.* 1911, **44**, 3423; Willstätter and Heidelberger, *ibid.* 1913, **46**, 517.
† *J. Org. Chem.* 1939, **3**, 603.
‡ B.I.O.S. Report, 1945, no. 137, item no. 22; Campbell, *Ann. Rep. Chem. Soc.* 1947, **44**, 120; Reppe, *Experientia*, 1949, **5**, 93.
§ *J. Amer. Chem. Soc.* 1947, **69**, 2068.
‖ See also Maccoll, *Nature*, 1946, **157**, 695.

THE AROMATIC COMPOUNDS

For aromatic stability, and high resonance energy, the molecule would have to be planar; in such a case, however, the angular valency angles would be 135° as against 120° for sp^2 hybridization characteristic of aromatic compounds. The experimental evidence indicates that the molecule is not planar, and that the bonds are not 'aromatic', but approximate to alternate single and double bonds.* It seems that the angular strain is sufficient to counteract the stabilizing influence of resonance.†

Of the possible bicyclic non-benzenoid aromatic hydrocarbons, both pentalene and heptalene have resisted attempts at preparation.‡

(XI)   (XII)

Certain derivatives of pentalene have been prepared. For example, Clar§ has described the preparation of aceanthreno-2':1'-1:2-aceanthrene, and other pentalene derivatives have been described by Brand and his co-workers.∥ 1:2:4:5-Dibenzopentalene (XIII) has been prepared by Blood and Linstead.¶ This compound behaves as a conjugated diene and has no aromatic stability.

(XIII)

The most important non-benzenoid aromatic compound is azulene (XIV). Azulene and its derivatives have considerable stability and form complexes with polynitro compounds as do

---

* Lippincott and Lord, *J. Amer. Chem. Soc.* 1946, **68**, 1868; Kaufman, Fankuchen and Mark, *Nature, Lond.*, 1948, **161**, 165; Pink and Ubbelohde, ibid. 1947, **160**, 502.
† Penney, *Proc. Roy. Soc.* A, 1934, **146**, 223.
‡ See, for example, Aspinall and Baker, *J. Chem. Soc.* 1950, p. 743.
§ *Ber. dtsch. chem. Ges.* 1939, **72**, 2134.
∥ See, for example, Brand and Hennig, *Chem. Ber.* 1948, **81**, 382, 387.
¶ *J. Chem. Soc.* 1952, p. 2263.

SOME PROPERTIES OF AROMATIC COMPOUNDS 57

most benzenoid compounds of similar complexity. It is interesting that the azulenes are all highly coloured—blue, violet and blue-green. Some derivatives occur naturally, and their isolation is made possible by the fact that they are soluble in concentrated aqueous phosphoric acid, from which solutions they separate on dilution with water.*

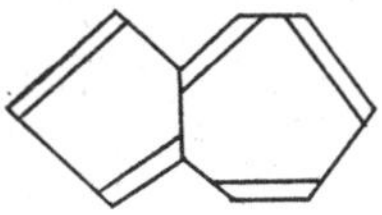

(XIV) 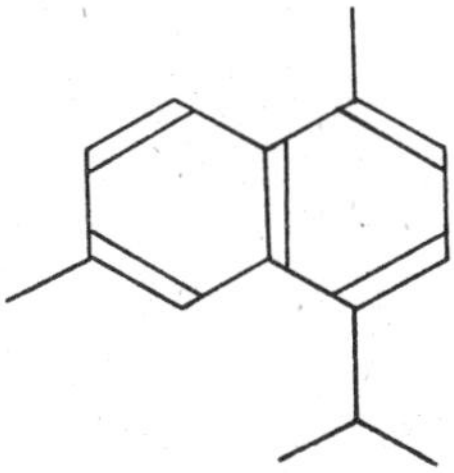   (XV) 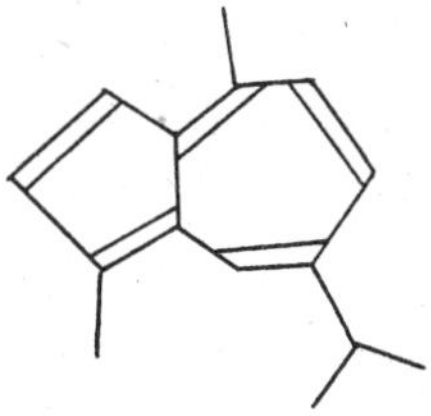   (XVI)

The resonance energy of azulene is considerable, although less than that of naphthalene. Perrottet, Taub and Briner† determined the heats of combustion of two isomeric compounds, cadalene (XV) and guaiazulene (XVI), and found that the former value is 29·5 kcal./mole lower than the latter. The resonance energy of azulene is therefore 45·5 kcal./mole as compared with naphthalene, 75 kcal./mole. Similarly, Heilbronner and Wieland‡ determined the resonance energy of azulene to be 46 kcal./mole. This resonance energy is considerable, and although little attention has yet been paid to the reactions of azulene, there is no doubt that it is 'aromatic'.

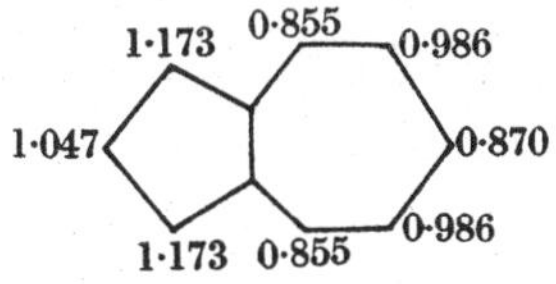

(XVII)

As explained in Chapter 2, the π-electron density at each annular carbon atom in benzene (and in all other unsubstituted benzenoid aromatic substances) is unity. With azulene and other

* Sherndal, *J. Amer. Chem. Soc.* 1915, **37**, 167, 1537.
† *Helv. chim. acta*, 1940, **23**, 1260.   ‡ Ibid. 1947, **30**, 947.

non-benzenoid aromatic compounds, however, the π-electron densities are sometimes greater than, and sometimes smaller than, unity. The figures for azulene are given in structure (XVII).*

These calculated figures imply that azulene should have a dipole moment, and this has been confirmed experimentally. It must be admitted, however, that the agreement between the calculated and observed dipole moments is far from satisfactory.[†]

TROPOLONE

Closely related to the non-benzenoid aromatic hydrocarbons is the aromatic system tropolone (I), or cycloheptatrienolone. Dewar[‡] postulated the existence of this stable system in order to account for the properties of stipitatic acid (II). Its occurrence in several complex molecules, such as purpurogallin (III) and colchicine (IV), now seems to be established or (in the latter case) to be fairly certain.[§] Various derivatives of tropolone have been prepared, and several independent syntheses of the parent aromatic ring system itself have also been published.[‖]

The chemistry of tropolone has not yet been extensively investigated, but aromatic character has certainly been established. Dewar[¶] calculates that tropolone should have a resonance energy comparable with that of benzene, and this has been confirmed experimentally. The resonance energy obtained from thermochemical data is 28·6 kcal./mole.** It cannot be hydrogenated over a palladium catalyst, so the double bonds are not olefinic in character. Moreover, it couples with *para*-toluenediazonium salts,[††] and is nitrated.[‡‡]

---

* Brown, *Trans. Faraday Soc.* 1948, **44**, 984; Pullman and Berthier, *C.R. Acad. Sci., Paris*, 1948, **227**, 677.
† Berthier and Pullman, ibid. 1949, **229**, 761.
‡ *Nature, Lond.*, 1945, **155**, 50.
§ Loudon, *Ann. Rep. Chem. Soc.* 1948, **45**, 187.
‖ Cook, Gibb, Raphael and Somerville, *Chem. and Ind.* 1950, p. 426; *J. Chem. Soc.* 1951, p. 503; Haworth and Hobson, *Chem. and Ind.* 1950, p. 441; *J. Chem. Soc.* 1951, p. 561; Doering and Knox, *J. Amer. Chem. Soc.* 1950, **72**, 2305; 1951, **73**, 828.
¶ *Nature, Lond.*, 1950, **166**, 790.
** Cook, Gibb, Raphael and Somerville, *J. Chem. Soc.* 1951, p. 503.
†† Cook *et al.* loc. cit.
‡‡ Haworth and Jefferies, *Chem. and Ind.* 1950, p. 841.

HETEROCYCLIC AROMATIC COMPOUNDS CONTAINING SIX-MEMBERED RINGS

Not all aromatic compounds are hydrocarbons. In 1849 Anderson[*] described the isolation of pyridine, C_5H_5N, and it was soon recognized that this substance is an analogue of benzene. Korner and Dewar[†] in 1869 proposed the hexagonal structure (I) having alternate double and single bonds, and the subsequent history of the attempts to assign an adequate structural formula to this molecule closely parallels that relating to benzene. Riedel[‡] proposed a structure (II) similar to the Dewar structure for benzene, and von Pechmann and Baltzer[§] suggested a centric formula (III) corresponding to the Armstrong-Baeyer benzene formula. Later a structure of the Thiele type (IV) was proposed.

There is no doubt that pyridine has aromatic character. It is even more resistant to oxidation than benzene. Picoline can be oxidized to picolinic acid with potassium permanganate, and quinoline (V) and *iso*quinoline (VII) are preferentially oxidized

[*] *Trans. Roy. Soc. Edinb.* 1849, **16**, 123.
[†] See Dobbin, *J. Chem. Educ.* 1934, **11**, 596.
[‡] *Ber. dtsch. chem. Ges.* 1883, **16**, 1609.
[§] *Ibid.* 1891, **24**, 3144.

60 THE AROMATIC COMPOUNDS

in the benz-rings to give quinolinic acid (VI) and cinchomeronic acid (VIII) respectively.

The unsaturated nature of pyridine is illustrated by the fact that it is readily reduced to piperidine; but its reactions are predominantly of the substitution type. In this respect it is less 'reactive' than benzene, but resembles nitrobenzene very closely indeed. Pyridine can be nitrated, but very severe conditions are required, and the same is true of sulphonation. It can be brominated or chlorinated in the vapour phase over charcoal or pumice at 300°,* and it also undergoes other typical substitution reactions.

(I) Korner-Dewar (II) Riedel (III) von Pechmann (IV) Thiele

(V) → (VI)

(VII) → (VIII)

Although there are certain important differences to be considered, the solution to the 'pyridine problem' closely resembles that already detailed for benzene.

Unlike carbon, which has four valency electrons, nitrogen has five, and in most of its compounds it increases this number to form a stable octet. In ammonia (IX) the nitrogen is linked to

* Den Hertog and Wibaut, *Rec. trav. chim. Pays-Bas*, 1932, **51**, 381, 940; Wibaut and Bickel, ibid. 1939, **58**, 994.

SOME PROPERTIES OF AROMATIC COMPOUNDS

three hydrogen atoms and the compound is saturated with respect to this atom, leaving a 'lone pair' of electrons which are not involved in bonding. Ammonia is not saturated with respect to hydrogen ions, however, and the ammonium ion (X) is formed by the addition of a proton. The lone pair of electrons may also be used to form a linkage with an element, such as oxygen, which requires only two electrons to complete its octet. This is the case in compounds of the type $R_3\equiv N \rightarrow O$ (XI). The most important difference between benzene and pyridine is that the latter forms salts with all strong acids. Pyridine therefore resembles ammonia in having two electrons available for salt formation. This lone pair is also available for donation to another atom requiring two electrons to complete its octet, for, under suitable conditions, pyridine gives pyridine-N-oxide.

$$
\begin{array}{ccc}
\text{H} & \left[\text{H}\right]^{+} & R \\
\bullet\circ & \bullet\circ & \bullet\circ \quad\times\times \\
\text{H} \; {}^{\bullet}_{\circ}\text{N}{}^{\bullet}_{\bullet} & \text{H} \; {}^{\bullet}_{\circ}\text{N}{}^{\bullet}_{\bullet}\; \text{H} & R \; {}^{\bullet}_{\circ}\text{N}{}^{\bullet}_{\bullet}\text{O}{}^{\times}_{\times} \\
\bullet\circ & \bullet\circ & \bullet\circ \quad\times\times \\
\text{H} & \text{H} & R \\
\text{(IX)} & \text{(X)} & \text{(XI)}
\end{array}
$$

Bearing this in mind, the electronic structure of pyridine can be visualized. Each carbon atom has four valency electrons, three of which must be involved in the formation of the σ-bonds. The nitrogen atom has five valency electrons. Of these, two are involved in the formation of the σ-bonds, and two form the 'lone pair'. Six electrons remain, one from each carbon atom and one from the nitrogen atom (XII).

(XII)

These six electrons constitute the 'aromatic sextet' and must interact with one another to form π-bonds in the same manner as in benzene. Just as in benzene, therefore, there are no single bonds and no double bonds. The bonds are all of intermediate character; they may be considered as hybrid bonds, and pyridine may be considered as a hybrid of all the possible reasonable valence-bond structures (XIII)–(XVII). The resonance (or delocalization) energy has been determined experimentally from the heat of combustion, and there is no doubt that the resonance stability is of the same order as that for benzene. The electronic symmetry cannot be as perfect as in benzene, for nitrogen is more electronegative (see Chapter 6) than carbon and has a tendency to attract more than its share of electrons. The C—C bonds cannot therefore be all equivalent as is the case with benzene.

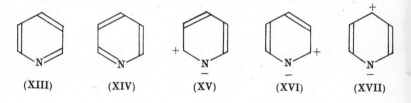

(XIII)　　　　(XIV)　　　　(XV)　　　　(XVI)　　　　(XVII)

This lack of electronic symmetry is also of importance in connexion with the basicity of pyridine. In the pyridinium ion the positive charge on the nitrogen atom still further reduces the electronic symmetry and hence the resonance energy. Some resonance energy is therefore lost in the conversion of pyridine into its ion, and for this reason pyridine is considerably less basic than ammonia.

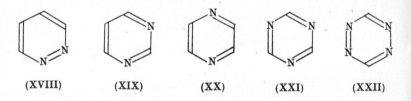

(XVIII)　　　　(XIX)　　　　(XX)　　　　(XXI)　　　　(XXII)

The other nitrogen-containing analogues of benzene, such as pyridazine (XVIII), pyrimidine (XIX), pyrazine (XX), triazine (XXI) and tetrazine (XXII), are all built up in the same way, and are likewise aromatic in character. Each annular carbon

SOME PROPERTIES OF AROMATIC COMPOUNDS

atom and each nitrogen atom must contribute one electron to the common pool of six π-electrons. Compared with benzene, however, all these compounds are deficient in electronic symmetry, because nitrogen is more electronegative than carbon.

The polynuclear heterocyclic compounds bear the same relationship to pyridine and the other nitrogen-containing analogues of benzene as do the polycyclic aromatic hydrocarbons to benzene. The reactions of quinoline, *iso*quinoline, acridine, phenazine, phenanthridine, and so on, are predominantly substitution reactions. The differences in properties as compared with the corresponding hydrocarbons can always be ascribed to the nitrogen atom.

It is not commonly realized just how closely some of the properties of the azahydrocarbons do parallel those of the corresponding hydrocarbons. For example, the increase in 'reactivity' and in the stability of dihydro derivatives as the series benzene, naphthalene, anthracene, naphthacene, pentacene and hexacene is ascended is well known; but the corresponding increase in 'reactivity' and in the stability of the dihydro derivatives of related azahydrocarbons does not seem to have attracted much attention. In the 'azine' series, however, this has already been investigated in some detail. The dihydrides of pyrazine and of quinoxaline are apparently unstable. The dihydride (XXIII) of phenazine is moderately stable, for it oxidizes in air to phenazhydrin, a molecular compound of phenazine and dihydrophenazine. Reduction of 2:3-benzophenazine proceeds very readily and the resulting dihydride (XXIV) is quite stable in air, although it can be oxidized to the aromatic structure with dichromate.* The dihydride (XXV) of 2:3:6:7-dibenzophenazine, on the other hand, is so stable that it cannot be oxidized to the aromatic compound.

This work is paralleled by investigations of other linear pentacyclic compounds. All attempts to prepare the aromatic compound 5:7:12:14-tetrazapentacene (XXVI) have failed. The deep blue dihydro compound (XXVII), on the other hand, is quite stable and easily prepared. The positions of the two hydrogen atoms and the quinonoid-like structure of the molecule follow from the deep colour and absorption spectrum of this

* Hinsberg, *Liebigs Ann.* 1901, **319**, 257.

compound,* and it may be concluded that this quinonoid structure is more stable than the aromatic system.

The tendency to form quinonoid systems of this kind seems to be much stronger in nitrogen-containing heterocyclic systems

(XXIII)

(XXIV)

(XXV)

(XXVI)

(XXVII)

than is the case with the aromatic hydrocarbons. This tendency seems to find its origin in the ability of an annular nitrogen atom to form an imino group and so isolate itself from the conjugated system.

* Badger and Pettit, *J. Chem. Soc.* 1951, p. 3211.

SOME PROPERTIES OF AROMATIC COMPOUNDS

The difference may be illustrated with reference to the keto-enol tautomerism in both hydrocarbon and azahydrocarbon ring systems. In the hydrocarbon series it will be recalled that although phenol and naphthol exist predominantly in the enol forms, and react as such, both 9-hydroxyanthracene and anthrone can be isolated. This substance reacts in either of its two tautomeric forms depending on the conditions. 5-Hydroxynaphthacene, on the other hand, reacts entirely in the keto form.

In the azahydrocarbon series, the equilibrium is shifted towards the keto forms by virtue of the additional tautomerism of the lactam-lactim type. Even the 2- and 4-hydroxy derivatives of pyridine sometimes react in the keto (or lactam) forms. 2-Hydroxypyridine (XXVIII) reacts with phosphorus pentachloride to give 2-chloropyridine and with diazomethane to give 2-methoxypyridine (XXX); but with methyl iodide it gives N-methyl-2-pyridone (XXXI).

Reaction in the lactam form naturally implies a quinonoid-like structure for the remainder of the molecule. Such a structure cannot be assumed in the case of 3-hydroxypyridine, which therefore behaves entirely as an enol, or lactim.

The 2- and 4-hydroxyquinolines behave in a similar fashion to the 2- and 4-hydroxypyridines. The spectrophotometric evidence suggests that these compounds are predominantly ketonic (or lactamic) in nature. This seems to be the case with all hydroxyquinolines and *iso*quinolines in which the hydroxy group occupies the α- or γ-position, but all other hydroxy derivatives behave as enols.*

As the series of linear azahydrocarbons is ascended the keto form becomes even more stable. 5-Hydroxyacridine, or acridone, behaves entirely as a lactam. Other hydroxyacridines have also been investigated,† and it seems that the amount of each tautomer present in solution depends on the dielectric constant of the solvent. Thus 4-hydroxyacridine (XXXII) is a yellow solid which dissolves in benzene to give a yellow solution. In 90% aqueous alcohol, however, the solution is green and in 20% aqueous alcohol it is blue. These differences have been ascribed to the different quantities of the lactam form present.

(XXXII)   (XXXIII)

2-Hydroxyacridine was likewise shown to exist both in lactam (XXXV) and lactim (XXXIV) forms depending on the solvent. However, 1- and 3-hydroxyacridines, for which no such tautomeric pairs can be written, did not show any essential change with the dielectric constant of the solvent.

(XXXIV)   (XXXV)

Similar lactam-lactim tautomerism involving the isolation of one nitrogen atom from the conjugated system has been observed with 2-phenazinol (XXXVI)–(XXXVII), and in other cases.‡

* Ewing and Steck, J. Amer. Chem. Soc. 1946, **68**, 2181.
† Albert and Short, J. Chem. Soc. 1945, p. 760.
‡ Badger, Pearce and Pettit, ibid. 1951, p. 3204.

SOME PROPERTIES OF AROMATIC COMPOUNDS

Amino and alkyl derivatives of azahydrocarbons, especially the former, behave in exactly the same way, and confirm this tendency of the nitrogen to form an imino group.

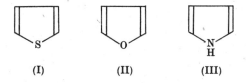

(XXXVI) (XXXVII)

HETEROCYCLIC AROMATIC COMPOUNDS CONTAINING FIVE-MEMBERED RINGS

In addition to the heterocyclic compounds containing six-membered rings, certain heterocyclic compounds containing five-membered rings also seem to possess aromatic character by virtue of their resonance energy and general properties. These compounds include thiophen (I), furan (II), pyrrole (III), and related compounds containing more than one hetero atom.

(I) (II) (III)

In many ways thiophen closely resembles benzene, except that it is much more reactive. Its physical properties are similar to those of benzene, and many thiophen derivatives closely resemble the corresponding derivatives of benzene. It undergoes a wide variety of substitution reactions, all with great ease, and under mild conditions.* Hypochlorous acid readily chlorinates thiophen, even when dissolved in benzene.† This pronounced reactivity has led to the description of thiophen (and some other substances) as 'super-aromatic' compounds. Reaction by addition occurs much more readily than with benzene, and thiophen

* Hartough, *J. Chem. Educ.* 1950, **27**, 500.
† Ardagh and Bowman, *J. Soc. Chem. Ind., Lond.*, 1935, **54**, 267 T.

derivatives are also much less stable than the corresponding derivatives of benzene. Cold concentrated nitric acid converts di- or tetrabromothiophen into dibromomaleic acid;[*] and both 2- and 3-hydroxythiophens are unstable and decompose on standing a few days even at ice temperature.

Furan, the oxygen analogue of thiophen, closely resembles the latter compound but is even more reactive. The nitration of furan, for example, occurs more readily than that of thiophen, which in turn occurs more readily than that of benzene.[†] Furthermore, 2-furyl-2-thienylketone (IV) gives 5-nitro-2-furyl-2-thienylketone (V) on nitration,[‡] also illustrating the enhanced reactivity of the furan nucleus.

(IV) → (V)

The furan nucleus is more reactive than benzene. This is illustrated by the fact that furylphenylketone is acylated and nitrated preferentially in the furyl nucleus.[§] Many other substitution reactions have also been observed. On the other hand, addition to furan also occurs very readily in certain circumstances, and, like thiophen, the ring system is relatively unstable. Furfural, for example, is readily oxidized to succinic acid with monopersulphuric acid; and the furan ring system is easily opened on hydrogenation.

Pyrrole is relatively unstable. It is rapidly destroyed by strong acids such as nitric or sulphuric, and is not a base in the usual sense of the word. As a matter of fact, it is a reasonably strong *acid*, and salts of some substituted pyrroles with bases are stable in aqueous solution.[ǁ] Like thiophen and furan, the pyrrole ring system is not very resistant to oxidation. Pyrrole itself gives maleimide, and many derivatives are oxidized in the same

---

[*] Angeli and Ciamician, *Ber. dtsch. chem. Ges.* 1891, **24**, 74, 1347.
[†] Rinkes, *Rec. trav. chim. Pays-Bas*, 1933, **52**, 1052.
[‡] Gilman and Young, *J. Amer. Chem. Soc.* 1934, **56**, 464.
[§] Ibid.
[ǁ] Fischer and Ernst, *Liebigs Ann.* 1926, **447**, 155.

SOME PROPERTIES OF AROMATIC COMPOUNDS

manner, with chromic oxide, nitric acid, lead dioxide, or even with bromine water (to give dibromomaleimide).*

Pyrrole and its derivatives are very reactive. They couple with aromatic diazonium compounds without difficulty, and halogenation also proceeds very readily. Tetraiodopyrrole is formed by treatment of pyrrole with iodine in potassium iodide.† The Gattermann synthesis, introducing an aldehyde group, also proceeds readily, as does the Friedel-Crafts reaction. The reactivity of pyrrole is so great that the Friedel-Crafts reaction sometimes proceeds even without a catalyst. 2-Acetylpyrrole, for example, has been prepared by heating pyrrole with acetic anhydride.‡

The aromaticity of pyrrole (VI) is easily explained if it is assumed that each carbon atom contributes one π-electron to the aromatic sextet, and the nitrogen atom contributes its 'lone pair'. This also explains the very weak basic character of pyrrole. As there is no lone pair available for salt formation, pyrrole can only form a salt at the expense of its aromatic character and resonance energy. On the other hand, the pyrrole anion is more symmetrical than pyrrole itself because the negative charge on the nitrogen reduces the electron affinity of this atom to a value comparable to that of carbon. This explains the acid character of pyrrole.

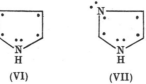

(VI)   (VII)

The aromatic compound imidazole (VII) is also of interest in this connexion because it is a mono-acidic base. This is reasonable enough if the nitrogen in the 1-position contributes its lone pair to the common pool of π-electrons, and the nitrogen in the 3-position contributes only *one* electron, leaving its lone pair available for salt formation.§

* Plancher and Cattadori, *R.C. Accad. Lincei*, 1904, **13**(i), 489; *Brit. Chem. Abstr.* 1904, **86**, 770.
† Ciamician and Silber, *Ber. dtsch. chem. Ges.* 1885, **18**, 1763.
‡ Ciamician and Dennstedt, *Gazz. chim. ital.* 1883, **13**, 455.
§ Turner, *J. Amer. Chem. Soc.* 1949, **71**, 3472.

According to Schomaker and Pauling,* pyrrole has a resonance energy of about 31 kcal./mole, in spite of the fact that the classical structure (X) is the only one which can be written for it in which each carbon atom forms four bonds with adjacent atoms. The most reasonable interpretation of the resonance energy would seem to be that pyrrole is a resonance hybrid of the structures (VIII)–(XII).†

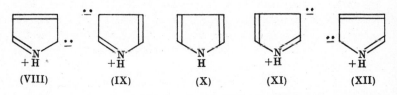

The electronic structure of thiophen (XIII) is evidently very similar to that of pyrrole, the sulphur atom contributing two electrons to the aromatic sextet.

(XIII)

These two electrons appear to resemble the π-electrons in a —CH=CH— group of an aromatic molecule very closely indeed,

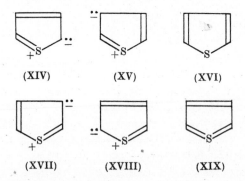

---

* *J. Amer. Chem. Soc.* 1939, **61**, 1769.
† Ingold, *J. Chem. Soc.* 1933, p. 1120; Pauling and Sherman, *J. Chem. Phys.* 1933, **1**, 606.

SOME PROPERTIES OF AROMATIC COMPOUNDS

and according to Longuet-Higgins* this similarity is accounted for by the *pd* hybridization of the sulphur orbitals. The resonance energy of thiophen appears to be of the order of 31 kcal./mole,† and it is reasonable to conclude that thiophen is also a resonance hybrid. The structures (XIV)–(XIX) are similar to those for pyrrole, with the addition of a further structure using five M orbits of the sulphur atom. Schomaker and Pauling assigned the following relative weights to the different types of structure: type (XVI), 70%; types (XIV) and (XVIII), 20%; types (XV), (XVII) and (XIX), 10%.

(XX)

Furan, the oxygen analogue of thiophen, also seems to have a similar electronic structure (XX), the oxygen contributing two electrons to the aromatic sextet. The resonance energy of furan is about 23 kcal./mole,† less than that of either pyrrole or thiophen; and a similar series of resonance structures (XXI)–(XXV) can be written.‡

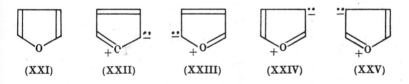

(XXI)   (XXII)   (XXIII)   (XXIV)   (XXV)

It is interesting to note that the various *six*-membered heterocyclic compounds containing one or more of the hetero atoms, S, O or NH, are *not* aromatic. Dioxadiene (XXVI), for example, is not aromatic. It does not undergo the usual reactions of aromatic compounds, and behaves as an unsaturated aliphatic ether comparable to vinyl ether.§ Similarly, pyran (XXVII), thiapyran, *para*-oxazine, *para*-thiazine and related compounds

---

* *Trans. Faraday Soc.* 1949, **45**, 173. † *J. Amer. Chem. Soc.* 1939, **61**, 1769.
‡ See Wheland, *The Theory of Resonance*, p. 62, John Wiley, New York, 1944.
§ Lappin and Summerbell, *J. Org. Chem.* 1948, **13**, 671.

are also aliphatic in character. The six-membered pyrylium *ion* (XXVIII), on the other hand, appears to be aromatic. Each carbon atom must contribute one π-electron, and the oxygen atom, having a positive charge, can also contribute one electron to the aromatic sextet.

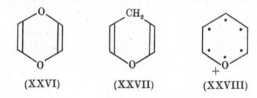

(XXVI)   (XXVII)   (XXVIII)

Apart from pyrrole, furan and thiophen, there is a large number of aromatic heterocyclic compounds which contain two or more hetero atoms in a five-membered ring. These include oxazole (XXIX), isoxazole (XXX), thiazole (XXXI), imidazole (XXXII) and pyrazole (XXXIII). There is a close analogy between the chemistry of thiazole and that of pyridine.*

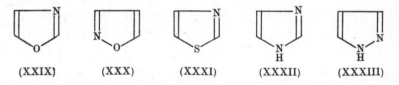

(XXIX)   (XXX)   (XXXI)   (XXXII)   (XXXIII)

In each case an aromatic sextet can be provided by the contribution of one electron from each carbon atom, one from each (basic) nitrogen atom, and two from each oxygen, sulphur or (imino-) nitrogen atom. Many of these ring systems are particularly stable, and resistant to oxidation. The same situation also holds with compounds containing *three* hetero atoms. For

(XXXIV)

* Ochiai and Nagasawa, *Ber. dtsch. chem. Ges.* 1939, **72**, 1470; Wibaut, ibid. 1939, **72**, 1708.

SOME PROPERTIES OF AROMATIC COMPOUNDS

example, on treatment with alkaline permanganate, the triazole (XXXIV) is oxidized at the methyl group, which is eliminated as carbon dioxide; and with acid permanganate, the phenyl group is similarly eliminated, to give the parent ring system.

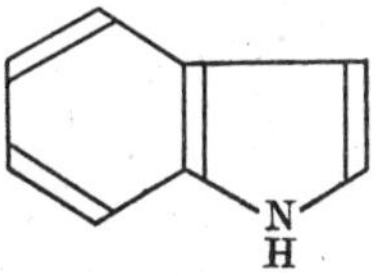

(XXXV) 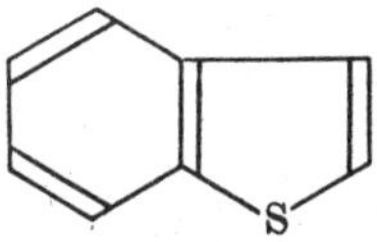   (XXXVI) 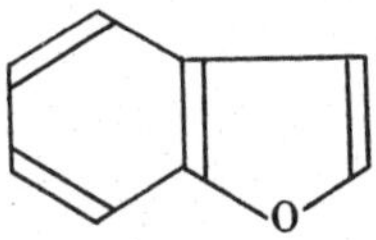   (XXXVII)

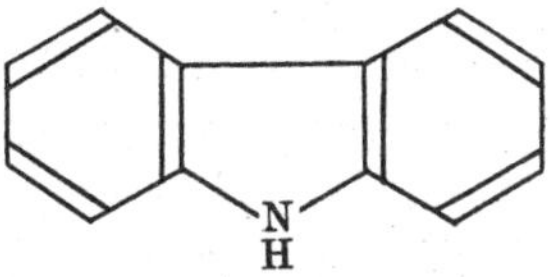

(XXXVIII) 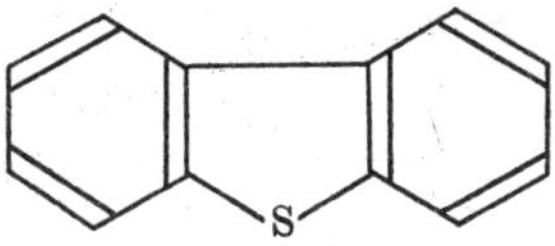   (XXXIX)

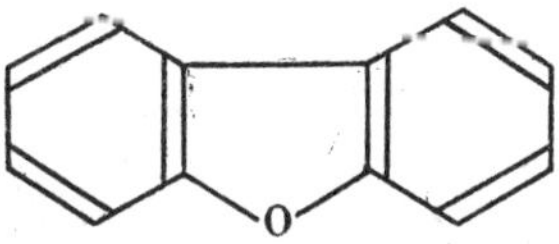

(XL)

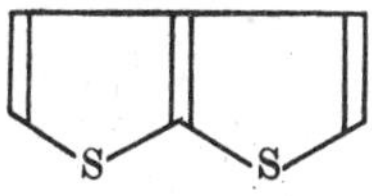

(XLI) 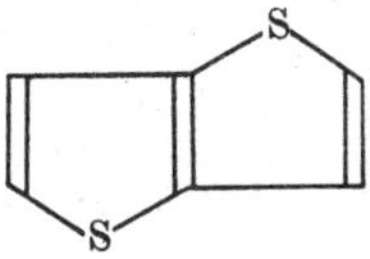   (XLII)

Polycyclic heterocyclic compounds can be built up in the same way as for the carbocyclic analogues. Indole (XXXV), thionaphthen (XXXVI) and benzofuran (XXXVII) are examples of bicyclic compounds, and carbazole (XXXVIII), dibenzothiophen (XXXIX) and dibenzofuran (XL) are examples of tricyclic compounds.

Alternatively, bicyclic and polycyclic compounds can be built up by the fusion of two or more heterocyclic rings, as, for example, in thiophthen (XLI) and thienothiophen (XLII).

In the pyrrole series an additional type of polycyclic aromatic compound can be built up from four pyrrole nuclei, extended conjugation being provided through α-methine groups. These compounds are the porphyrins, which have the basic structure (XLIII).

(XLIII)

(XLIV)

This stable ring system is of wide occurrence in nature, a metal atom being bound to the four nitrogen atoms. It occurs, for example, in haemin (XLIV) and in chlorophyll (XLV). It also occurs in the synthetic pigments known as the phthalocyanins.

SOME PROPERTIES OF AROMATIC COMPOUNDS

Closely related to pyrrole is indolizine (XLVI), which is also aromatic in character.* It undergoes a number of substitution reactions (including diazo coupling) and resembles pyrrole in giving the pine-chip reaction and in being very feebly basic. Each carbon atom must provide one electron, and the nitrogen must provide its lone pair, making a total of ten mobile electrons, as in naphthalene and azulene.

(XLV)

(XLVI)   (XLVII)

For similar reasons pyrimidazole (XLVII) must also be aromatic.

* Borrows and Holland, *Chem. Rev.* 1948, **42**, 611.

THE STRENGTH OF AROMATIC HETEROCYCLIC BASES

The most fundamental difference between the aromatic hydrocarbons and the nitrogen-containing heterocyclic aromatic compounds is that the latter are mostly weak bases, and dissolve in acids with the formation of salts.

According to the Lowry-Brönsted system,* a 'base' is a compound which acts as an acceptor of hydrogen nuclei. The reaction is

$$NH_3 + H^+ \rightleftharpoons NH_4^+,$$

or, generally, $\quad B + H^+ \rightleftharpoons BH^+.$

The acid dissociation constant, K_a, of a base is given by the expression

$$K_a = \frac{[B][H^+]}{[BH^+]}.$$

In practice it is usual to use pK_a, where $pK_a = -\log K_a$. The K_a for ammonia is found experimentally to be $5 \cdot 5 \times 10^{-10}$, so that the pK_a for this base is $9 \cdot 2$.

This conception has replaced the older idea of a base as a generator of hydroxyl ions. If a base is sufficiently strong, it may not merely accept the free hydrogen nuclei which have been driven away from the molecule of an acid, but may also remove them from compounds such as water in which they are only weakly held. By the older method, the basic dissociation constant K_b is derived from the expression

$$K_b = \frac{[OH^-][NH_4^+]}{[NH_4OH]}.$$

The pK_b value for ammonia is $4 \cdot 8$. If the pK_b value for a base is subtracted from the ionic product of water (14 at 25°), the pK_a value is obtained.

* Lowry, *Chem. and Ind.* 1923, **42**, 43; Brönsted, *Rec. trav. chim. Pays-Bas*, 1923, **42**, 718; Bell, *Quart. Rev. Chem. Soc., Lond.*, 1947, **1**, 113; Albert, *Chem. and Ind.* 1947, p. 51.

SOME PROPERTIES OF AROMATIC COMPOUNDS

The pK_a system has the advantage that it is very similar to that of pH. The very weak bases have pK_a 0–4, the weak bases 4–7, the moderately strong 7–10, and the very strong bases, such as sodium hydroxide and guanidine, have higher values.

The aliphatic amines usually have pK_a values between 10 and 11, and the saturated heterocyclic systems have comparable strengths. Piperidine, for example, has pK_a 11·1. On the other hand, the *aromatic* heterocyclic bases are much weaker, pyridine having the value pK_a 5·3. The explanation is probably that, in the pyridinium ion, the positive charge on the nitrogen increases the electron affinity, thereby reducing both the electronic symmetry and the resonance energy. This decrease in resonance energy reduces the stability of the salts and hence the basicity.*

TABLE 3·2. *Strengths of aromatic heterocyclic bases*[†]

Compound	pK_a (in water)	pK_a (in 50% alcohol)
Pyridine	5·23	—
Quinoline	4·94	—
*iso*Quinoline	5·14	—
Acridine	5·60	—
5:6-Benzoquinoline	5·15	3·90
6:7-Benzoquinoline	5·05	3·84
7:8-Benzoquinoline	4·25	3·15
Phenanthridine	—	3·30
3:4-Benzacridine	4·70	4·16
2:3-Benzacridine	—	4·52
1:2-Benzacridine	—	3·45
Pyridazine	2·33	—
Pyrimidine	1·30	—
Pyrazine	0·6	—
Cinnoline	2·70	—
Phthalazine	3·47	—
Quinazoline	3·51	—
Quinoxaline	~0·8	—
Phenazine	1·23±0·10	—
ortho-Phenanthroline	—	4·27
meta-Phenanthroline	—	3·11
para-Phenanthroline	—	3·12

The strengths of numerous aromatic heterocyclic bases have been determined by Albert, Goldacre and Phillips, and the results for the unsubstituted ring systems are given in table 3·2.

---

* Dewar, *The Electronic Theory of Organic Chemistry*, p. 186, Oxford University Press, 1949.
† After Albert, Goldacre and Phillips, *J. Chem. Soc.* 1948, p. 2240.

THE AROMATIC COMPOUNDS

All the aromatic heterocyclic compounds related to pyridine have been shown to be weak bases, with pK_a values of approximately 5·0. The nitrogen 'lone pair' is therefore less available for salt formation in such compounds than is the case with the aliphatic bases. Additional benzene rings have relatively little effect, for quinoline, acridine and 6:7-benzoquinoline all have approximately the same strength as pyridine. On the other hand, the introduction of a second nitrogen atom into a six-membered ring greatly reduces the basic strength. Pyrazine, pyrimidine and pyridazine, for example, are all much weaker bases than pyridine; and cinnoline, quinazoline, phthalazine and quinoxaline are all much weaker than quinoline. Similarly, phenazine is much weaker than acridine. The reason for this is simply that the second nitrogen is electron-attracting. It attracts the lone pair of the first nitrogen, thereby reducing its availability for salt formation.

(I)   (II)
(III)   (IV)

It is of some interest that phenanthridine (I) is a relatively weak base, weaker than pyridine or acridine. *ortho*-Phenanthroline (II), *meta*-phenanthroline (III) and *para*-phenanthroline (IV) are also weak bases.

Pyrrole is a very weak base, pK_a 0·4, and indole is no stronger. In the five-membered series of aromatic heterocyclic bases, however, the insertion of another nitrogen in the ring is base-

SOME PROPERTIES OF AROMATIC COMPOUNDS

strengthening, pyrazole having pK_a 2·53, glyoxaline 7·03, and benzimidazole 5·33. This is because the unshared electrons of only one nitrogen are required for the formation of the aromatic sextet, those of the other nitrogen being free for salt formation.

In experimental work it is frequently useful to know the extent of ionization of a base at a given pH value. In all cases the percentage ionization may be readily determined from the fundamental equation, namely, $K_a = [B][H^+]/[BH^+]$, and the table calculated by Albert[*] is reproduced for reference.

TABLE 3·3. *To find the percentage ionization of a base at any given pH value*

$pH - pK_a$	% base ionized	$pH - pK_a$	% base ionized
−3·0	99·9	3·0	0·1
−2·0	99·0	2·0	1·0
−1·5	97	1·5	3
−1·2	94	1·2	6
−1·0	91	1·0	9
−0·9	89	0·9	11
−0·8	86	0·8	14
−0·7	83	0·7	17
−0·6	80	0·6	20
−0·5	76	0·5	24
−0·4	71	0·4	29
−0·3	67	0·3	33
−0·2	61	0·2	39
−0·1	56	0·1	44
0	50	0	50

MOLECULAR COMPLEXES WITH AROMATIC COMPOUNDS

One of the most striking properties of aromatic compounds, and especially of the aromatic hydrocarbons, is their ability to form a large range of molecular compounds.

These complexes are of various types. The largest group is that formed between aromatic hydrocarbons, amines, etc., on the one hand, and polynitro compounds on the other. Picric acid and *sym.*-trinitrobenzene are probably the most widely used components of the second type, but many other polynitro compounds, including styphnic acid, dinitrobenzoic acid, dinitro-

[*] *Chem. and Ind.* 1947, p. 51.

anthraquinone and trinitrofluorenone, have also received considerable attention. Complexes between naphthalene and well over fifty different nitro compounds have already been described in the literature.*

These complexes between aromatic hydrocarbons and polynitro compounds are usually formed in the ratio 1:1, but this is not always the case. Ratios of 1:2, 2:1 and 3:2 are sometimes observed, and remarkably little change in the structure of either component sometimes suffices to change the ratio. For example, naphthalene forms a complex with picric acid (2:4:6-trinitrophenol) in the ratio 1:1, but that with 2:4:6-trinitrophenetole is in the ratio 1:2. The complex with 2:4:6-trinitro-3-*tert*.-butylphenol is also in the ratio 1:2, but with 1:2:4:6-tetranitrobenzene is in the ratio 3:2.* Similarly, *sym*.-trinitrobenzene forms 1:1 complexes with naphthalene, anthracene, phenanthrene, acenaphthene and chrysene; but that with stilbene is in the ratio 2:1, and with fluorene 3:2. Sudborough† suggested that the molecular ratio is largely dependent on the number of 'independent' aromatic nuclei present in the hydrocarbon moiety. That is, 1:1 complexes are formed if the hydrocarbon (anthracene, chrysene, naphthalene, benzene) has only one 'independent' ring system, but 1:2 complexes are formed if two 'independent' aromatic ring systems are present (stilbene, diphenylethane, etc.). In many cases this classification is valid, but there are so many exceptions that the generalization can have little significance. In this connexion it is worth noting that although the majority of the methylbenzanthracenes form monopicrates, that from 1'-methyl-1:2-benzanthracene is a dipicrate.‡

The complexes between aromatic hydrocarbons and polynitro compounds are all highly coloured, usually yellow, red or purple, and the depth of colour seems to parallel their stability. Benzene itself forms an unstable yellow picrate, but naphthalene picrate is a deep yellow and is relatively stable. Anthracene picrate is red and is quite stable. In general, the complexes prepared from methyl- and amino-anthracenes are more stable and more highly

---

* Elsevier's *Encyclopaedia of Organic Chemistry*, **12B**, 38.
† *J. Chem. Soc.* 1916, **109**, 1339.
‡ Cook and Robinson, ibid. 1938, p. 505.

coloured than those prepared from the corresponding derivatives of benzene or of naphthalene.* Naphthacene does not form a picrate which can be isolated, but solutions of naphthacene and picric acid are red, indicating that complex formation does take place. Possibly the sparing solubility of the hydrocarbon precludes the isolation of the complex, and this factor often seems to be a limiting condition in the formation of these molecular compounds.

Substituents in the hydrocarbon moiety have been shown to have a profound effect on the stability of the resulting complexes. Thus nitrobenzene does not form a complex with polynitro compounds, and halogen substituents also decrease the tendency to form stable complexes. On the other hand, methyl and amino groups increase the stability of the resulting complexes. This is illustrated by the fact that although benzene gives a yellow unstable picrate, the corresponding complex from hexamethylbenzene is red in colour and is stable. Similarly, the picrates of the tetramethyl-naphthalenes are orange-red or red, whereas that from naphthalene itself is yellow.

With polycyclic compounds, such as 1:2-benzanthracene, it is possible to observe differences in the stability and colour of the complexes with the position of the substituent. The complexes formed from 9:10-dimethyl-1:2-benzanthracene and its further alkylated derivatives, for example, are dark red or purple in colour, but other dimethyl and monomethyl derivatives of benzanthracene form less stable picrates, which are also lighter in colour.

The stability of the complex also seems to be governed by the number of electron-attracting groups in the polynitro compound. Nitrobenzene does not form complexes with aromatic hydrocarbons, but *meta*-dinitrobenzene and *sym*.-trinitrobenzene do so, and so does 1-nitro-3:5-dicarbomethoxybenzene (I). In general, Bennett and Wain[†] found that complex formation is favoured by the presence of carbomethoxy, cyano, and sulphonyl chloride substituents in the nitro compound, or by an increase in the number of nitro groups. Methyl substituents in the polynitro compound were found to have the opposite effect.

* Cadre and Sudborough, ibid. 1916, **109**, 1349.
† *J. Chem. Soc.* 1936, p. 1108.

Complex formation therefore seems to be favoured by the presence of electron-releasing groups in the first component and by electron-attracting groups in the nitro compound.

(I) (II) (III)

It might be asked whether the nitro group itself is required, or whether any concentration of electron-attracting substituents would suffice. This seems to be the case, for certain anhydrides form complexes which seem to be of the same type as the hydrocarbon-polynitro compound group. Tetrachlorophthalic anhydride (II), tetrabromophthalic anhydride and mellitic trianhydride (III) are typical of the anhydrides which form complexes with aromatic hydrocarbons. Tetrachlorophthalic anhydride forms yellow complexes with naphthalene, phenanthrene and anthracene. Mellitic trianhydride forms an orange-red complex with naphthalene, and even forms complexes with such 'deactivated' aromatic compounds as pyridine.[*] Similarly, Bennett and Wain[†] found that *sym.*-tricyanobenzene and trimesic acid trichloride both form highly coloured complexes with aromatic amines. The complexes formed between quinols, phenolic ethers and aromatic hydrocarbons on the one hand and quinones (such as chloranil) on the other also seem to be of the same general type.[‡] The deep green complex between benzoquinone and hydroquinone, viz. quinhydrone, is a special example.

Many suggestions have been put forward as to the precise nature of the union between the aromatic compounds and the polynitro compounds.

It has often been suggested that covalent bonds are formed

---

[*] Mustafin, *J. Gen. Chem., Moscow*, 1947, **17**, 560.
[†] *J. Chem. Soc.* 1936, p. 1108.
[‡] Pfeiffer, *Liebigs Ann.* 1914, **404**, 1.

between the two components,* but it now seems to be established that the distances between the component molecules are too great for ordinary covalent bond formation.† In this connexion the hypothesis of Brackmann‡ is interesting. Brackmann ascribes complex formation to 'complex resonance' and postulates a type of covalent bonding the character of which lies in between that of a single bond and no bond at all.

Pfeiffer§ held that complex formation is due to the mutual saturation of 'residual valencies'. Briegleb,‖ however, regarded complexes as 'polarization aggregates' maintained by electrostatic forces between polar molecules (nitro groups) and electric dipoles induced in polarizable hydrocarbons and other aromatic compounds,¶ and this hypothesis has received considerable support.

More recently, it has been suggested by Weiss** that the complex molecule is essentially ionic in character, being formed from the two components by an electron transfer from the aromatic component (donor, A) to the polynitro compound (acceptor, B), according to the net reaction $A + B \rightleftharpoons [A]^+[B]^-$.

In other words, the complex between naphthalene and *sym.*-trinitrobenzene is formed by the transfer of an electron from naphthalene to *sym.*-trinitrobenzene. Similarly, the complex (IV) between quinone and hydroquinone is formed by the transfer of an electron from the hydroquinone to the quinone. Again, the formation of phenazhydrins (V) and related compounds†† is ascribed to the transfer of an electron from the dihydro compound (e.g. dihydrophenazine) to the heterocyclic aromatic compound (phenazine).

This theory has much to commend it. It offers a reasonable explanation for the colour of the complexes, and it explains the effects of substituents both on the donor and acceptor components.

* Sudborough, *J. Chem. Soc.* 1901, **79**, 522; Moore, Shepherd and Goodall, ibid. 1931, p. 1447.
† Powell and Huse, *Nature, Lond.*, 1939, **144**, 77.
‡ *Rec. trav. chim. Pays-Bas*, 1949, **68**, 147.
§ *Organische Molekülverbindungen*, Enke, Stuttgart, 1927.
‖ *Zwischenmolekulare Kräfte und Molekülstruktur*, Stuttgart, 1937.
¶ See also Gibson and Loeffler, *J. Amer. Chem. Soc.* 1940, **62**, 1324.
** *J. Chem. Soc.* 1942, p. 245; 1943, p. 462; 1944, p. 464.
†† Clemo and McIlwain, *J. Chem. Soc.* 1934, p. 1991; Blout and Corley, *J. Amer. Chem. Soc.* 1947, **69**, 763; Badger, Seidler and Thomson, *J. Chem. Soc.* 1951, p. 3207.

No considerable measure of agreement has yet been reached on these theories,* but a few experimental observations appear to be especially noteworthy.

Weiss† found that naphthalene-picric acid and naphthalene-trinitrobenzene complexes have a small but measurable conductivity when dissolved in sulphur dioxide, supporting the ionic theory.

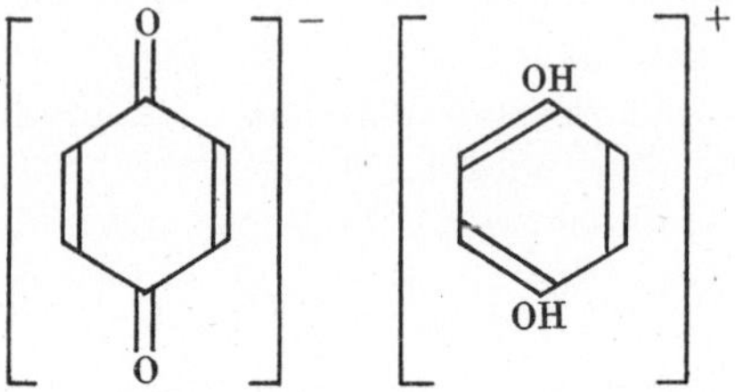

(IV)

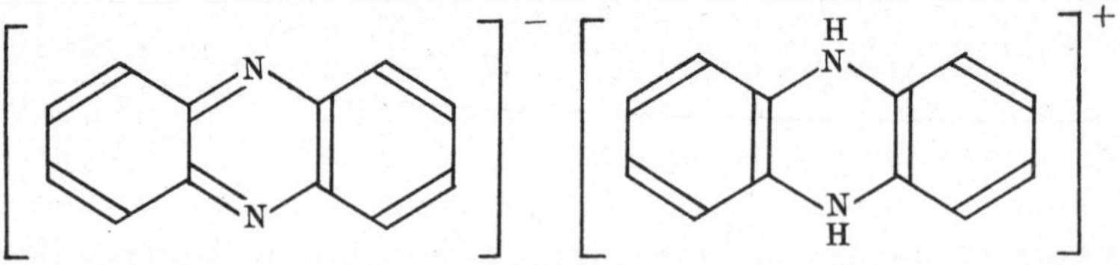

(V)

Secondly, Baddar and Mikhail‡ found that the magnetic susceptibility of naphthalene picrate does not differ widely from the additive value for naphthalene and picric acid. This seems to support the assumption that the components are linked together by weak electrostatic forces.

Thirdly, although X-ray crystallographic investigations of complexes have not, for various reasons, provided much information of value, Abrahams and Robertson§ have obtained some interesting results with *para*-nitroaniline. In this structure, in addition to rather weak hydrogen bridges connecting the oxygen atoms of the nitro group to the amine groups of adjoining molecules, there is a very close approach (2·7–3·0 Å.) between

---

* See Chemical Society Discussions, *Chem. and Ind.* 1938, 57, 512; 1942, p. 453.
† *J. Chem. Soc.* 1942, p. 245; 1943, p. 462; 1944, p. 464.
‡ Ibid. 1949, p. 2927. § *Acta Cryst.* 1948, 1, 252.

SOME PROPERTIES OF AROMATIC COMPOUNDS

one of the oxygen atoms of the nitro group and three of the aromatic carbon atoms of an adjoining molecule. Abrahams and Robertson suggest that this is an intermolecular attraction of a new type, which may also be present in molecular complexes of the hydrocarbon-polynitro compound type.

In addition to the complexes of the above type, many aromatic compounds form complexes with a great variety of *inorganic* compounds. The latter are usually polychloro compounds, the best known being aluminium chloride, antimony trichloride, antimony pentachloride and stannic chloride. Some of these complexes are colourless, but many are highly coloured. Naphthacene forms a green complex with antimony pentachloride in the ratio of 2:1, and with stannic chloride in the ratio 1:2. Highly coloured molecular compounds with antimony pentachloride and (in some cases) stannic chloride have also been obtained with such diverse hydrocarbons as 1:2-benzanthracene, 9:9'-dianthryl, 1:12-benzoperylene and 2:3:10:11-dibenzoperylene.*

It is surprising that complexes of this type have received so little detailed attention, especially as the aromatic hydrocarbon-aluminium chloride complexes are of such practical importance.[†] Brown, Pearsall and Eddy[‡] have, however, studied the interaction of aluminium chloride and toluene. No complex formation was observed at $-80°$, but on the addition of hydrogen chloride the aluminium chloride dissolved, forming a brilliant green solution. One mole of hydrogen chloride was absorbed per mole of aluminium chloride, and although the free acid, $HAlCl_4$, does not exist independently, it seems that it is this acid which forms the complex with the very weakly basic aromatic compound. In other words, the complex is probably ionic in type, a proton being donated to the aromatic component, as in (VI).

Aromatic hydrocarbons dissolve in anhydrous hydrogen fluoride[§] and in concentrated sulphuric acid (but paraffins do not in general dissolve) to give highly coloured solutions which are probably of the same type.

---

* Brass and Fanta, *Ber. dtsch. chem. Ges.* 1936, **69**, 1; Brass and Tengler, ibid. 1931, **64**, 1650, 1654.

† Thomas, *Anhydrous Aluminium Chloride in Organic Chemistry*, Reinhold, New York, 1941.

‡ *J. Amer. Chem. Soc.* 1950, **72**, 5347.

§ Klatt, *Z. anorg. Chem.* 1937, **234**, 189.

86 THE AROMATIC COMPOUNDS

In addition to these two main classes of molecular compounds some examples of the 'clathrate' type have also been found. Benzene itself forms a pale violet crystalline complex with nickel cyanide and ammonia[*] which has been shown by X-ray diffraction methods to consist of sheets of polymerized nickel cyanide

$$\left[\begin{array}{c} CH_3 \\ \bigcirc_+ \\ H\ H \end{array} \xleftrightarrow{\text{(etc.)}} \begin{array}{c} CH_3 \\ \bigcirc_+ \\ H\ H \end{array} \right] \left[AlCl_4^- \right]$$

(VI)

with two NH_3 groups on alternate nickel atoms, and a molecule of benzene imprisoned in this 'cage',[†] as in fig. 3·2.

Homologues of benzene do not form similar compounds, presumably because their molecular dimensions are too large for them to fit into the cage. As saturated compounds do not form complexes either, it is possible to devise a method for the complete purification of benzene which is based on the formation of this clathrate complex.[‡]

Thiophen, furan and pyrrole form similar complexes, but aniline and phenol appear to be the only simple derivatives of benzene which do so.

Other examples of complexes of the clathrate type are those from hydroquinone and sulphur dioxide[§] and those from hydroquinone and the rare gases.[‖]

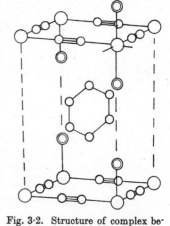

Fig. 3·2. Structure of complex between benzene, nickel cyanide and ammonia. (From Powell and Rayner, *Nature*, Lond., 1949, **163**, 566.)

---

[*] Hofmann and Höchtlen, *Ber. dtsch. chem. Ges.* 1903, **36**, 1149.
[†] Powell and Rayner, *Nature*, Lond., 1949, **163**, 566.
[‡] Evans, Ormrod, Goalby and Staveley, *J. Chem. Soc.* 1950, p. 3346.
[§] Palin and Powell, ibid. 1947, p. 208; 1948, pp. 571, 815.
[‖] Powell, ibid. 1950, pp. 298, 300, 468.

SOME PROPERTIES OF AROMATIC COMPOUNDS

Finally, brief mention may be made of the fact that evidence is accumulating that the halogens form complexes with aromatic compounds.* Silver ions also seem to form complexes with several aromatic hydrocarbons.†

OXIDATION-REDUCTION POTENTIALS OF QUINONES

It is of some interest to inquire into the reason for the pronounced tendency of many aromatic compounds to give addition rather than substitution products, and in this connexion the interpretation of the oxidation-reduction potentials of the corresponding quinones is of importance.‡

All aromatic quinones have oxidation-reduction or redox potentials, E^0, characteristic of the reaction (I) → (II), which are measures of their 'oxidizing power'.

The value of E^0 for *para*-benzoquinone is 0·71 V., and that for α-naphthaquinone, 0·48 V. In other words, α-naphthaquinone has less tendency than *para*-benzoquinone to undergo reduction; it is more stable than benzoquinone, and is less prone to change to the aromatic structure. This is explained by the fact that one of the ethylenic type double bonds in benzoquinone,

---

* Benesi and Hildebrand, *J. Amer. Chem. Soc.* 1949, **71**, 2703; Cromwell and Scott, ibid. 1950, **72**, 3825; Mulliken, ibid. 1950, **72**, 600; Keefer and Andrews, ibid. 1950, **72**, 4677.
† Andrews and Keefer, ibid. 1949, **71**, 3644; 1950, **72**, 5034.
‡ Clar, *Ber. dtsch. chem. Ges.* 1940, **73**, 104; *Aromatische Kohlenwasserstoffe*, 2nd ed., Springer-Verlag, Berlin, 1952; Fieser in Gilman's *Organic Chemistry*, **1**, 117; Branch and Calvin, *The Theory of Organic Chemistry*, p. 303, Prentice-Hall, New York, 1941; Waters, *J. Chem. Soc.* 1948, p. 727.

which largely contribute to the reactivity of the quinone groups, is, in naphthaquinone, incorporated in a benzenoid ring and rendered comparatively inert. In anthraquinone, both double bonds are incorporated in benzenoid rings and the potential is correspondingly smaller, namely, 0·15 V. The higher members of this series have even less reactive quinones: naphthacenequinone is converted into a 'vat' (i.e. an alkaline solution of the quinol) only with difficulty; and pentacenequinone does not form a vat.

Dihydro compounds have the same nuclear bond systems as the corresponding quinones, and the redox potentials may therefore be taken as measures of the stability of the dihydrides. The enhanced stability of 1:4-dihydronaphthalene as compared with 1:4-dihydrobenzene is therefore understandable. Similarly, the dihydrides of anthracene, naphthacene, pentacene, etc., are even more stable, the stability increasing with the number of linear benzene rings.

The same argument can be applied to the *ortho*-quinones. Phenanthraquinone has a lower redox potential than β-naphthaquinone, and this in turn is lower than the potential for *ortho*-benzoquinone. That is, the tendency for these quinones to undergo reduction increases in the order phenanthraquinone, β-naphthaquinone and *ortho*-benzoquinone. Dihydrophenanthrene is therefore more stable than 1:2-dihydronaphthalene, and this in turn is more stable than 1:2-dihydrobenzene. In other words, the tendency towards *addition* rather than *substitution* reactions decreases in the order: phenanthrene, naphthalene and benzene. The latter has very little tendency to form addition compounds.

The same explanation may again be offered, namely, that the reactivity of the quinone is largely determined by the adjacent double bonds. In β-naphthaquinone one of the double bonds is incorporated in a benzenoid ring and rendered largely inactive, and in phenanthraquinone both double bonds are essentially benzenoid. The way in which the redox potentials of *para*-benzoquinone and *ortho*-benzoquinone are affected by the fusion of further benzene or naphthalene rings is indicated in tables 3·4 and 3·5.

The determination of the redox potentials has also been extended to some heterocyclic quinones, and it is therefore

SOME PROPERTIES OF AROMATIC COMPOUNDS

TABLE 3·4. *Redox potentials of* para-*quinones**

Quinone	Redox potential		Quinone	Redox potential	
	E^0 (experimental value) (volts)	E^0 (corrected value) (volts)		E^0 (experimental value) (volts)	E^0 (corrected value) (volts)
(benzoquinone)	0·71	0·71	(dibenz anthraquinone)	0·27	0·42
(phenanthrenequinone)	0·53	0·61	(benzanthraquinone)	0·23	0·36
(naphthoquinone)	0·48	0·53	(anthraquinone)	0·15	0·25
(naphthacenequinone)	0·40	0·45			

* After Branch and Calvin, *The Theory of Organic Chemistry*, see p. 312.

TABLE 3·5. *Redox potentials of* ortho-*quinones**

Quinone	Redox potential		Quinone	Redox potential	
	E^0 (experimental value) (volts)	E^0 (corrected value) (volts)		E^0 (experimental value) (volts)	E^0 (corrected value) (volts)
o-benzoquinone	0·81	0·71	anthracene-1,2-quinone	0·49	0·42
phenanthrene-3,4-quinone	0·66	0·59	chrysene-quinone	0·47	0·45
phenanthrene-1,2-quinone	0·62	0·57	phenanthrene-9,10-quinone	0·46	0·41
1,2-naphthoquinone	0·58	0·51	naphthacene-quinone	0·43	0·38

* After Branch and Calvin, *The Theory of Organic Chemistry*, see p. 312.

possible to determine the effect of certain hetero atoms on the stability of the dihydrides.

4:7-Thionaphthenquinone (VI), 5:8-quinolinequinone (IV) and 5:8-*iso*quinolinequinone (V) are all very similar to α-naphthaquinone (III), and the redox potentials have all been determined.*

The potentials of all three heterocyclic quinones are somewhat higher than that for α-naphthaquinone, indicating increased reactivity of the quinones, and reduced stability of the corresponding dihydrides as compared with 1:4-dihydronaphthalene. Similarly, thiophanthrenquinone (VIII) has a higher potential than its analogue, anthraquinone (VII), again indicating reduced stability of the dihydro heterocyclic system as compared with 9:10-dihydroanthracene.†

* Fieser and Kennelley, *J. Amer. Chem. Soc.* 1935, **57**, 1611; Fieser and Martin, ibid. 1935, **57**, 1840.
† Fieser and Ames, ibid. 1927, **49**, 2604.

Some other heterocyclic analogues of anthraquinone have also been investigated. The pyrazole analogues (IX), (X) were found to have almost the same potentials as anthraquinone itself, but the triazoles (XI), (XII) gave higher values. The potential of the unsubstituted triazole analogue (XII) was found to differ considerably from that of the methyl derivative (XI), and a different bond structure was therefore suggested.*

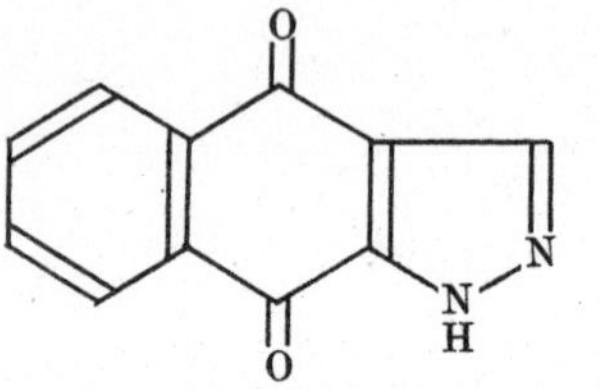

(IX) 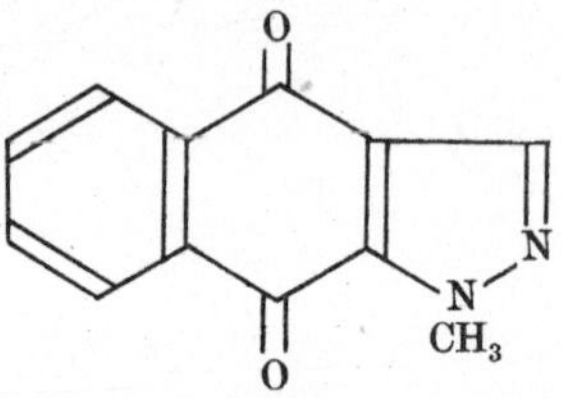   (X)

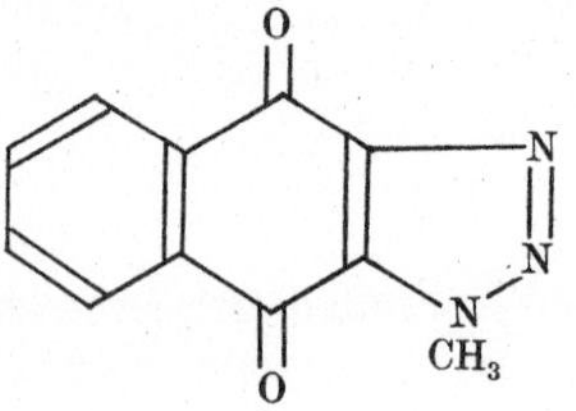

(XI) 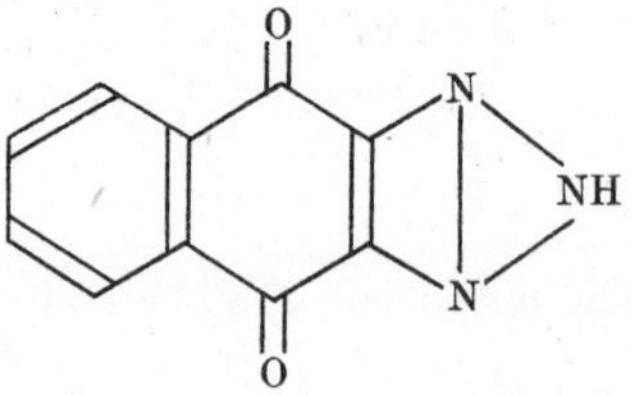   (XII)

In the *ortho*-quinone series, the potential of α-naphthaquinolinequinone (XIII) in 0·1 N-HCl was found to be 0·112 V. higher than that of 9:10-phenanthraquinone (XIV), but the variation of the potential of some nitrogen-containing quinones with increasing acidity of the solution indicates that the difference between the potentials would be less in a solution of such acidity that no dissociation could occur. Naphthatriazole-4:5-quinone (XV) also has a higher potential than phenanthraquinone.†

The redox potential of 2:3-thionaphthenquinone (XVI) is of special interest, being much lower than that of β-naphthaquinone (XVII).‡ This is indicative of the fact that the thiophen ring is much more prone to pass into the quinonoid or dihydride

---

* Fieser and Martin, *J. Amer. Chem. Soc.* 1935, **57**, 1844.
† Fieser and Ames, ibid. 1927, **49**, 2604.
‡ Ibid.; Conant and Fieser, ibid. 1924, **46**, 1858.

condition than is benzene. That is, 2:3-dihydrothionaphthen is more stable than 1:2-dihydronaphthalene.

In a critical survey of the properties of quinones, Branch and Calvin* have pointed out that steric repulsion between a phenolic hydrogen and an adjacent hydrogen atom may lead to a redox potential which is lower than would otherwise be the case. They

(XIII) (XIV) (XV)

(XVI) (XVII)

(XVIII) (XIX)

estimate the steric repulsion correction for the system (XVIII) to be about 0·025 V., and for the system (XIX) to be about 0·050 V. They therefore suggest the use of 'corrected' redox potentials. These corrected values are obtained by adding the above corrections to the values determined experimentally.

Similarly, in *ortho*-quinones the hydroquinone is, to some extent, stabilized by hydrogen bond formation between the

* *The Theory of Organic Chemistry*, pp. 303 et seq.

hydroxy groups, as in (XX). In order to compare the redox potentials of *ortho-* and *para-*quinones, therefore, Branch and Calvin subtract 0·100 V. (the difference between *ortho-* and *para-*benzoquinone) from the potential of each *ortho-*quinone.

(XX)

When these corrections are made, there is a linear relationship between the redox potential and the ratio

$$\frac{N_H - N_Q}{N_H + N_Q},$$

where N_H and N_Q are the numbers of possible valence-bond forms of the hydroquinone and the quinone respectively.

Carter* has given the above relationship between the corrected potential of a quinone and the number of structures in the hydroquinone (N_H) and quinone (N_Q) in the logarithmic form

$$E^0 \text{ (corr.)} = 0·26 + 0·65 \ln \frac{N_H}{N_Q},$$

and from this equation he has derived another, from which the resonance energy of aromatic hydrocarbons can be calculated. This second equation may be expressed in the form

$$R/\beta = 0·6D + 1·5 \ln N - 1,$$

where R is the resonance energy, D is the number of 'double bonds', N is the number of Kekulé structures, and β is the resonance integral.

The essentially empirical nature of this equation must be emphasized. Nevertheless, taking β to be 20·0 kcal., resonance energies may be derived (table 3·6) which are in good agreement with those obtained by other methods.

* *Trans. Faraday Soc.* 1949, **45**, 597.

TABLE 3·6. *Calculated resonance energies of polycyclic hydrocarbons*

Hydrocarbon	D	N	Resonance energy (kcal./mole)
Naphthalene	5	3	$20 \times 3·65 = 73$
Anthracene	7	4	$5·28 = 105·6$
Phenanthrene	7	5	$5·61 = 112·2$
Naphthacene	9	5	$6·81 = 136·2$
Chrysene	9	8	$7·52 = 150·4$
Pyrene	8	6	$6·49 = 129·8$
Triphenylene	9	9	$7·70 = 154$
Perylene	10	9	$8·30 = 166$
Pentacene	11	6	$8·29 = 165·8$
Hexacene	13	7	$9·72 = 194·4$
Heptacene	15	8	$11·10 = 222$

The relationship between the redox potential of a quinone and the tendency of the corresponding hydrocarbon to undergo addition has already been emphasized. Furthermore, Clar[*] has pointed out that the quinones corresponding to the very reactive hydrocarbons have very little reactivity, and vice versa. The series benzene, naphthalene, anthracene, naphthacene and pentacene is characterized by a very marked and progressive increase in 'reactivity', and in the tendency to undergo addition rather than substitution reactions, as the number of benzene rings increases. The series of *para*-quinones, *para*-benzoquinone, α-naphthaquinone, anthraquinone, naphthacenequinone and pentacenequinone, on the other hand, is characterized by an equally marked and progressive *decrease* in reactivity as the series is ascended.

There can be little doubt that there is a close relationship between the redox potential of a quinone and the velocity of an addition reaction to the corresponding hydrocarbon. (It must be admitted, however, that strongly polarizing reagents may introduce additional factors, especially in *substituted* hydrocarbons.) There is, for example, a very great similarity between the addition of bromine to the 9:10 bond of phenanthrene (XXI) to give the dibromide (XXII), and the oxidation of phenanthra-

[*] *Ber. dtsch. chem. Ges.* 1940, **73**, 104; *Aromatische Kohlenwasserstoffe*, 2nd ed., Springer-Verlag, Berlin, 1952.

hydroquinone (XXIII) to phenanthraquinone (XXIV).* It is noteworthy that the substituents in the 2- and 3-positions of phenanthrene which were found to decrease the free energy of bromine addition (retard the reaction) are precisely those which were earlier† found to decrease the free energy of oxidation of

(XXI) → (XXII)

(XXIII) → (XXIV)

hydroquinones (increase the potential). Similarly, the substituents which increased the free energy of bromine addition also increased the free energy of oxidation of the hydroquinones. Moreover, although the free-energy change in the bromine addition is much less than in the oxidation, Fieser and Price found that a given substituent has nearly the same *proportionate* influence in each case.

It seems reasonable, therefore, to conclude that there is a correlation between the velocity of an addition to a suitable aromatic double bond and the magnitude of the redox potential of the corresponding *ortho*-quinone. Other evidence bearing on this subject is given in Chapter 5.

* Fieser and Price, *J. Amer. Chem. Soc.* 1936, **58**, 1838.
† Fieser and Fieser, ibid. 1935, **57**, 491.

CHAPTER 4

ADDITION REACTIONS OF AROMATIC COMPOUNDS

BENZENE AND ITS HETEROCYCLIC ANALOGUES

It is often held to be a characteristic of aromatic compounds that they are generally *substituted* by reagents which tend to *add* to ethylenic compounds. Nevertheless, most aromatic compounds do undergo addition reactions under suitable conditions, and this fact is of importance in all studies of the nature of aromaticity.

In the presence of suitable catalysts, benzene readily adds three molecules of hydrogen, and it also adds three molecules of ozone (see Chapter 5). Furthermore, as early as 1825, Michael Faraday obtained a crystalline hexachloro*cyclo*hexane ('benzene hexachloride') by passing chlorine into benzene in sunlight. Although it is not intended to discuss the mechanism of the addition at this stage, it may be mentioned that the reaction is essentially different from the halogenation of benzene in the presence of catalysts such as ferric chloride (Chapter 7). The *photochemical* halogenation of benzene apparently proceeds by a chain reaction involving halogen free-radicals, and hydrogen halide is not eliminated (I)–(IV).

The reaction has often been investigated, and in 1912 van der Linden[*] established the existence of four isomers of hexachloro*cyclo*hexane, termed α, β, γ and δ. A fifth isomer, ε, has since been discovered. Of recent years the reaction has achieved considerable commercial importance, for the γ isomer has potent insecticidal properties.[†] The commercial preparation contains 70 % of the α isomer, 5 % of the β, and about 12 % of the γ, and it also contains small quantities of more highly chlorinated products.[‡]

The unsaturated character of benzene and of other aromatic hydrocarbons has also been demonstrated by their reactions

---
[*] *Ber. dtsch. chem. Ges.* 1912, **45**, 231.
[†] Slade, *Endeavour*, 1945, **4**, 148; *Chem. and Ind.* 1945, p. 314.
[‡] Ramsey and Patterson, *J. Ass. Off. Agric. Chem., Wash.*, 1946, **29**, 337.

with fluorine (usually in the presence of certain catalysts). Benzene gives perfluoro*cyclo*hexane, C_6F_{12}, by substitution and addition, as well as several fragmentation products. Other aromatic substances behave similarly, anthracene yielding $C_{14}F_{24}$.[*] In this connexion it is also of interest that hexachlorobenzene reacts with BrF_3 to give $C_6F_6Br_2Cl_4$ by addition and substitution.[†]

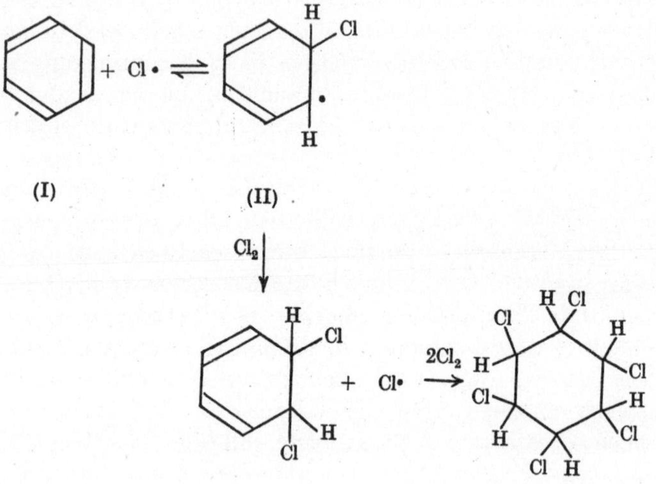

(I)   (II)

(III)   (IV)

Pyridine (V) shows less tendency than benzene to undergo addition reactions, but the addition of *hydrogen* to pyridine proceeds very readily, more readily than to benzene. It is reduced (to piperidine (VI)) not only by hydrogen and nickel[‡] and by hydrogen and platinum,[§] but also chemically, with sodium and ethanol.[||] In this connexion it is also noteworthy

[*] Fukuhara and Bigelow, *J. Amer. Chem. Soc.* 1941, **63**, 2792; Cady, Grosse, Barber, Burger and Sheldon, *Industr. Engng Chem.* 1947, **39**, 290.
[†] For a review of fluorine chemistry see Smith, *Ann. Rep. Chem. Soc.* 1947, **44**, 86.
[‡] Adkins, Kuick, Farlow and Wojcik, *J. Amer. Chem. Soc.* 1934, **56**, 2425; Burrows and King, ibid. 1935, **57**, 1789.
[§] Zelinsky and Borisoff, *Ber. dtsch. chem. Ges.* 1924, **57**, 150; Hamilton and Adams, *J. Amer. Chem. Soc.* 1928, **50**, 2260.
[||] Ladenburg, *Ber. dtsch. chem. Ges.* 1884, **17**, 388; *Liebigs Ann.* 1888, **247**, 51; Marvel and Lazier, *Organic Syntheses*, Coll. vol. 1, 2nd ed. p. 99.

AROMATIC COMPOUNDS—ADDITION REACTIONS 99

that quinoline (VII) is reduced to the 1:2:3:4-tetrahydro derivative (VIII), and that reduction of the benz-ring occurs only under more vigorous conditions.

(V) (VI) (VII) (VIII)

(IX) (X) (XI)

Some of the five-membered heterocyclic analogues of benzene are very reactive and show considerable aliphatic character in that addition products are obtained more easily than with benzene.

Catalytic hydrogenation of thiophen is not practicable, as this substance acts as a catalyst poison, and Raney nickel serves to desulphurize thiophen and its derivatives. Certain chemical methods of reduction also remove the sulphur. This is the case with hydriodic acid at 140°. On the other hand, thiophen-2:5-dicarboxylic acid is reduced to a tetrahydro derivative with a mild reducing agent such as sodium amalgam,* and the reduction of thiophen with sodium and methyl alcohol in liquid ammonia at $-40°$ yields a mixture of 2:3- and 2:5-dihydrothiophen.†

Thiophen also reacts with chlorine to produce both addition and substitution products in proportions dependent on the reaction conditions and the ratio of chlorine to thiophen.‡ Two of the six possible geometrical isomers of 2:3:4:5-tetrachlorothiolane (IX) have been isolated in this reaction, as well as a pentachlorothiolane (X) and a hexachlorothiolane (XI). Indeed,

---

* Ernst, *Ber. dtsch. chem. Ges.* 1886, **19**, 3274.
† Birch and McAllan, *Nature, Lond.*, 1950, **165**, 899.
‡ Coonradt and Hartough, *J. Amer. Chem. Soc.* 1948, **70**, 1158.

in a series of chlorinations under different conditions, Coonradt, Hartough and Johnson[*] have identified *all nine* possible substitution and addition products.

With furan, substitution reactions occur much more readily than with benzene, and addition products are also obtained very much more easily. Furan is hydrogenated over Raney nickel catalyst (100–150°/100–150 atm.), and it is reduced by hydrogen in the presence of palladium. The halogens also add to furan at low temperatures. Clauson-Kaas[†] has obtained satisfactory evidence that bromine adds to furan (XII) by a 1:4 addition to yield 2:5-dibromo-2:5-dihydrofuran (XIII).

(XII)   (XIII)   (XIV)

In acetic acid containing potassium acetate, bromine and furan gave 2:5-diacetoxy-2:5-dihydrofuran (XIV) in good yield, and the same addition product was obtained by the action of lead tetra-acetate on furan.[‡] Furthermore, Hill and Hartshorn[§] obtained a hexabromodihydrofuran (probably (XV)) from tetrabromofuran and bromine. Similarly, Moureu, Dufraisse and Johnson[∥] obtained a tetrabromo addition-compound (XVI) from furylacrylic acid by the action of bromine in chloroform at −15°; and Gilman and Wright[¶] obtained a dibromo addition-compound (XVII) with ethyl furylacrylate.

As a matter of fact, the presence of 1:4 addition products of furan has been demonstrated after interaction with a large number of reagents, including bromocyanogen,[**] oxygen,[††] nitric acid in acetic anhydride,[‡‡] osmium tetroxide and hydrogen peroxide[§§] and with per-acids.[∥∥]

[*] *J. Amer. Chem. Soc.* 1948, **70**, 2564.   [†] *Acta chem. scand.* 1947, **1**, 379.
[‡] Clauson-Kaas, *Kgl. danske vidensk. Selsk., Mat. fys. Medd.*, 1947, **24**, no. 6.
[§] *Ber. dtsch. chem. Ges.* 1885, **18**, 448.
[∥] *Ann. Chim. (Phys.)*, 1927, **7**, 8.   [¶] *J. Amer. Chem. Soc.* 1930, **52**, 3349.
[**] Klopp and Wright, *J. Org. Chem.* 1939, **4**, 142.
[††] Schenck, *Naturwissenschaften*, 1943, **31**, 387; *Ber. dtsch. chem. Ges.* 1944, **77**, 661.
[‡‡] Clauson-Kaas and Fakstorp, *Acta chem. scand.* 1947, **1**, 210.
[§§] *Ibid.* 1947, **1**, 216.
[∥∥] *Ibid.* 1947, **1**, 415.

AROMATIC COMPOUNDS—ADDITION REACTIONS

The third five-membered heterocyclic analogue of benzene is pyrrole, which is a very reactive aromatic compound. Many substitution reactions take place with very great facility, and certain addition reactions also take place very easily. Pyrrole is reduced by a number of chemical methods as well as by catalytic procedures. It is, for example, reduced to pyrrolidine (XX) with red phosphorus and hydriodic acid[*] as well as with hydrogen and platinum[†] and with hydrogen and nickel;[‡] but reduction with zinc dust and acetic or hydrochloric acid gives 2:5-dihydropyrrole (XIX) by 1:4 addition to the 'diene'.[§]

It is also of some interest that triphenylmethyl adds to pyrrole to give 2:5-*bis*-(triphenylmethyl)pyrroline.[‖]

The related aromatic compounds, pyrazole (XXI) and imidazole (XXII), are more resistant to reduction than is pyrrole. Pyrazole is not reduced by sodium and boiling amyl alcohol,[¶] and imidazole is not reduced with zinc and hydrochloric acid, nor by phosphorus and hydriodic acid.[**]

---

[*] Knorr and Rabe, *Ber. dtsch. chem. Ges.* 1901, **34**, 3491.
[†] Putokhin, *J. Soc. phys.-chim. russe*, 1930, **62**, 2216; *Chem. Abstr.* 1931, **25**, 3995.
[‡] Signaigo and Adkins, *J. Amer. Chem. Soc.* 1936, **58**, 1122.
[§] Ciamician, *Ber. dtsch. chem. Ges.* 1904, **37**, 4244.
[‖] Conant and Chow, *J. Amer. Chem. Soc.* 1933, **55**, 3475.
[¶] Buchner, Fritsch, Papendieck and Witter, *Liebigs Ann.* 1893, **273**, 214.
[**] Wyss, *Ber. dtsch. chem. Ges.* 1877, **10**, 1365.

1:2 ADDITION REACTIONS

The reactions of naphthalene are mainly of the substitution type, but, as early as 1833, Laurent* obtained 1:2:3:4-tetrachloro-1:2:3:4-tetrahydronaphthalene ((I), 'naphthalene tetrachloride') by passing chlorine over solid naphthalene at room temperature. Improved conditions for the reaction were devised by Faust and Saame,† and Leeds and Everhart‡ found that sunlight favoured the formation of the addition compound.

Similarly, Orndorff and Moyer§ obtained 1:2:3:4-tetrabromo-1:2:3:4-tetrahydronaphthalene ((II), 'naphthalene tetrabromide') in 3% yield by the bromination of naphthalene. Here again it seems that the reaction must be predominantly photochemical in character, for the yield has been increased to 30% by carrying out the bromination under strong ultra-violet irradiation.‖

The same type of addition compound has also been obtained in the anthracene field. For example, 9:10-dichloroanthracene

* *Ann. chim. (Phys.)*, 1833, **52**, 275; 1835, **59**, 196.
† *Liebigs Ann.* 1871, **160**, 66.
‡ *J. Amer. Chem. Soc.* 1880, **2**, 205.
§ *Amer. Chem. J.* 1897, **19**, 262.
‖ Sampey, Cox and King, *J. Amer. Chem. Soc.* 1949, **71**, 3697.

AROMATIC COMPOUNDS—ADDITION REACTIONS

has been chlorinated to 1:2:3:4:9:10-hexachloro-1:2:3:4-tetrahydroanthracene, and 1:2:3:4:9:10-hexabromo-1:2:3:4-tetrahydroanthracene (III) has been obtained by the addition of bromine to 9:10-dibromoanthracene. It is of some interest that this addition compound has been obtained in two stereoisomeric forms. One form was obtained by exposing 9:10-dibromoanthracene to bromine vapour at ordinary temperatures, without the aid of a solvent.* The other isomer was obtained by Meyer and Zahn† by treating 9:10-dibromoanthracene with bromine in a little chloroform.

(IV)　　　(V)　　　(VI)

In all these cases it seems likely that the mechanism involves the 1:2 addition of chlorine or bromine atoms; but none of these reactions has yet been adequately studied.

The addition of bromine to phenanthrene has, however, received more attention.

The fact that bromine adds to the 9:10 positions of phenanthrene (IV) to give 9:10-dibromo-9:10-dihydrophenanthrene ((V), 'phenanthrene dibromide') has been known since the discovery of the hydrocarbon.‡ The addition product (V) is of moderate stability, but it loses hydrogen bromide on prolonged standing at room temperature, or more rapidly on warming, to give 9-bromophenanthrene (VI). Chlorine adds to phenanthrene in the same way, and 'phenanthrene dichloride' is even more stable than the dibromide.

* Anderson, *Liebigs Ann.* 1862, **122**, 303; Grandmougin, *C.R. Acad. Sci., Paris*, 1921, **173**, 1176.
† *Liebigs Ann.* 1913, **396**, 166.
‡ Fittig and Ostermayer, *Ber. dtsch. chem. Ges.* 1872, **5**, 933; *Liebigs Ann.* 1873, **166**, 361; Graebe, *Ber. dtsch. chem. Ges.* 1872, **5**, 861, 968; *Liebigs Ann.* 1873, **167**, 131.

Using this reaction as a model, it has often been suggested that most aromatic substitution reactions proceed by a process of addition followed by elimination.* Analysis of the kinetics of the phenanthrene-bromine reaction has, however, shown this hypothesis to be erroneous. Moreover, there is now overwhelming evidence in other cases that the reactions of addition and substitution proceed by independent mechanisms.†

The addition of bromine to pure phenanthrene is probably almost entirely photochemical. According to Kharasch, White and Mayo,‡ in the absence of oxygen and other catalytic impurities, there is no detectable addition of bromine to phenanthrene in the dark, in 24 hr. The matter requires further investigation, however, for Sampey, Cox and King§ apparently detected considerable addition under similar experimental conditions.

When light is not rigidly excluded, however, the reaction of bromine and pure phenanthrene in carbon tetrachloride solution at 25° is certainly one of pure addition, no hydrogen bromide being formed even after several days. The reaction proceeds with measurable velocity, and is reversible. Measurements of the position of the equilibrium and the rate of its establishment have been carried out. Certain substances, such as diphenylamine and tetrabromohydroquinone, were found to inhibit the reaction, and Price‖ therefore suggested that it proceeds by a chain mechanism, the actual reagent being bromine atoms. This mechanism is illustrated in formulae (VII)–(IX).

In the presence of certain catalysts, however, the reaction is considerably modified. Small amounts of iodine, aluminium chloride, antimony pentachloride, phosphorus trichloride, phosphorus pentachloride or of stannic chloride, for example, were found to catalyse the formation of hydrogen bromide (and of 9-bromophenanthrene) in appreciable amounts within an hour or two, although in the absence of such catalysts none was formed in several days. These catalysts therefore promote the

---

* See Fieser, 'Theory of structure and reactions of aromatic compounds', in Gilman, *Organic Chemistry*, 2nd ed. p. 117, John Wiley, New York, 1943.
† See, for example, Gillespie and Millen, *Quart. Rev. Chem. Soc.* 1948, **2**, 277.
‡ *J. Org. Chem.* 1938, **2**, 574.
§ *J. Amer. Chem. Soc.* 1949, **71**, 3697.
‖ Ibid. 1936, **58**, 1834.

AROMATIC COMPOUNDS—ADDITION REACTIONS

substitution reaction, and the way in which they do this is clearly of the greatest importance.*

Price[†] was able to show that no hydrogen bromide is evolved from phenanthrene dibromide when it is allowed to stand, in carbon tetrachloride solution, with iodine. On the addition of a little bromine, however, hydrogen bromide was slowly evolved. It seems clear, therefore, that although catalytic amounts of iodine do promote the formation of hydrogen bromide from solutions containing phenanthrene, bromine and phenanthrene dibromide, this hydrogen bromide does not arise directly by decomposition of the phenanthrene dibromide. This evidence shows that the addition-elimination theory of aromatic substitution is erroneous, and it must be concluded that the catalysts promote the substitution of phenanthrene by facilitating the elimination of hydrogen bromide in another way.

(VII)   (VIII)   (IX)

The effect of the usual bromination catalysts is almost certainly to promote the formation of bromine cations, or to facilitate electrophilic reaction by means of the formation of a polarized complex of type $Br^{\delta+} \ldots BrI_2^{\delta-}$ (see also Chapter 7). In this case the reaction would be represented as in (X)–(XIII).[‡]

In other words, both the substitution reaction and the (polar) addition reaction proceed through the common intermediate (XI). A similar intermediate is involved in the substitution of

---

* It is of some interest that these catalysts may either promote or inhibit the rate of the *addition* reaction (Price, *J. Amer. Chem. Soc.* 1936, **58**, 2101). Thus iodine was found to retard the addition reaction, but antimony pentachloride greatly accelerated it.

† Ibid. 1936, **58**, 2101.

‡ Price and Arntzen, ibid. 1938, **60**, 2835.

benzene, and the relative ease with which such an intermediate either loses a proton to give the substitution product, or adds an anion to give the addition product, seems to depend on the double-bond character of the original bond.*

(X)   (XI)   (XII)

(XIII)

1:4 ADDITION REACTIONS

Reactions of this type are relatively common with anthracene (I) and its derivatives, the addition taking place at the reactive *meso* or 9:10 positions. As early as 1876, Perkin[†] obtained anthracene dibromide, or 9:10-dibromo-9:10-dihydroanthracene (II), by bromination at a low temperature, and the dichloride

* Compare Price, *Mechanisms of Reactions at Carbon-Carbon Double Bonds*, pp. 35 et seq., Interscience, 1946.
† *Chem. News*, 1876, **34**, 145; *Bull. Soc. chim. Paris*, 1877, **27**, 464.

AROMATIC COMPOUNDS—ADDITION REACTIONS 107

was obtained in a similar manner. The reaction is reversible, and if the addition compound is treated with ice-cold hydriodic acid, or with substances such as phenol which take up bromine very readily, anthracene is regenerated. When anthracene is treated simultaneously with pyridine and bromine,* addition takes place with the formation of a pyridinium salt (III); and the same salt is also obtained when anthracene dibromide is treated with pyridine.

Anthracene dibromide slowly evolves hydrogen bromide (to give 9-bromoanthracene) at room temperature, but the reaction is not complete even after three days.†

Addition compounds of this nature can clearly be either *cis* or *trans*, and either isomer may be formed. For example, 1:5-dichloroanthracene gave 1:5:9:10-tetrachloro-9:10-dihydro-anthracene (IV) on chlorination, and this was shown to have a *cis* configuration by measurement of its dipole moment.‡ 9:10-Diphenylanthracene also gave a *cis* dichloride. On the other hand, the addition of chlorine to 1:8-dichloroanthracene is a purely *trans* reaction, for the dipole moment of the resulting dichloride (V) was found to be smaller than that of the parent compound.

---

* Barnett and Cook, *J. Chem. Soc.* 1921, **119**, 901.
† Ibid. 1924, **125**, 1084.
‡ Bergmann and Weizmann, *J. Amer. Chem. Soc.* 1938, **60**, 1801.

It has also been shown that nitric acid reacts with anthracene by addition, although the resulting addition compound (VI) is only stable enough to be isolated in the form of an ester. Meisenheimer and Connerade * obtained the acetate, nitrite and chloride of the addition compound.

(IV)   (V)

(VI)   (VII)

Meisenheimer† has also shown that anthracene reacts with nitrogen dioxide by addition, to give 9:10-dinitro-9:10-dihydroanthracene (VII). Other compounds of this type were prepared by Barnett,‡ who found that substituents either in the *meso* positions or in the benz-rings have very little effect on the ease of reaction, although they have a pronounced influence on the stability of the resulting addition compound.

Another reaction of interest is the addition of alkali metals to anthracene and derivatives. Sodium reacts to give 9:10-disodio-9:10-dihydro derivatives; and lithium adds in the same way. These alkali-metal addition compounds are very highly coloured (intense blue or purple) and they are very reactive.§ When

---

* *Liebigs Ann.* 1904, **330**, 133; compare Dimroth, *Ber. dtsch. chem. Ges.* 1901, **34**, 219.
† *Liebigs Ann.* 1902, **323**, 205; 1904, **330**, 133.
‡ *J. Chem. Soc.* 1925, **127**, 2040.
§ Bachmann and Pence, *J. Amer. Chem. Soc.* 1937, **59**, 2339.

AROMATIC COMPOUNDS—ADDITION REACTIONS

treated with carbon dioxide they give the corresponding acids; when treated with alkyl halides they give the corresponding dialkyl dihydro derivatives; and when treated with alcohol the alkali metal is replaced by hydrogen, giving the dihydro derivative.

The addition of sodium to anthracene itself seems to be a *cis* reaction, giving the addition compound (VIII), for alkylation with methyl iodide gave a 9:10-dimethyl-9:10-dihydroanthracene (IX) which is almost certainly of *cis* configuration.*

(VIII)   (IX)

Many reductions can be classified as 1:4 additions; but it is usually difficult to determine the exact mechanism. The reactions are often complicated by the fact that 1:2 addition also occurs; moreover, isomerism of dihydro derivatives occurs quite frequently. Furthermore, it is only in rather special cases that reduction stops at the dihydro stage.

Benzene can be reduced under many different conditions, usually to *cyclo*hexane. On the other hand, although naphthalene is reduced to tetralin both by many chemical and catalytic methods of reduction, it can be reduced to 1:4-dihydronaphthalene with sodium and alcohol, under suitable conditions. Anthracene can be reduced quite readily by both chemical and catalytic methods to 9:10-dihydroanthracene. In this connexion it is interesting that the reduction of 9:10-dimethylanthracene with sodium and alcohol yields a mixture of the *cis*- and *trans*-9:10-dihydro derivatives.†

More vigorous methods of hydrogenation can convert naphthalene and anthracene into higher hydrides. With anthracene, 1:2:3:4-tetrahydroanthracene and *sym.*-octahydroanthracene

---
* Badger, Goulden and Warren, *J. Chem. Soc.* 1941, p. 18; Badger, Jones and Pearce, ibid. 1950, 1700.
† Ibid.

can be obtained. Moreover, the hydrogenation of anthracene, or of *sym*.-octahydroanthracene, over nickel catalysts at high temperatures, gives a perhydroanthracene; and hydrogenation over Adams's platinum oxide catalyst gives an isomeric perhydroanthracene.

The complexity of the reduction process can be illustrated by a consideration of the results obtained by the hydrogenation of pyrene.* With a molybdenum-sulphur-carbon catalyst 1:2-dihydropyrene and 1:2:6:7-tetrahydropyrene were obtained. Hydrogenation in ethanol over copper chromite gave a mixture of two hexahydropyrenes; but when Raney nickel was used as catalyst a decahydropyrene was formed. Finally, when the reduction was carried out in *cyclo*hexane over Raney nickel, the completely reduced compound, hexadecahydropyrene, was obtained.

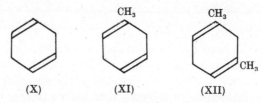

(X)   (XI)   (XII)

One of the most interesting of all the 1:4 addition reactions with aromatic compounds is the reduction process using sodium in liquid ammonia containing alcohol. In this way benzene has been reduced to 1:4-dihydrobenzene (X), toluene has been reduced to 2:5-dihydrotoluene (XI), and *meta*-xylene has given 2:5-dihydro-*meta*-xylene (XII).†

This reaction, which has many synthetic applications both in aliphatic and aromatic chemistry, has been reviewed in detail.‡

Sodium in liquid ammonia behaves as a solution of sodium cations and solvated electrons in equilibrium with sodium atoms,§ and it seems that the first step in the reduction process is the addition of two electrons to the aromatic ring system, to give

---

* Coulson, *J. Chem. Soc.* 1937, p. 1298; Cameron, Cook and Graham, ibid. 1945, p. 286; Campbell, *Ann. Rep. Chem. Soc.* 1947, 44, 130.

† Birch, *J. Chem. Soc.* 1944, p. 430; Wibaut and Haak, *Rec. trav. chim. Pays-Bas*, 1948, 67, 85; Haak and Wibaut, ibid. 1950, 69, 1382.

‡ Birch, *Quart. Rev. Chem. Soc.* 1950, 4, 69.

§ Kraus, *J. Amer. Chem. Soc.* 1921, 43, 749.

AROMATIC COMPOUNDS—ADDITION REACTIONS

a bivalent mesomeric anion. The reduction is then completed by the subsequent addition of two protons.

As Birch has pointed out, the two charges cannot reside on the same carbon atom, or on carbon atoms which are *meta* to one another. The two charges must therefore be disposed among alternate carbon atoms, as in (XIII), one charge being distributed between the carbon atoms at the corners of each triangle.

(XIII)

(XIV) → (XV)

(XVI) → (XVII)

The effects of substituents on the nature of the reduction product can then be evaluated. With *ortho*-methylanisole, for example, one electron must be distributed between the 1:3:5 positions, and the other between the 2:4:6 positions. However, as the electron-repulsive power of a methoxy group is greater than that of a methyl group, the unsubstituted 3- and 5-positions must have the highest charge densities. The first proton would therefore be expected to add at either of these positions (to give (XIV) or (XVI)) and subsequent further protonization would give the two dihydrides ((XV) and (XVII)) obtained experimentally.

ELIMINATION REACTIONS IN DIHYDROAROMATIC SYSTEMS

The disubstituted dihydro derivatives of anthracene, phenanthrene and of other aromatic compounds are of varying stability. Some compounds of this nature are stable at room temperature; others decompose to give the aromatic structure in a few days; others are too unstable to be isolated at room temperature. The re-establishment of the completely aromatic structure is usually favoured by heat, and many catalysts may also be used to facilitate the transformation.

In nearly every case the elimination and re-establishment of the aromatic structure can take place in at least two different ways. For example, anthracene dibromide (I) can be expected either to eliminate bromine with the formation of anthracene (II), or to eliminate hydrogen bromide with the formation of 9-bromoanthracene (III). In point of fact, both reactions may be observed under suitable conditions. When treated with ice-cold hydriodic acid, or with compounds such as phenol which take up bromine very readily, anthracene is formed; but when anthracene dibromide is warmed, 9-bromoanthracene is the main product, and a little anthracene and 9:10-dibromoanthracene are also obtained.* Furthermore, when anthracene dibromide is treated with pyridine, the dipyridinium salt is obtained, but on warming this passes to anthranylpyridinium bromide.

* Barnett and Cook, *J. Chem. Soc.* 1924, **125**, 1084.

The method of carrying out the elimination reaction therefore has a profound effect on its course. In addition to the above experiments, Barnett, Cook and Matthews* have also shown that decomposition by heating in a solvent can often give results which differ from those obtained by warming in the absence of a solvent. This is seen in table 4·1, which records the products obtained by carrying out the decomposition (i) in the absence of solvent and (ii) with either xylene or cymene as solvent.

TABLE 4·1. *Decomposition of disubstituted dihydroanthracenes by heating alone or in the presence of a solvent*

9:10-Dihydroanthracene derivative	Anthracene derivative obtained by	
	Heating alone	Boiling with xylene (X) Boiling with cymene (C)
1:5:9:10-Tetrachloro-	1:5:9-Trichloro-	1:5-Dichloro- (C)
1:8:9:10-Tetrachloro-	1:8:10-Trichloro-	1:8:10-Trichloro- (X)
1:5-Dichloro-9:10-dibromo-	—	1:5-Dichloro- (X)
1:8-Dichloro-9:10-dibromo-	1:8-Dichloro-10-bromo-	Mixture (X)
1:5:9:10-Pentachloro-	1:5:9:10-Tetrachloro-	1:5:9-Trichloro- (C)
1:8:9:10:10-Pentachloro-	1:8:9:10-Tetrachloro-	1:8:10-Trichloro- (X

It appears that heating in a solvent favours elimination of a molecule of halogen, while heating alone favours elimination of hydrogen halide.

In many cases the 9:10-addition product is unsymmetrical, so that elimination can take additional alternative courses.

In the case of 9-chloro-10-nitro-9:10-dihydroanthracene (IV), for example, the re-establishment of the aromatic structure could take place either by loss of hydrogen chloride to give 9-nitroanthracene (V), or by loss of nitrous acid with the production of 9-chloroanthracene (VI). Experimentally, it is found that the former reaction takes place exclusively.

Similarly, the corresponding 9-acetoxy-10-nitro-9:10-dihydroanthracene (VII) decomposes exclusively by elimination of acetic acid, with the formation of 9-nitroanthracene (VIII). The acetate (VII) can be recrystallized from benzene without decomposition, but is quantitatively converted into nitroanthracene

* *Rec. trav. chim. Pays-Bas*, 1926, **46**, 68.

by cold pyridine,* and the alternative elimination of nitrous acid to give 9-acetoxyanthracene (IX) does not occur.

In the same way, the intermediate dichloride (XI) obtained by the action of chlorine on 9-bromoanthracene might be

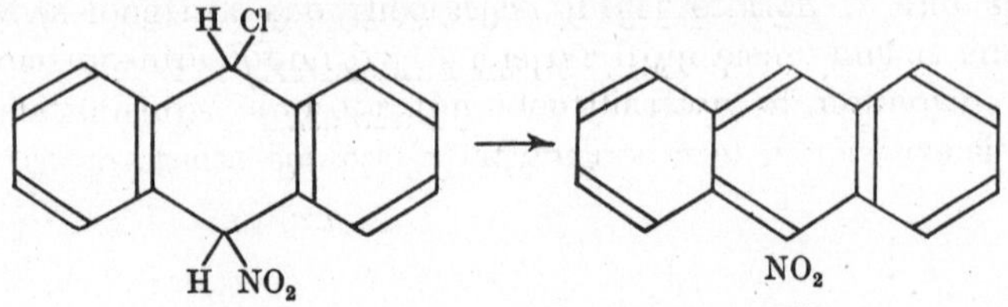

(IV) → (V)

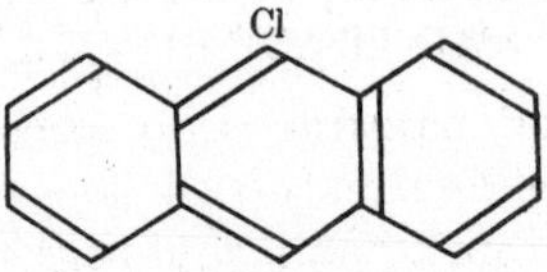

(VI)

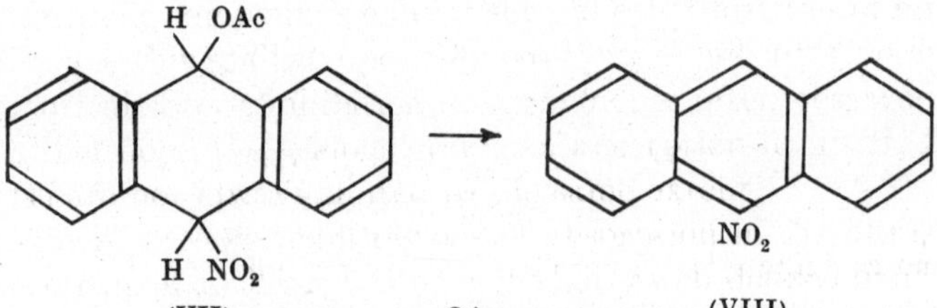

(VII) → (VIII)

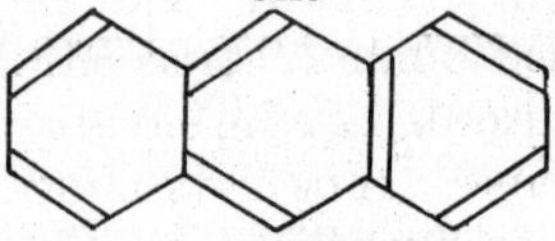

(IX)

expected to give either 9:10-dichloroanthracene (XII) or 9-bromo-10-chloroanthracene (X). In this case, only chlorobromoanthracene (X) was isolated, hydrogen chloride being eliminated.†

Again, the action of nitric acid on 9-bromoanthracene might be expected either (i) to give nitroanthranol ((XIII), which

* Barnett, Cook and Grainger, *J. Chem. Soc.* 1922, **121**, 2059; Barnett, Cook and Matthews, ibid. 1923, **123**, 1994.
† Barnett and Cook, ibid. 1924, **125**, 1084.

would at once tautomerize to nitroanthrone) by elimination of hydrogen bromide from the intermediate addition compound

(XIV), or (ii) to form 9-bromo-10-nitroanthracene (XV) by elimination of water. In this case both reactions seem to take place, the former predominating. It is also noteworthy that when 9-nitroanthracene is brominated, 9:10-dibromoanthracene is obtained, evidently by elimination of nitrous acid from the intermediate addition complex.*

* Barnett, Cook and Grainger, ibid. 1922, **121**, 2059.

CHAPTER 5

THE AROMATIC 'DOUBLE' BOND*

BOND FIXATION IN AROMATIC COMPOUNDS

It seems to be universally accepted that the two *ortho* positions in any mono-substituted benzene derivative are entirely equivalent in every way. This is in accordance with the modern theoretical views on the structure of benzene (see Chapter 2), but for many years this equivalence between the two *ortho* positions was interpreted in terms of the mobility of the double bonds, as postulated by Kekulé.[†] The double bonds in phenol, for example, were thought to 'oscillate' between the positions shown in structures (I) and (II).

(I) ⇌ (II)

(III)   (IV)   (V)

By simple extension of this theory it would be expected that the double bonds in naphthalene would oscillate between the positions shown in structures (III), (IV) and (V). As early as 1893, however, Marckwald[‡] called attention to the remarkable difference between the two positions *ortho* to the functional group in 2-substituted naphthalene derivatives such as 2-naphthol and 2-naphthylamine. In all such cases the 1-position

* The argument in this chapter closely follows that in the author's review article, *Quart. Rev. Chem. Soc., Lond.*, 1951, **5**, 147.
† *Liebigs Ann.* 1872, **162**, 77.
‡ *Ibid.* 1893, **274**, 331; 1894, **279**, 1.

is found to be reactive and the 3-position to be inert, or nearly so;* and it was suggested by Marckwald that the double bonds in naphthalene are not mobile, as in benzene, but that they are 'fixed' in the positions shown in the symmetrical Erlenmeyer formula (III) for this substance. According to this theory, therefore, the 2-position is always linked to the 1-position by a double bond, and to the 3-position by a single bond, so that the difference in reactivity of the two positions *ortho* to a substituent in the 2-position is at once understandable.

Of course, it is now known that pure double and pure single bonds do not exist in aromatic compounds, but that all C—C bonds are of hybrid character. This is not to say, however, that all the C—C bonds in aromatic compounds are identical in character. Considerable evidence has now accumulated that the individual bonds in di- and polycyclic systems show pronounced differences in 'reactivity', and this evidence is summarized in the following sections.

CYCLIZATION EXPERIMENTS

In his classical papers on the bond structure of naphthalene, Marckwald[†] pointed out that the Skraup reaction with 2-naphthylamine (I), glycerol, sulphuric acid, and an oxidizing agent, gives the angular benzoquinoline (II) exclusively. In other words, the cyclization occurs exclusively into the 1- rather than into the 3-position. Marckwald interpreted this result as indicating the presence of a double bond in the 1:2 but not in the 2:3 position of naphthalene. This tendency to form the angular rather than the alternative linear benzoquinoline is so great that with 1-bromo- and 1-nitro-2-naphthylamines the substituent in the 1-position is eliminated during the reaction, 5:6-benzoquinoline (II) being formed.

On the other hand, 1-chloro-2-naphthylamine yields a considerable amount of the expected chloroazanthracene (III), and angular cyclization is by no means the invariable rule.[‡]

---

* See Bell and Hunter, *J. Chem. Soc.* 1950, p. 2903, for some recent experiments illustrating this point.
† *Liebigs Ann.* 1893, **274**, 331; 1894, **279**, 1.
‡ Fries, Walter and Schilling, ibid. 1935, **516**, 285.

Huisgen* has recently found that a large number of 1-substituted derivatives of 2-naphthylamine may be induced to undergo cyclization to give the linear derivative under suitable conditions. For example, 1-methyl-2-acetamidonaphthalene (IV) gave 8-methyl-6:7-benzoquinoline (V) by the Skraup reaction, and 2-acetacetamido-1-methylnaphthalene (VI) was

cyclized by treatment with sulphuric acid at room temperature to 2-hydroxy-4:8-dimethyl-6:7-benzoquinoline (VII). The corresponding bromo derivative reacted in the same way to give 2-hydroxy-4-methyl-8-bromo-6:7-benzoquinoline. Furthermore, the product (IX) of the interaction of 1-methyl-2-naphthylamine (VIII) and ethyl acetoacetate gave 2:8-dimethyl-4-hydroxy-6:7-benzoquinoline (X) on heating.

Similarly, Johnson and Mathews† obtained 2:4-dimethyl-6:7-benzoquinoline (XII) by cyclization of 4-(2′-naphthylimino-)-

---

* *Liebigs Ann.* 1948, **559**, 101.
† *J. Amer. Chem. Soc.* 1944, **66**, 210; see also Clemo and Legg, *J. Chem. Soc.* 1947, p. 545.

pentan-2-one (XI), itself formed from 2-naphthylamine and acetylacetone.

In spite of these exceptions, cyclization to form the angular product is very much more common, and occurs with a great

number of compounds, including heterocyclic derivatives. For example, Boggust and Cocker* have shown that the Skraup reaction with 6-aminobenzothiazole leads to the angular product, pyridino-(3':2':4:5)-benzothiazole (XIII), and not the

* Ibid. 1949, p. 355.

linear product (XIV). Similarly, the crotonate (XV) gives the angular structure (XVI) and not the linear derivative (XVII).

Very similar results have been obtained by experiments on the formation of cyclic ketones by intramolecular acylation. The cyclization of β-2-naphthylpropionic acid and of γ-2-naphthylbutyric acid proceeds to give the angular, (XVIII) and (XIX), rather than the linear structures. That is, the cyclization takes place at the 1-position.*

Other experiments have been carried out using propionic and butyric acid derivatives of the tricyclic and other polycyclic aromatic hydrocarbons. Thus γ-2-anthrylbutyric acid cyclizes to give the angular product (XX),† and β-3-phenanthrylpropionic acid also gives an angular ketone (XXI).‡

(XX) (XXI)

This cyclization of β-3-phenanthrylpropionic acid is of some interest, for the corresponding butyric acid (XXII) cyclizes in the *other* direction to give a keto-tetrahydrobenzanthracene (XXIII).§ In this case the steric effect of the angular benz-ring is evidently the controlling factor, for the Skraup reaction with 3-aminophenanthrene proceeds in the expected direction to give the azabenzophenanthrene (XXIV).‖

Other examples of the influence of steric factors are also known. γ-8-Methyl-2-naphthylbutyric acid (XXV, $R=H$) and γ-5:6:7:8-tetramethyl-2-naphthylbutyric acid (XXV, $R=CH_3$) both give anthracene derivatives (XXVI, $R=H$ or CH_3), and not phenanthrene derivatives on cyclization. In these cases the 8-methyl group apparently offers steric hindrance to cyclization at the 1-position.¶

In a few instances the direction of cyclization is influenced by the reagent. The most striking example is probably the

* Schroeter, Müller and Huang, *Ber. dtsch. chem. Ges.* 1929, **62**, 645; Haworth, *J. Chem. Soc.* 1932, p. 1125; Bachmann and Edgerton, *J. Amer. Chem. Soc.* 1940, **62**, 2219.
† Cook and Robinson, *J. Chem. Soc.* 1938, p. 505; Fieser and Heymann, *J. Amer. Chem. Soc.* 1941, **63**, 2333.
‡ Bachmann and Kloetzel, ibid. 1937, **59**, 2207.
§ Haworth and Mavin, *J. Chem. Soc.* 1933, p. 1012.
‖ Mosettig and Krueger, *J. Org. Chem.* 1938, **3**, 317.
¶ Haworth and Sheldrick, *J. Chem. Soc.* 1934, p. 1950; Hewett, ibid. 1940, p. 293.

cyclization of γ-2-phenanthrylbutyric acid (XXVIII). With hydrogen fluoride, 8-keto-5:6:7:8-tetrahydro-1:2-benzanthracene (XXIX) was obtained in excellent yield. When the cyclization was effected, however, by the Friedel-Crafts stannic chloride

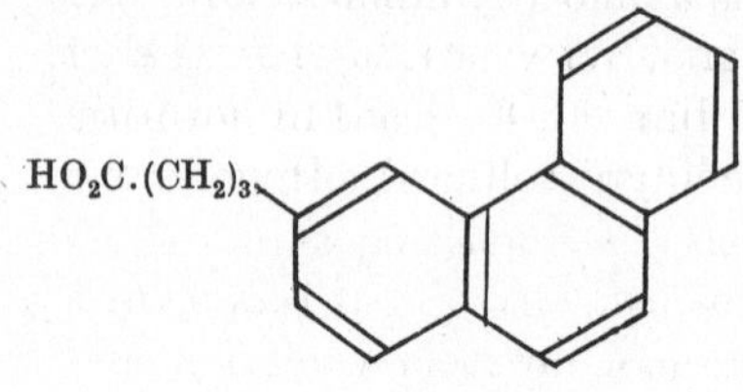

(XXII)

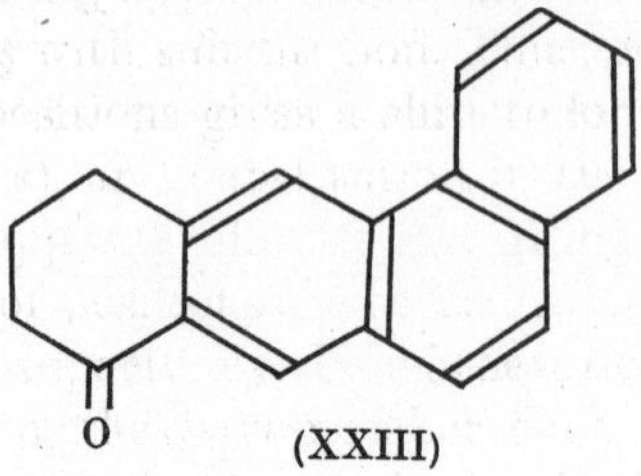

(XXIII)

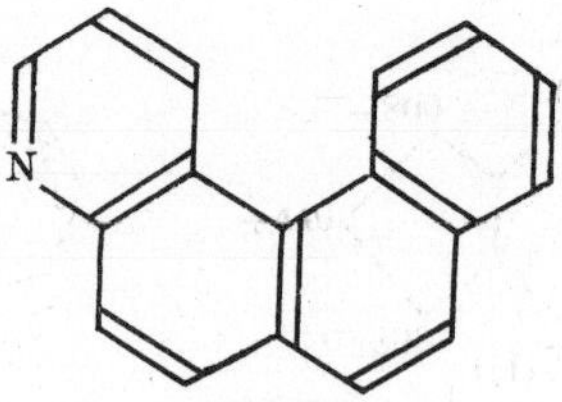

(XXIV)

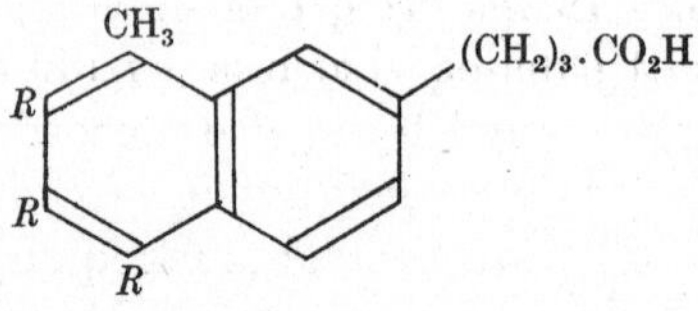

(XXV)

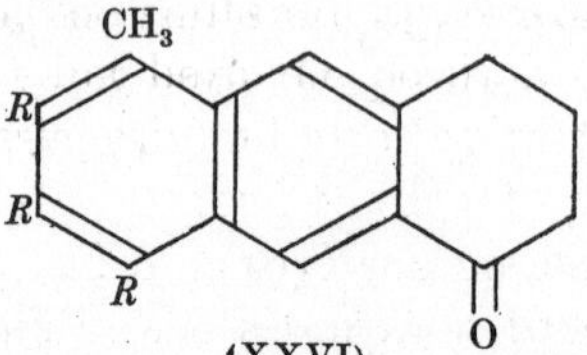

(XXVI)

method, the isomeric ketone, 6-keto-3:4:5:6-tetrahydrochrysene (XXVII), was obtained, also in excellent yield. Under other conditions, mixtures of the two products were obtained,* and both the solvent and the temperature of the reaction were found to influence the proportion of each ketone.

In conclusion, therefore, it seems that in the absence of special factors (such as steric hindrance) the cyclization of 2-substituted naphthalenes usually occurs into the 1- rather than into the 3-position, and in this respect the original observations of

* Fieser and Johnson, *J. Amer. Chem. Soc.* 1939, **61**, 1647: Bachmann and Struve, *J. Org. Chem.* 1939, **4**, 456.

THE AROMATIC 'DOUBLE' BOND

Marckwald have been confirmed by most of the subsequent work. On the other hand, these experiments, although indicating the superior reactivity of the 1-position, do not *necessarily* indicate that this superior reactivity is caused by the presence of a double bond in the 1:2 position. This problem is discussed in the final section of this chapter.

THE DIAZO COUPLING AND HALOGENATION OF PHENOLS

Phenols may be considered as enolic systems, and it has often been supposed* that the reactions of phenols are similar to and take place by the same mechanisms as those of aliphatic enols. Diazo compounds and the halogens are known to react with aliphatic enols in such a way that the substituent is introduced at the adjacent carbon atom which forms part of the enolic unit, and it was therefore considered feasible to use the diazo coupling and halogenation reactions to demonstrate the presence or otherwise of double bonds in phenols.

* See Fieser, in Gilman's *Organic Chemistry*, vol. 1, chapter 2, 2nd ed., John Wiley, New York, 1943.

A *para*-alkylphenol is brominated first in one of the positions *ortho* to the hydroxy group, and then in the other. If the reaction does involve an enolic double bond, it must be supposed that the second substitution is preceded by a migration of the double bond. With 2-naphthol, however, the situation is different. The first substituent enters the 1-position, but further halogenation of 1-chloro-(or bromo-)2-naphthol (I, III) does not give the 3-substituted derivative as would be expected if the double bond in the 1:2 position can migrate to the 2:3 position. A keto-halogenide (II) is formed.

(I) → (II) ← (III)

The reaction appears to be quite general, for ketones of this type have been isolated following halogenation of 1-methyl-2-naphthol, as well as the 1-halogeno-2-naphthols.*

Similarly, if the solution is made sufficiently and progressively alkaline, phenol couples three times with diazo compounds to form 2:4:6-trisazophenols. Coupling at both *ortho* positions may again be explained readily enough on the basis of oscillating double bonds and reaction with an enolic system.

When 2-naphthol is treated with diazotized amines, coupling takes place in the 1-position, and certain types of substituents, if present in this position, are displaced during the reaction; but if the 1-position is blocked by a stable substituent, such as a methyl group, no coupling takes place. For example, 1-methyl-2-naphthol does not couple, even with very reactive diazo compounds such as diazotized picramide.† Interpreted on the basis of the assumptions mentioned above, this appears to indicate that the double bond evidently present in the 1:2 position cannot migrate to the 2:3 position.

---

* Zincke, *Ber. dtsch. chem. Ges.* 1888, **21**, 3378, 3540; Fries *et al.*, ibid. 1906, **39**, 435; 1908, **41**, 2614; *Liebigs Ann.* 1930, **484**, 245; Fries, ibid. 1927, **454**, 121; Fieser in Gilman's *Organic Chemistry*, **1**, 151.
† Bell and Hunter, *J. Chem. Soc.* 1950, p. 2903.

Investigations of this type have been carried out for a number of compounds. Ruggli and Courtin* found that 2:7-dihydroxynaphthalene (IV) couples with diazo compounds in the 1:8 positions, and not the 1:6 positions, as might be expected if naphthalene has an unsymmetrical distribution of double bonds as in formula (V).

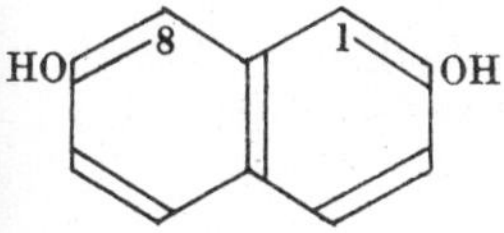

(IV) 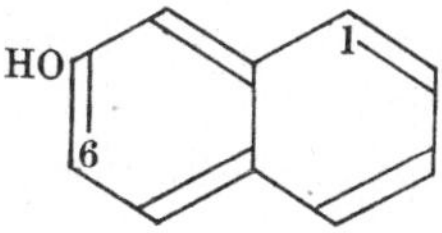 (V) 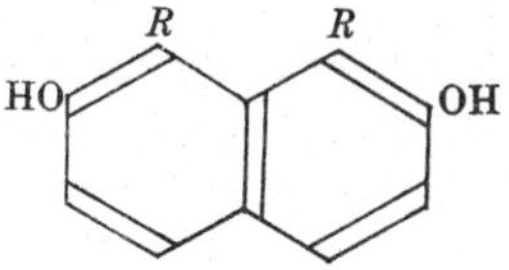 (VI)

Fieser and Lothrop† found further support for the symmetrical Erlenmeyer formula in the observation that 1:8-dialkyl-2:7-dihydroxynaphthalenes (VI) and 1:5-dialkyl-2:6-dihydroxynaphthalenes do not couple, even with particularly active diazo components, and they also interpreted their results as indicating that naphthalene differs from benzene in the matter of the mobility of the bonds.

The same technique has been applied to the study of double-bond distribution and bond fixation in anthracene. Fries, Walter and Schilling‡ showed that 2:6-dihydroxyanthracene (VII) is brominated in the 1- and 5-positions, and Fieser and Lothrop§ found that derivatives of 2:6-dihydroxyanthracene having substituents in the 1- and 5-positions do not couple with diazotized amines. According to the classical view, therefore, it would seem that there is a double bond in the 1:2 but not in the 2:3 position of anthracene.

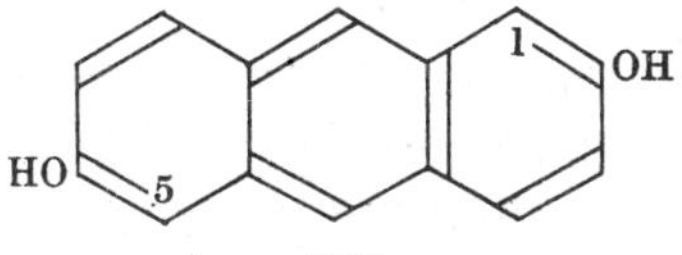

(VII) 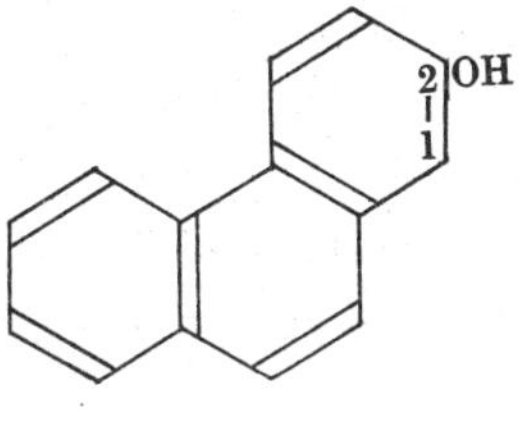 (VIII)

* *Helv. chim. acta*, 1932, **15**, 110. † *J. Amer. Chem. Soc.* 1935, **57**, 1459.
‡ *Liebigs Ann.* 1935, **516**, 248.   § *J. Amer. Chem. Soc.* 1936, **58**, 749.

In the same way, 2-phenanthrol (VIII) couples with diazotized amines in the 1-position, and the presence of an alkyl group at C_1 entirely prevents the coupling. It seems, therefore, that the 1:2 linkage is a double bond, and the 2:3 linkage a single bond. Similarly, the reactions of 3-phenanthrol indicate the presence of a double bond in the 3:4 position.*

Once again, however, it must be admitted that this method of probing for the presence of double bonds is open to serious objection, especially as it is based on certain assumptions with regard to the mechanisms of the reactions.

THE CLAISEN REARRANGEMENT

The Claisen rearrangement involves the conversion of alkyl ethers of enols, and of certain phenols, into C-alkyl derivatives by heating to a high temperature. It may be illustrated with O-allylacetoacetate (I), which rearranges to (II)† when distilled at atmospheric pressure in the presence of ammonium chloride.

$$\begin{array}{cc} O.CH_2.CH=CH_2 & \quad CH_2.CH=CH_2 \\ | & \quad \overset{O}{\underset{\|}{}} \; | \\ CH_3C=CH.CO_2Et & \quad CH_3C.CH.CO_2Et \\ (I) & (II) \end{array}$$

$$\begin{array}{cc} O.CH_2.CH=CH_2 & OH \\ \text{[benzene ring]} \longrightarrow & \text{[benzene ring]}\text{--}CH_2.CH=CH_2 \\ (III) & (IV) \end{array}$$

Allyl ethers of phenols (III) rearrange to give the corresponding *ortho*-allylphenol (IV), except when both *ortho* positions are already substituted. In such cases the *para*-allylphenol is usually formed.

---

* Fieser and Young, *J. Amer. Chem. Soc.* 1931, **53**, 4120.
† Claisen, *Ber. dtsch. chem. Ges.* 1912, **45**, 3157.

From a study of the rearrangement with a number of ethers having α- or γ- substituents on the allyl group it has been shown that the reaction is almost invariably accompanied by inversion, so that the mechanism probably involves a cyclic transition state, and may be represented as follows:

Both double bonds are essential to this mechanism, which seems to be confirmed by the experimental results.*

For this reason the reaction appears to provide an additional method for the study of double-bond distribution in aromatic systems other than benzene. The classical example is β-naphthol allyl ether (V), which is smoothly rearranged, in very good yield, to 1-allyl-2-naphthol (VI) on heating. If the 1-position is blocked, no rearrangement takes place even though the 3-position is free,†‡ but there is no *inherent* reluctance for a rearrangement to give a β-substituted derivative, for 1-allyloxynaphthalene rearranges to 2-allyl-1-naphthol as expected.

The only reasonable explanation for the failure of 1-substituted-2-allyloxynaphthalenes to undergo rearrangement is that although the 1:2 bond can function as a double bond, the 2:3 bond cannot do so, but is best represented as a single linkage.

* Tarbell, *Organic Reactions*, 2, 1, John Wiley, New York, 1944.
† Claisen, *Ber. dtsch. chem. Ges.* 1912, 45, 3157.
‡ Tarbell, *Organic Reactions*, 2, 1.

Further evidence of this nature is provided by the fact that while 2:6-diallyloxynaphthalene rearranges to give 1:5-diallyl-2:6-dihydroxynaphthalene in very good yield, 1:5-diallyl-2:6-diallyloxynaphthalene does not rearrange.* Again, 2-allyloxyphenanthrene (VII) rearranges to 1-allyl-2-phenanthrol (VIII), and 3-allyloxyphenanthrene (IX) to 4-allyl-3-phenanthrol (X); but 1-allyl-2-allyloxyphenanthrene does not rearrange.† 2:6-Diallyloxyanthracene rearranges as expected to give the 1:5-diallyl derivative, but 1:5-dimethyl-2:6-diallyloxyanthracene gives only decomposition products.‡

Similar results have been obtained in a few heterocyclic systems. 7-Allyloxyquinoline (XI) rearranges to 8-allyl-7-hydroxyquinoline, but 7-allyloxy-8-allylquinoline does not rearrange, indicating the presence of a double bond between C_7 and C_8, but not between C_6 and C_7.§ 2-Methyl-4-allyloxyquinoline (XII) rearranges to 3-allyl-2-methyl-4-hydroxyquinoline‖

---
* Fieser and Lothrop, *J. Amer. Chem. Soc.* 1935, **57**, 1459.
† Fieser and Young, ibid. 1931, **53**, 4120
‡ Fieser and Lothrop, ibid. 1936, **58**, 749.
§ Ochiai and Kokeguti, *J. Pharm. Soc. Japan*, 1940, **60**, 271; *Chem. Abstr.* 1941, **35**, 458.
‖ Mander-Jones and Trikojus, *J. Amer. Chem. Soc.* 1932, **54**, 2570.

and 8-allyloxyquinoline gives 7-allyl-8-hydroxyquinoline.* 8-Allyloxy-7-allylquinoline rearranges to give a *para* rearrangement product, namely, 4:7-diallyl-8-hydroxyquinoline.*

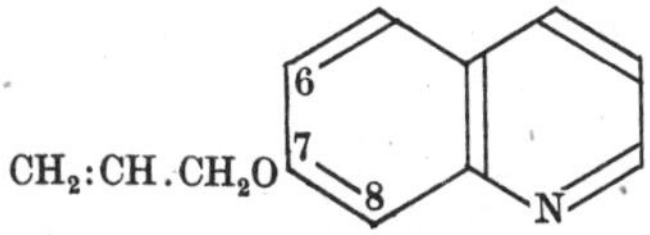

(XI) 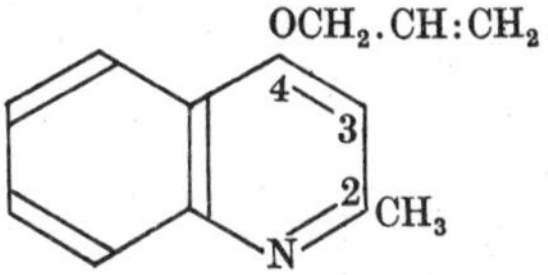 (XII)

The experiments with derivatives of benzothiazole are of considerable interest, as they appear to indicate only partial bond fixation. 6-Allyloxy-2-methylbenzothiazole rearranges to give a mixture of 7-allyl-6-hydroxy-2-methylbenzothiazole (20 parts) and 5-allyl-6-hydroxy-2-methylbenzothiazole (1 part). Moreover, 6-allyloxy-5-allyl-2-methylbenzothiazole (XIII), and 6-allyloxy-7-allyl-2-methylbenzothiazole (XV), both rearrange to give the same compound, 5:7-diallyl-6-hydroxy-2-methylbenzothiazole (XIV).†

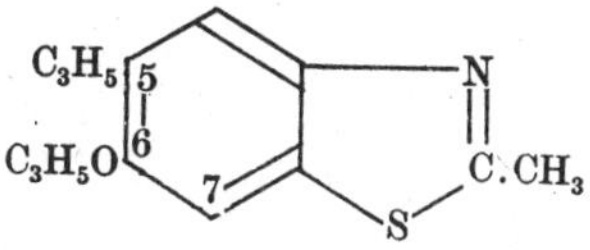

(XIII) 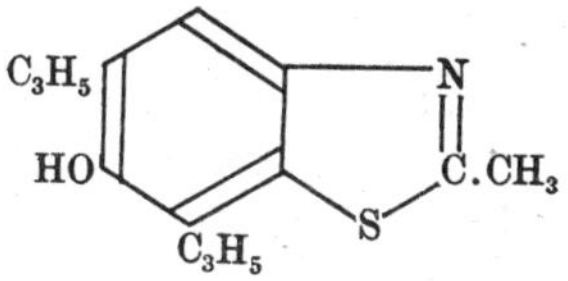 (XIV)

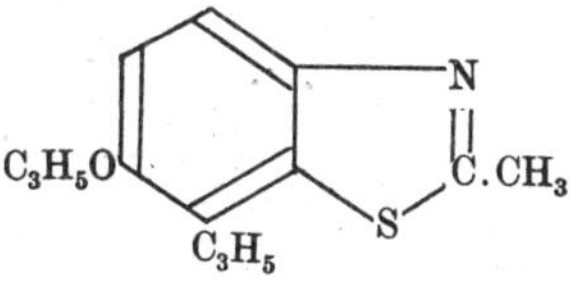

(XV)

* Mander-Jones and Trikojus, *J. Roy. Soc. N.S.W.* 1932, **66**, 300; *Chem. Abstr.* 1933, **27**, 1350.
† Ochiai and Nisizawa, *Ber. dtsch. chem. Ges.* 1941, **74**, 1407; *Chem. Abstr.* 1942, **36**, 5475.

CHELATION AND BOND FIXATION

In 1934 Baker* pointed out that the ability to form a six-membered chelate ring containing co-ordinately linked hydrogen seems to be dependent on the presence of a double bond between the carbon atoms bearing the hydroxy and acetyl groups. For example, chelation occurs in *ortho*-hydroxyacetophenone (I) and in the enolic form of acetylacetone (II), but not, apparently, in the saturated compound (III).†

It would seem, therefore, that the formation of a chelate ring can be used to demonstrate the presence of a double bond. If naphthalene has the symmetrical arrangement of double bonds, then 1-hydroxy-2-acetylnaphthalene (IV) and 2-hydroxy-1-acetylnaphthalene (V) should behave as chelated compounds, and 2-hydroxy-3-acetylnaphthalene (VI) should be non-chelated.

Baker and Carruthers‡ detected chelation in all three compounds, and the results seemed inconclusive, but by examination of the infra-red spectra Hunsberger§ has recently shown that the chelation in the 1:2- and 2:1-disubstituted naphthalenes is of

---

* J. Chem. Soc. 1934, p. 1684.
† Freymann and Heilmann, C.R. Acad. Sci., Paris, 1944, **219**, 415.
‡ J. Chem. Soc. 1937, p. 479.
§ J. Amer. Chem. Soc. 1950, **72**, 5626.

THE AROMATIC 'DOUBLE' BOND

about equal strength but is considerably greater than in the isomeric 2:3-disubstituted naphthalene. This at least supports the view that the 1:2 bond differs in character from the 2:3 bond.

Other evidence of a like nature also supports this conclusion. The infra-red spectra of the nitronaphthylamines show that there is considerable hydrogen bonding in 2-nitro-1-naphthylamine (VII), and also in 1-nitro-2-naphthylamine (VIII), but that hydrogen bonding does not occur in 3-nitro-2-naphthylamine (IX).*

Furthermore, Calvin and Melchior† observed that the *metallic* chelates from 3-hydroxy-2-naphthaldehyde are less stable than those derived from 1-hydroxy-2-naphthaldehyde or from 2-hydroxy-1-naphthaldehyde.

ELECTRON DISPLACEMENTS AND BOND FIXATION

Further evidence that the 1:2 bond of naphthalene differs from the 2:3 bond has been obtained by studying the properties of suitably substituted derivatives. McLeish and Campbell‡ pointed out that if naphthalene has a double bond in the 1:2 position then the bromine atom in 1-bromo-2-nitronaphthalene (I) and in 2-bromo-1-nitronaphthalene (II) should be 'reactive'. In the same way, if naphthalene has a single bond in the 2:3 position, then the bromine atom in 3-bromo-2-nitronaphthalene (III) should be 'unreactive'. These expectations were confirmed experimentally, and it seems that the latter compound does not undergo the polarization necessary to 'loosen' the bromine.

* Hathway and Flett, *Trans. Faraday Soc.* 1949, **45**, 818.
† *J. Amer. Chem. Soc.* 1948, **70**, 3273.
‡ *J. Chem. Soc.* 1937, p. 1103.

THE AROMATIC COMPOUNDS

A somewhat similar method was used by Mills and Smith* in their study of quinoline and *iso*quinoline. They pointed out that the nitrogen atom in *iso*quinoline activates a methyl group in the 1-position, but not in the 3-position. 1-Methyl*iso*quinoline (IV), for example, condenses with benzaldehyde, but 3-methyl-*iso*quinoline (V) does not react in this fashion. The activation can only be brought about by the transmission of electronic effects *via* a double bond, and Mills and Smith therefore concluded that there is a double bond in the 1:2 position of *iso*-quinoline, and a single bond in the 2:3 position. Similarly, quinaldine (2-methylquinoline (VI)) also undergoes condensations with arylaldehydes,† apparently indicating the presence of a double bond in the 1:2 position.

The effects of substituents on the strengths of acids and amines are also of interest. The effects always seem to be greater when relayed through a 1:2 bond of naphthalene than through a 2:3 bond, and this again appears to indicate the presence of a double bond in the 1:2 but not in the 2:3 position. Thus, *ortho*-nitroaniline is a much weaker base than aniline because the nitro group reduces the availability of the 'lone pair' for salt formation. Similarly, 1-nitro-2-naphthylamine is much weaker than 2-naphthylamine (see table 5·1). On the other hand,

---

* *J. Chem. Soc.* 1922, **121**, 2724.
† Henrich, *Ber. dtsch. chem. Ges.* 1899, **32**, 668.

3-nitro-2-naphthylamine is only *slightly* weaker than the parent amine, as the nitro group cannot exert its full electron-attracting effect through the 2:3 bond.*

TABLE 5·1. *Strengths of* ortho-*substituted bases*

Compound	pK_a
Aniline	4·6
ortho-Nitroaniline	−0·2
2-Naphthylamine	4·2
1-Nitro-2-naphthylamine	−1·0
3-Nitro-2-naphthylamine	3·0

Again, Bergmann and Hirshberg† have shown that a chlorine atom in the 3-position has a much smaller influence on the strength of 2-naphthoic acid than one in the 1-position. This also indicates that the electronic influence of the substituent is transmitted much more strongly through the 1:2 bond than through the 2:3 bond, and confirms the different nature of the two linkages.

THE MILLS-NIXON EFFECT

It was long held that the angle (α) between the single bonds in molecules of the type (I) is the same as that between the carbon valencies of methane ($109\frac{1}{2}°$), and that the angle between each single bond and the double linkage is (by difference) $125\frac{1}{4}°$. This being so, Mills and Nixon‡ pointed out that if benzene has a Kekulé structure, the external valencies cannot be directed from the centre of the hexagon, but must lie, as indicated in (II), in directions alternately to either side of these.

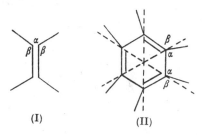

(I)   (II)

* Bryson, *Trans. Faraday Soc.* 1949, **45**, 257.
† *J. Chem. Soc.* 1936, p. 331.   ‡ *Ibid.* 1930, p. 2510.

THE AROMATIC COMPOUNDS

Although it was recognized that the inclusion of the system (1) in a cyclic molecule would affect the magnitude of the angles, it was maintained that the angle α would certainly be smaller than the angle β. In these circumstances, Mills and Nixon reasoned that a *five*-membered ring could be fused on to a benzene ring with very little distortion of the valency angles *only if it included two of the smaller angles*; that is, if it was attached to two carbon atoms joined by a *single* bond. On the other hand, Mills and Nixon suggested that a *six*-membered ring could be attached, without distortion of valency angles, only to two carbon atoms which are joined by a *double* bond.

(favoured) (strained)
(III) (IV)

(favoured) (strained)
(V) (VI)

(VII) (VIII) (IX)

It was thought that the net effect would be to stabilize or 'fix' the double bonds of indane (hydrindene) in the positions shown in formula (III), and that the double bonds of tetralin would be stabilized as in (V). The alternative distribution of double bonds ((IV) and (VI)) would involve strain, and the compounds would not, therefore, exist in these forms.

In order to test these predictions, Mills and Nixon investigated the products of diazo coupling and of bromination of 5-hydroxyindane (VII), and of 6-hydroxytetralin (VIII).

Substitution was found to occur in the positions marked, apparently indicating the presence of double bonds in the positions required by the theory. The evidence was, however, inconclusive, for 3:4-dimethylphenol (IX) was also found to react in the same way as 5-hydroxyindane. Mills and Nixon were forced to conclude that 5-hydroxyindane and 3:4-dimethylphenol have the 'normal' arrangement of double bonds, but that 6-hydroxytetralin has an 'abnormal' arrangement brought about by the attachment of the tetramethylene chain.

The evidence cited by Mills and Nixon has been considerably weakened by the recent work of Parkes,* who has found that the relative reactivities of the two positions *ortho* to the hydroxy group in 3:4-dialkylphenols are dependent on the alkyl groups. Nevertheless, the original results were striking and the subject attracted very considerable attention.

Fieser and Lothrop[†] extended the original work by studying the effect of blocking the reactive positions with methyl groups, a procedure which the same authors had used in studies of bond fixation in naphthalene. To test the degree of fixation in indane derivatives, they examined 5-hydroxy-6-methylindane (X) and 5-hydroxy-4:7-dimethylindane (XI). The latter coupled very readily, as expected, but even (X) coupled, provided the reaction was carried out in weakly alkaline solution. Similarly, *both* the 6-hydroxytetralins, namely, (XII) and (XIII), were found to couple readily, and it was concluded that bond fixation, if any, is not comparable with that observed in naphthalene. Moreover, it was also found that the allyl ethers of the hydroxyindanes (X) and (XI) both undergo the Claisen rearrangement to allyl phenols without difficulty, again indicating lack of bond fixation.

Baker[‡] attempted to apply the study of chelation between hydroxy and acetyl groups to the Mills-Nixon effect. He had previously shown[§] that chelation between these groups in *ortho*-hydroxyacetophenones, and in β-hydroxyketones, depends upon

---
* *J. Chem. Soc.* 1948, p. 2143.
† *J. Amer. Chem. Soc.* 1936, **58**, 2050; 1937, **59**, 945; Lothrop, *ibid.* 1940, **62**, 132.
‡ *J. Chem. Soc.* 1937, p. 476.   § *Ibid.* 1934, p. 1684.

the presence of a double bond between the carbon atoms bearing these groups. When applied to the acetylhydroxyindanes (XIV) and (XV), however, the method gave inconclusive results, for both compounds exhibited chelation.

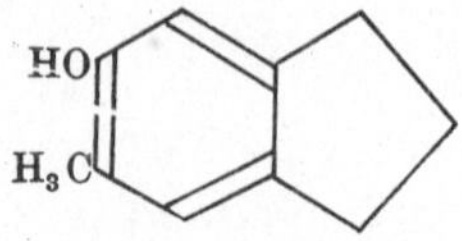

(X) 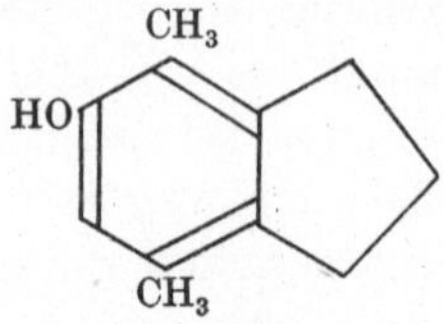   (XI)

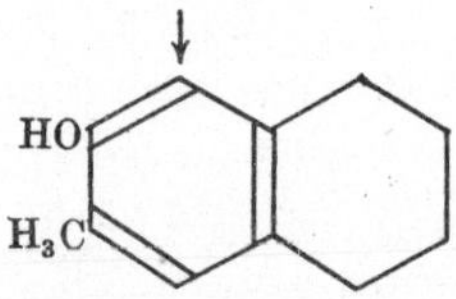

(XII) 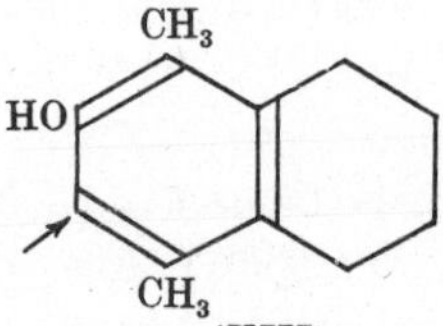   (XIII)

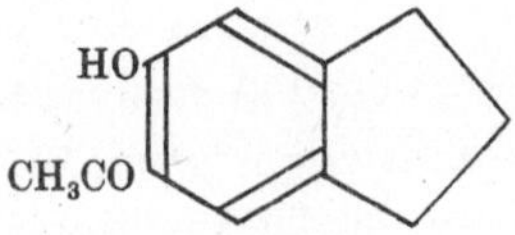

(XIV) 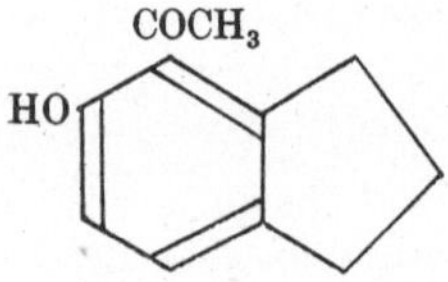   (XV)

Other investigators have studied the bond structures of indane and of tetralin by examining the dipole moments and Br—Br distances in 5:6-dibromoindane, 6:7-dibromotetralin and related compounds. Electron-diffraction experiments, for example, showed the Br—Br distances in dibromo-indane, -tetralin, -*ortho*-xylene, and -benzene to be identical (3·39 ± 0·02 Å.), indicating free resonance in all compounds, with no bond fixation.* This result invalidated the provisional conclusion of Sidgwick and Springall† that the magnitude of the dipole moment of dibromoindane indicated bond fixation in accordance with the views of Mills and Nixon, for this conclusion was based

* Kossiakoff and Springall, *J. Amer. Chem. Soc.* 1941, **63**, 2223; Springall, *Chem. and Ind.* 1943, 21, 149.
† *J. Chem. Soc.* 1936, p. 1532.

on the assumption that the Br—Br distance in this compound is abnormally large.*

Finally, Dolliver, Gresham, Kistiakowsky and Vaughan[†] have found that the heats of hydrogenation of alkylbenzenes and of indane are essentially identical, entirely precluding bond fixation.

There can be little doubt, therefore, that the bond fixation in indane and tetralin as originally postulated by Mills and Nixon does not exist. It must be remembered that the original reasoning was based on the assumption that the angle (α) between the two single bonds in ethylene and its derivatives is $109\frac{1}{2}°$. It is now known to be nearer $120°$,[‡] and the argument as originally put forward is therefore entirely untenable. Sutton and Pauling[§] gave a simple theoretical treatment of the problem, but this work must also be disregarded, as it also assumed that the natural angle between single bonds attached to double-bonded carbon atoms is $109\frac{1}{2}°$. In view of this, Longuet-Higgins and Coulson[‖] have re-examined the problem and concluded that any differences in the bonds are very small indeed, and, in any case, in the opposite direction to those commonly postulated.

Although there seems to be no bond fixation in indane and in tetralin as suggested by Mills and Nixon, there is ample evidence that the attachment of saturated rings to benzene does result in a distortion of the normal valency angles. It has been pointed out that a distortion of the angles α and β in (II) from the normal aromatic value of $120°$ would have a big effect on the steric influence of methylene groups adjacent to the ring. The steric effect of a five-membered ring would be smaller than that of a six-membered ring. This fact has been conclusively demonstrated by a number of methods.[¶]

Contrary to the views of Mills and Nixon, however, this distortion of the valency angles does not bring about bond fixation, and the problem becomes one of finding an alternative

---

* See also Springall, Hampson, May and Spedding, ibid. 1949, p. 1524.

† *J. Amer. Chem. Soc.* 1937, **59**, 831.

‡ Penney, *Proc. Roy. Soc.* A, 1937, **158**, 306; Thompson, *Trans. Faraday Soc.* 1939, **35**, 697; Gallaway and Barker, *J. Chem. Phys.* 1942, **10**, 88; see also Coulson, *Quart. Rev. Chem. Soc., Lond.*, 1947, **1**, 149.

§ *Trans. Faraday Soc.* 1935, **31**, 939. ‖ Ibid. 1946, **42**, 756.

¶ Arnold and Rondestvedt, *J. Amer. Chem. Soc.* 1945, **67**, 1265; 1946, **68**, 2176; Arnold and Craig, ibid. 1948, **70**, 2791; 1950, **72**, 2728; Arnold and Richter, ibid. 1948, **70**, 3505; Orchin and Golumbic, ibid. 1949, **71**, 4151.

explanation of the fact that while 5-hydroxyindane couples and brominates in the 6-position, 6-hydroxytetralin reacts in the 5-position. The most likely explanation seems to be that the activating influence of a saturated five-membered ring is not identical with that of a six-membered ring, and that the π-electron density on the 5-position of 6-hydroxytetralin exceeds that on the 7-position, but that the π-electron density on the 6-position of 5-hydroxyindane exceeds that on the 4-position.*

REACTIONS WITH DIAZOACETIC ESTER

The aliphatic diazo compounds, such as diazomethane and diazoacetic ester, react with a variety of different types of organic compounds. With unsaturated compounds (I) they often react by addition to double bonds. The primary products are usually pyrazolines (II), which are often decomposed at moderately high temperatures, with elimination of nitrogen and the formation of *cyclo*propane derivatives (III).

$$R.CH=CH.R' + N_2CHR'' \longrightarrow \begin{array}{c} R.CH - CHR \\ | \quad\quad | \\ N \quad\quad CHR'' \\ \diagdown N \diagup \end{array}$$

(I) (II)

$$\longrightarrow \begin{array}{c} R.CH - CHR' \\ \diagdown \quad\diagup \\ CH \\ | \\ R'' \end{array} + N_2$$

(III)

The reaction often offers a suitable method for the synthesis of such compounds.† In some cases, however, nitrogen may be

* Berthier and Pullman, *Bull. Soc. chim. Fr.* 1950, **17**, 88; see also Waters, *J. Chem. Soc.* 1948, p. 727; Wheland, *J. Amer. Chem. Soc.* 1942, **64**, 900; Pullman, *Bull. Soc. chim. Fr.* 1947, **14**, 337.
† Hancox, *J. Aust. Chem. Inst.* 1949, **16**, 282.

eliminated from the reagent *before* addition, and the reacting molecule may be the radical :$CH\!-\!CO_2Et$.*

Under suitable conditions diazoacetic ester also reacts with aromatic 'double' bonds. In this case, however, the reaction proceeds only with difficulty and the yields are seldom good, or even satisfactory. The reaction of diazoacetic ester with benzene and its derivatives was first investigated by Buchner and his co-workers. In a long series of papers† it was clearly demonstrated that the primary reaction product is the ester of a norcaradiene acid (IV). This acid, which is isomeric with phenylacetic acid (VI), was called pseudophenylacetic acid. It formed a tetrabromide by addition, as expected, and its structure was established by the formation of 1:2:3-*cyclo*propanetricarboxylic acid on permanganate oxidation. The primary products of type (IV) were also shown to undergo a variety of rearrangements, especially at moderately high temperatures, leading to derivatives of *cyclo*heptatriene carboxylic acid (V), phenylacetic acid (VI) and (in the case of methyl-substituted benzenes) hydrocinnamic acid. For example, toluene reacted with diazoacetic ester to give ethyl 4-methylnorcaradiene carboxylate. Hydrolysis of this ester with boiling 30% sulphuric acid gave 4-methyl*cyclo*heptatriene carboxylic acid, and hydrolysis of 4-methylnorcaradienecarboxylamide gave a small quantity of *para*-tolylacetic acid.‡

(IV) (V) (VI)

With naphthalene, diazoacetic ester adds to the 1:2 bond, to give ethyl benzonorcaradiene carboxylate (VII),§ but, unlike the similar products obtained by addition to benzene derivatives, this compound could not be induced to rearrange. The addition

---

* Compare Buchner and Hediger, *Ber. dtsch. chem. Ges.* 1903, **36**, 3502.
† *Ber. dtsch. chem. Ges.* 1885, **18**, 2377; ibid. 1901, **34**, 982; ibid. 1903, **36**, 3502, 3509; *Liebigs Ann.* 1908, **358**, 1; 1910, **377**, 259; *Ber. dtsch. chem. Ges.* 1920, **53**, 865.
‡ Buchner and Feldmann, ibid. 1903, **36**, 3509.
§ Buchner and Hediger, ibid. 1903, **36**, 3502.

compound (VII) gave a crystalline dibromide (VIII) by addition, and the dibromide of the amide was found to react with boiling water to form bromohydroxybenzonorcarene carboxylic acid and dihydroxybenzonorcarene carboxylic acid (IX).

The structure of the addition product (VII) was confirmed not only by these reactions but also by the fact that the acid was oxidized by alkaline permanganate to 1-(2'-carboxyphenyl-)2:3-*cyclo*propanedicarboxylic acid (X) and, finally, to 1:2:3-*cyclo*-propanetricarboxylic acid.

With phenanthrene, diazoacetic ester was found to add, with elimination of nitrogen, to the 9:10 bond, to give the ethyl ester of dibenzonorcaradiene carboxylic acid (XI). As with the naphthalene addition product, this acid (XI) was found to be remarkably stable. It did not undergo rearrangement when heated with alkali even under quite vigorous conditions. Oxidation with alkaline permanganate gave 1-(2'-carboxyphenyl-)2:3-*cyclo*propanedicarboxylic acid (X), and the isolation of this substance established the structure of the acid (XI).*

* Drake and Sweeney, *J. Org. Chem.* 1946, **11**, 67; Cook, Dickson and Loudon, *J. Chem. Soc.* 1947, p. 746.

THE AROMATIC 'DOUBLE' BOND

Some attempts have recently been made to apply the reaction to other polycyclic aromatic compounds. For example, anthracene gave an addition product which is, almost certainly, represented by (XII);* pyrene probably gave (XIII), and 1:2-benzanthracene probably (XIV).† It would seem, therefore,

(XII)

(XIII)

(XIV)

* At first sight it may seem surprising that diazoacetic ester adds only once to each molecule of benzene, naphthalene and anthracene, for the remaining bonds in the ring attacked would appear to be ethylenic in character. However, as early as 1916, Robinson (*J. Chem. Soc.* 1916, **109**, 1042) postulated the conjugation of *cyclo*propane with ethenoid groups, and evidence in favour of this concept has accumulated. Recently, van Volkenburgh, Greenlee, Derfer and Boord (*J. Amer. Chem. Soc.* 1949, **71**, 3595) have produced evidence for the conjugation in vinyl*cyclo*propane itself (see also van Volkenburgh, Greenlee, Derfer and Boord, ibid. 1949, **71**, 172; Derfer, Greenlee and Boord, ibid. 1949, **71**, 175; Mariella, Peterson and Ferris, ibid. 1948, **70**, 1494; Rogers, ibid. 1947, **69**, 2544; Allen and Boyer, *Canad. J. Res.* 1933, **9**, 159; Carr and Burt, *J. Amer. Chem. Soc.* 1918, **40**, 1590; Kishner, *J. Soc. phys.-chim. russe*, 1911, p. 1163; Walsh, *Nature, Lond.*, 1947, **159**, 165). This also appears to be the case in the present compounds. In this connexion it is interesting that hydrogenation of the compound (presumably (XII)) derived from anthracene gives a hexahydro derivative, a compound which probably corresponds to the symmetrical octahydro derivative obtained by hydrogenation of anthracene itself.

† Clar, 'Reichsamt Wirtschaftsausbau', *Chem. Ber.*, Prüf-Nr O15 (PB 52017), 859–78 (1942); *Chem. Abstr.* 1947, **41**, 6553; Badger, Cook and Gibb, *J. Chem. Soc.* 1951, p. 3456.

that, although the mechanism of the reaction is entirely unknown, the reagent does seek out the most reactive 'double' bond.

The application of the diazoacetic ester reaction to indane and its derivatives is of considerable interest. On the one hand, many azulenes have been synthesized in this manner, and, secondly, the Mills-Nixon effect has very frequently been cited in an attempt to determine the structure of the products. It is doubtful, however, whether any of the work so far carried out is of any value in assessing the reality or otherwise of bond fixation in indanes, because steric hindrance appears to play such a large part in the addition of diazoacetic ester to aromatic double bonds.

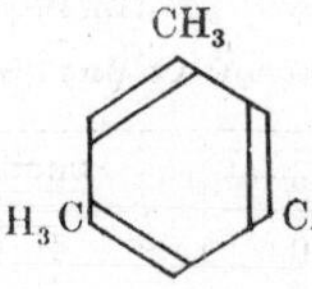

(XV) 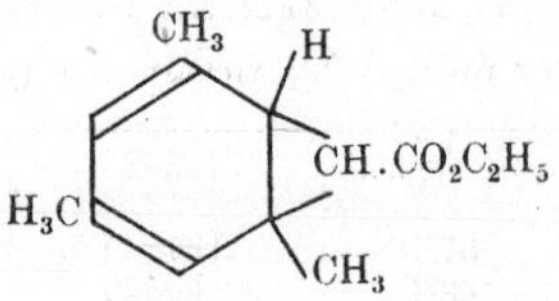   (XVI)

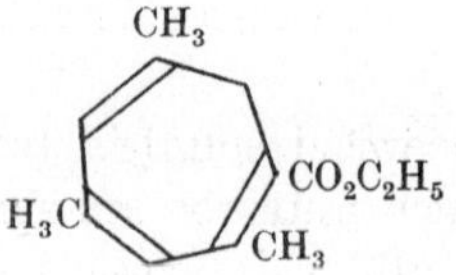

(XVII) 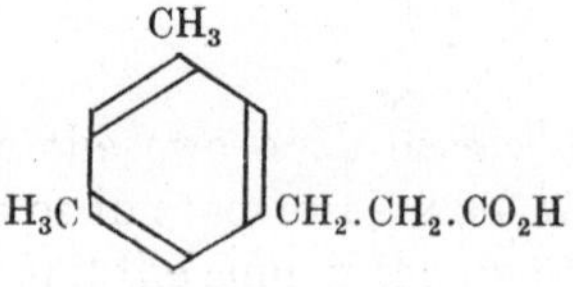   (XVIII)

Buchner, in his pioneer work, investigated the course of the reaction with a number of alkylated benzenes, and concluded that the steric hindrance of such alkyl groups is very considerable. The addition was always found to involve *only* unsubstituted carbon atoms, and a norcaradiene derivative with a quaternary carbon atom was never produced, or at least could never be isolated.

With certain polyalkyl-benzenes an unsubstituted double bond is not always available. Mesitylene (XV) is such a compound, and in this case no bicyclic compound (XVI) could be detected after reaction with diazoacetic ester. The main product was found to be the trimethyl*cyclo*heptatriene acid (XVII), and smaller amounts of 3:5-dimethylhydrocinnamic acid (XVIII)

THE AROMATIC 'DOUBLE' BOND

and mesitylacetic acid were also formed.* Again, no unsubstituted 'double' bond is available in durene, and in this case the only product isolated was shown to be 2:4:5-trimethylhydrocinnamic acid.†

In *substituted* compounds, therefore, the reagent does not always seem to add to double bonds. Although compounds of type (XVI) have nearly always been postulated as intermediates, there is no evidence for their formation, and the so-called rearrangement products may be formed directly from unstable transition complexes. Indeed, this would seem to be the more likely course. The formation of compounds of type (XVI) involves forcing the methyl group out of the plane of the ring, and there is little doubt that there would be considerable resistance to this movement, leading to preferential rupture of the molecule with the formation of compounds of types (XVII) and (XVIII), etc.

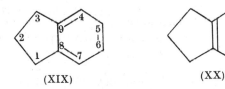

(XIX)   (XX)

An interesting attempt to apply diazoacetic ester to the study of the bond structure of indane was carried out by Arnold.‡ Indane can, theoretically, react either in the form (XIX) or in the form (XX). The former structure (XIX) is that favoured by Mills and Nixon.§ For steric reasons already outlined above, it is unlikely that there will be any addition to either the 4:9, the 8:9 or the 7:8 bonds, and only the 4:5, 5:6 and 6:7 bonds would seem to be likely to take part in the reaction. Assuming the compound reacts in the form (XIX), as expected by the Mills-Nixon effect, then the intermediate (XXI) would be formed which, on rearrangement, reduction by the Bouveault-Blanc method, and dehydrogenation, would give 6-methylazulene (XXIII).

* Buchner and Schottenhammer, *Ber. dtsch. chem. Ges.* 1920, **53**, 865.
† Smith and Tawney, *J. Amer. Chem. Soc.* 1934, **56**, 2167.
‡ *Chem. Ber.* 1947, **80**, 123.   § *J. Chem. Soc.* 1930, p. 2510.

On the other hand, reaction in the form (XX) would give 5-methylazulene. In point of fact, Arnold isolated only 5-methylazulene and concluded that indane does *not* react in the form (XIX) as postulated by Mills and Nixon, but in the alternative form (XX). The evidence, however, was not conclusive. Supposing the 4:5, 5:6 and 6:7 bonds to be *equally* reactive, then on a simple statistical basis there is twice the chance that indane will react in the form (XX) as in the form (XIX), for the simple reason that the 4:5 and 6:7 bonds are equivalent. Furthermore, the initial condensation product had a wide boiling range, and, after reduction, was separated into three fractions. The first of these gave 5-methylazulene in reasonable yield, and the third in only poor yield. The evidence does not therefore exclude the possibility that the 4:5, 5:6 and 6:7 bonds were all involved in the reaction, leading to two products, only one of which was isolated.

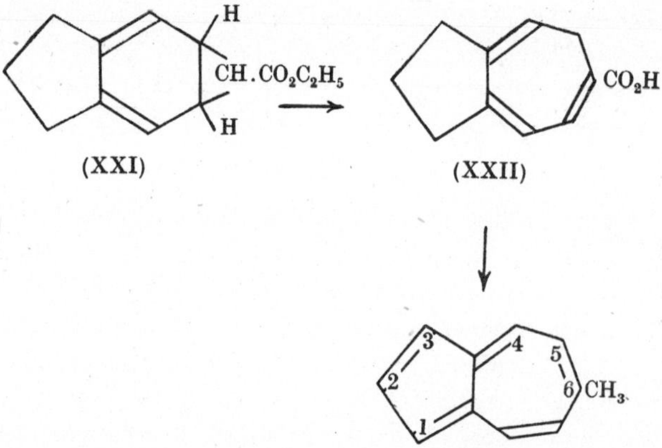

(XXI) → (XXII)

↓

(XXIII)

This conclusion is supported by the recent work of Plattner, Fürst, Müller and Somerville,* who have re-examined the addition of diazoacetic ester to indane. The crude azulene carboxylic acid was found to be a mixture of the 5- and 6-carboxylic acids, the former predominating. With 2-methylindane, however, more of the 6- isomer was obtained.

* *Helv. chim. acta*, 1951, **34**, 971.

The diazoacetic ester method was first applied to the synthesis of azulenes by St Pfau and Plattner.* 2-*iso*Propyl-4:7-dimethylindane (XXIV) was heated with the ester, and the product (XXV) saponified and then heated with palladium-charcoal. In this way, vetivazulene, or 2-*iso*propyl-4:8-dimethylazulene (XXVI), was obtained.

Similarly, Plattner and Wyss[†] prepared 4:8-dimethylazulene from 4:7-dimethylindane, and Plattner, Fürst and Schmid[‡] obtained 1:3:4:8-tetramethylazulene from 1:3:4:7-tetramethylindane. In all these cases there can be little doubt that the addition is forced to take place at the 5:6 bond *not* by any bond fixation, but by the steric hindrance offered by the presence of the two methyl groups at the 4- and 7-positions.

When applied to indanes having a methyl substituent at the 5- or 6-position, the results are again consistent with the view that the reagent adds to an unsubstituted bond, if that is possible. For example, 5-methylindane (XXVII) is known to give 5-methylazulene (XXIX), almost certainly because the 6:7 bond is the only unsubstituted bond available. Complex considerations involving a quaternary carbon atom and the Mills-Nixon

---

* *Helv. chim. acta*, 1939, **22**, 202.
† Ibid. 1941, **24**, 483. ‡ Ibid. 1945, **28, 1647**.

effect are not necessary to explain this result.* 5-Methyl-2-*iso*propylindane is also attacked predominantly at the 6:7 bond.† Another example is provided by the fact that 1:3:5-trimethylindane gives 1:3:5-trimethylazulene‡ in this synthesis.

(XXVII) (XXVIII) (XXIX)

In some dimethylindanes, no unsubstituted double bond is available for the addition, and here the evidence is consistent with the view that the first addition of the reagent is to the most reactive *centre*, and that the adjacent bond is then ruptured and re-formed to give the seven-membered ring. In this way, 1:5:7-trimethylindane (XXX) gave 1:6:8-trimethylazulene (XXXI) and not the 1:5:8-derivative (XXXII) as might be expected by the Mills-Nixon effect.§

(XXX) (XXXI) (XXXII)

On the other hand, with the analogous 1-*iso*propyl-4:6-dimethylindane (XXXIII), the 1-*iso*propyl group offers considerable steric hindrance at the 7-position and the initial addition must be at the 5-position. The product is, almost certainly, 1-*iso*propyl-4:7-dimethylazulene (XXXIV) and not 1-*iso*propyl-4:6-dimethylazulene (XXXV).∥

* Compare Plattner and Roniger, *Helv. chim. acta*, 1942, **25**, 590.
† Arnold and Spielmann, *Chem. Ber.* 1950, **83**, 28.
‡ Wagner-Jauregg and Hippchen, *Ber. dtsch. chem. Ges.* 1943, **76**, 694.
§ Wagner-Jauregg, Friess, Hippchen and Prier, ibid. 1943, **76**, 1157.
∥ Wagner-Jauregg, Arnold and Hüter, ibid. 1942, **75**, 1293; see also Plattner and Roniger. *Helv. chim. acta*, 1943, **26**, 905.

THE AROMATIC 'DOUBLE' BOND

The reaction of fluorene with diazoacetic ester has also been investigated, but it is not known whether the addition is to the 1:2, 2:3 or 3:4 bond. Any or all of these addition products would give the 1:2-benzazulene actually isolated.*

(XXXIII)   (XXXIV)   (XXXV)

The diazoacetic ester reaction has also been developed as a method for the synthesis of tropolones.† Treatment of 1:2-dimethoxybenzene, veratrole, with diazoacetic ester, for example, gave a dimethoxy*cyclo*heptatriene carboxylic ester which on oxidative hydrolysis gave β-carbethoxytropolone. In the same way, 1:2:4-trimethoxybenzene was converted into stipitatic acid.

It is also noteworthy that diazomethane reacts with benzene in a similar fashion. Oxidation of the product gave a small yield of tropolone.‡

THE ADDITION OF OZONE

Ozone adds very readily to ethylenic double bonds to form the so-called ozonides. The structure of these addition compounds appears to be entirely unknown, although certain speculations have been published.§ Indeed, the structure of ozone itself has not yet been satisfactorily confirmed. It would seem to be

---

* Treibs, *Naturwissenschaften*, 1946, **33**, 371; *Chem. Ber.* 1948, **81**, 38; Plattner, Fürst, Chopin and Winteler, *Helv. chim. acta*, 1948, **31**, 501; Horn, Nunn and Rapson, *Nature, Lond.*, 1947, **160**, 829; Nunn and Rapson, *J. Chem. Soc.* 1949, p. 825.
† Johnson, *Sci. Progr.* 1951, **39**, 495; Bartels-Keith and Johnson, *Chem. and Ind.* 1950, p. 677; Bartels-Keith, Johnson and Taylor, *ibid.* 1951, p. 337; *J. Chem. Soc.* 1951, p. 2352.
‡ Doering and Knox, *J. Amer. Chem. Soc.* 1950, **72**, 2305.
§ Long, *Chem. Rev.* 1940, **27**, 437.

important, however, that the decomposition of the ozonides either with water or by hydrogenation almost invariably results in the cleavage of the C—C bonds.* It is this circumstance which is largely responsible for the general utility of the reaction. The reagent has been used very extensively in degradations designed to establish the position of ethylenic double bonds by examining the fragments formed on scission of the molecule.

The ozonolysis of aromatic compounds was first investigated by Harries.† Ozone reacts with benzene only slowly, but a triozonide (I) was isolated as might be anticipated from the Kekulé formula. This triozonide gave glyoxal on decomposition. Similarly, naphthalene formed a diozonide (II), and as *ortho*-phthalaldehyde (III) and glyoxal were detected following decomposition of this product, the 1:2 and 3:4 bonds must be involved in the addition.‡

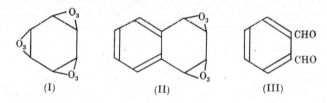

(I)   (II)   (III)

Other evidence has now accumulated that ozone, like diazoacetic ester, is a 'double-bond reagent', and that it adds to double bonds, but does not usually react with reactive centres. The most conclusive evidence is provided by the demonstration that ozone first attacks the 1:2 bond of pyrene (IV) and then the 6:7 bond. The usual ionic reagents such as Br^+, on the other hand, attack the most reactive *centres*, the 3-, 5-, 8- and 10-positions. That ozone does attack the 1:2 and 6:7 bonds of pyrene was clearly demonstrated by the isolation of 4-phenanthrene-aldehyde-5-carboxylic acid (V) and of diphenyl-2:2′:6:6′-tetra-aldehyde (VI) after decomposition of the mixture of ozonides.§

* The ozonides from acetylene and diphenylacetylene, however, yield glyoxal and benzil respectively.
† *Liebigs Ann.* 1905, **343**, 311; Harries and Weiss, *Ber. dtsch. chem. Ges.* 1904, **37**, 3431.
‡ Wibaut and Kampschmidt, *Proc. K. Akad. Wet. Amst.* 1950, **53**, 1109.
§ Vollmann, Becker, Corell and Streeck, *Liebigs Ann.* 1937, **531**, 1; Fieser and Novello, *J. Amer. Chem. Soc.* 1940, **62**, 1855.

This reaction has been applied to the synthesis of 4:5-dimethylphenanthrene.*

On the other hand, attempts to ozonize anthracene gave only anthraquinone,† but further work on this reaction is required.

(IV) (V) (VI)

It is also of some interest that Vollmann et al.‡ found that 1:9-benzanthrone (VII) gave anthraquinone-1-aldehyde (VIII) in 20% yield, together with a large amount of anthraquinone-1-carboxylic acid, indicating the addition of ozone to the bonds marked.

(VII) (VIII)

Similarly, fluoranthene (IX) was ozonized in about 30% yield to a mixture of fluorenone-1-aldehyde (X) and fluorenone-1-carboxylic acid.

* Newman and Whitehouse, ibid. 1949, **71**, 3664; Badger, Campbell, Cook, Raphael and Scott, *J. Chem. Soc.* 1950, p. 2326.
† Roitt and Waters, ibid. 1949, 3060.
‡ Vollmann, Becker, Corell and Streeck, *Liebigs Ann.* 1937, **531**, 1; Fieser and Novello, *J. Amer. Chem. Soc.* 1940, **62**, 1855.

THE AROMATIC COMPOUNDS

All these experiments are consistent with the view that ozone is a 'double-bond reagent', and that it seeks out the most reactive bond in an aromatic compound for attack. For this reason the reagent has been extensively used in studies of the fine structure of such substances.

(IX)   (X)

In 1932, Levine and Cole* used the reagent to examine the question of the existence or otherwise of two isomeric *ortho* disubstitution products of benzene. They reasoned that, if *ortho*-xylene exists in two isomeric forms, (XI) and (XII), then two distinct triozonides should be formed. Decomposition of the triozonide of (XI) would be expected to yield two molecules of methylglyoxal and one of glyoxal, but decomposition of the isomeric triozonide (from (XII)) would be expected to give two molecules of glyoxal and one of diacetyl. In fact, glyoxal, methylglyoxal and diacetyl were all identified in the mixture obtained by scission of the crude triozonides, and Levine and Cole therefore concluded that *ortho*-xylene does indeed exist in two isomeric configurations.

(XI) → 2Me.CO.CHO
       CHO.CHO

(XII) → Me.CO.CO.Me
        2CHO.CHO

* *J. Amer. Chem. Soc.* 1932, **54**, 338.

Haaijman and Wibaut* have more recently repeated and extended this work on the ozonolysis of *ortho*-xylene. By refining the technique it was possible to isolate the decomposition products as the oximes, and to determine their relative proportions. The mixture of oximes was found to consist of 21% of dimethylglyoxime, 34·2% of methylglyoxime and 43·9% of glyoxime, in good agreement with the calculated figures of 20, 35 and 45% if diacetyl, methylglyoxal and glyoxal are formed in the molecular ratios of 1:2:3.

The interpretation of the results is, of course, difficult. The conclusion of Levine and Cole cannot now be accepted, although at that time it seemed to be the simplest explanation. According to the modern view, *ortho*-xylene is a resonance *hybrid* of the two structures (XI) and (XII) which may react according to either structure. Indeed, Haaijman and Wibaut interpret their results on this basis.† In discussing this and other problems in ozonolysis, Kooyman and Ketelaar‡ have suggested that the initial process is the conversion of an aromatic linkage into an isolated double bond, i.e. two π-electrons are 'localized' by the approach of the reagent, the remaining disturbed system of π-electrons rearranging to form a structure of minimum energy. The ozonolysis of *ortho*-xylene may therefore be depicted as follows. The hybrid molecules (XIII) are converted into the *reacting* molecules (XIV a) and (XIV b) by the approach of the ozone molecules, and the mono-ozonides (XV a) and (XV b) are formed. These mono-ozonides, having two ethylenic double bonds, are rapidly attacked further to give the triozonides, which, on decomposition, yield the products as already enumerated.

It is rather surprising that steric hindrance seems to play little or no part in this reaction when it seems to be the controlling factor in the analogous addition of diazoacetic ester to double bonds. Nevertheless, this appears to be the case, for the polyalkylbenzenes, including hexamethylbenzene, are more readily ozonized than benzene itself. Van Dijk§ has carried out velocity

---

* *Rec. trav. chim. Pays-Bas*, 1941, **60**, 842.
† But, as Campbell (*Ann. Rep. Chem. Soc.* 1947, **44**, 127) has said, 'experimental data first used to substantiate one theory lose some of their evidential value when later applied for the same purpose to a second and fundamentally different theory'.
‡ *Rec. trav. chim. Pays-Bas*, 1946, **65**, 859.
§ *Ibid.* 1948, **67**, 945.

measurements of the ozonization of benzene, toluene, *meta*-xylene, mesitylene and hexamethylbenzene, and found that the reactivity of these substances increases in the order named.

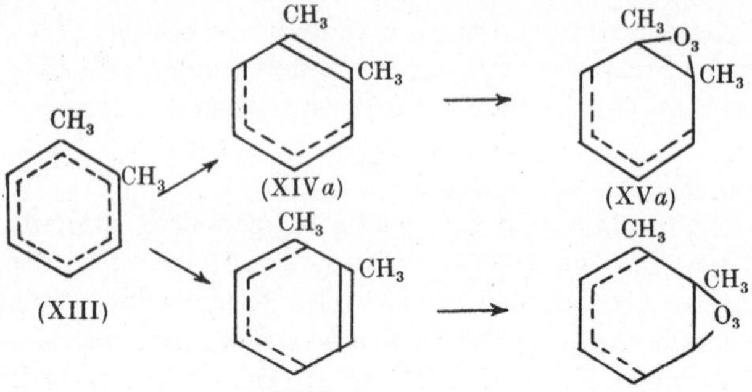

CHO.CHO
$CH_3.CO.CO.CH_3$
$NH_2.CHO$

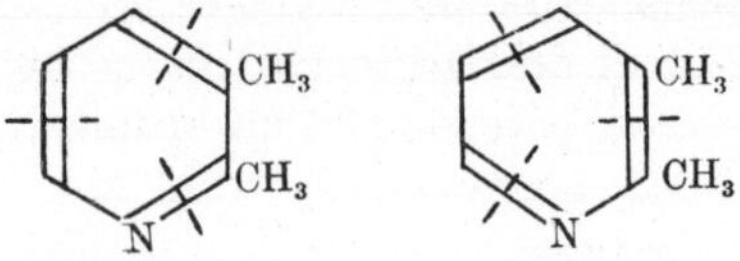

$CH_3.CO.CHO$
CHO.CHO
$CH_3.CONH_2$

Wibaut has extended these ozonolysis experiments to the study of other aromatic, including heterocyclic, compounds. Ozonolysis of 2:3-dimethylpyridine and of 2:3:4-trimethylpyridine,* and subsequent decomposition of the reaction products with water, was found to give glyoxal, methylglyoxal and diacetyl, as well as formic acid, oxalic acid and ammonia. This may be taken as evidence that 2:3-dimethylpyridine reacts according to the two structures (XVI) and (XVII).

The acid amides ($NH_2.CHO$ and $CH_3.CONH_2$) have not yet been detected amongst the decomposition products, and to this extent the mechanism proposed is still hypothetical.

* Wibaut and Kooyman, *Rec. trav. chim. Pays-Bas*, 1944, **63**, 231; 1946, **65**, 141; see also Shive *et al. J. Amer. Chem. Soc.* 1942, **64**, 909; 1946, **68**, 2144; Lochte, Crouch and Thomas, ibid. 1942, **64**, 2753.

THE AROMATIC 'DOUBLE' BOND

With naphthalene and its derivatives, the interpretation of the results is even more difficult. Wibaut and van Dijk* studied the ozonization of 2:3-dimethyl- and of 1:4-dimethylnaphthalene. The ozonides formed from 2:3-dimethylnaphthalene were found to yield glyoxal, methylglyoxal and diacetyl on decomposition, the molecular ratio of diacetyl to methylglyoxal being 10:1. Similarly, decomposition of the ozonides from 1:4-dimethylnaphthalene gave glyoxal and methylglyoxal. Naphthalene has three resonance structures of the Kekulé type, the symmetrical Erlenmeyer-Graebe structure (XVIII) and the two unsymmetrical Erdmann structures (XIX) and (XX). The identification of methylglyoxal among the products of decomposition of 2:3-dimethylnaphthalene ozonides was first interpreted* as evidence that naphthalene reacts partly in accord with the structure (XX), a result which seemed to be at variance with most other chemical work, which indicates the presence of reactive bonds in the $\alpha\beta$ positions only. This interpretation was, however, revised by Kooyman,† who pointed out that the diozonides formed by addition of two molecules of ozone to one ring have an intact benzene ring which is still capable of ordinary resonance and which may be expected to react according to either of its Kekulé structures.

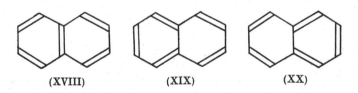

(XVIII)   (XIX)   (XX)

The most satisfactory mechanism for the addition of ozone to naphthalene and its derivatives would therefore seem to be as follows. The approach of a molecule of ozone first localizes the π-electrons in one $\alpha\beta$ bond, to give the *reacting* molecule (XXII) which gives the diozonide (XXIII) as originally postulated by Harries and Weiss.‡

---

* *Rec. trav. chim. Pays-Bas*, 1946, **65**, 413.
† Ibid. 1947, **66**, 201.
‡ *Liebigs Ann.* 1905, **343**, 311; Harries and Weiss, *Ber. dtsch. chem. Ges.* 1904, **37**, 3431.

THE AROMATIC COMPOUNDS

The unattacked ring in the diozonide (XXIII) is aromatic in character, and on the approach of a further molecule of ozone it may be expected to be converted into either the reacting molecule (XXIV a) or (XXIV b). The former would lead to the pentozonide (XXV a) and the latter to the isomeric pentozonide

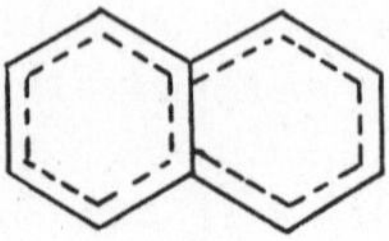

(XXI) 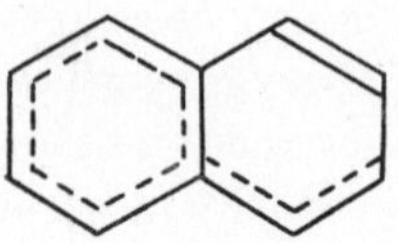   (XXII) 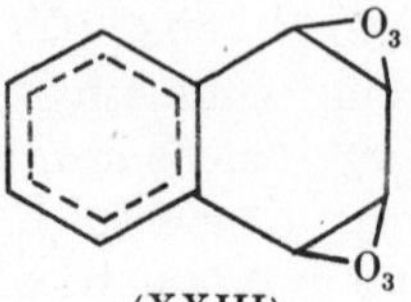   (XXIII)

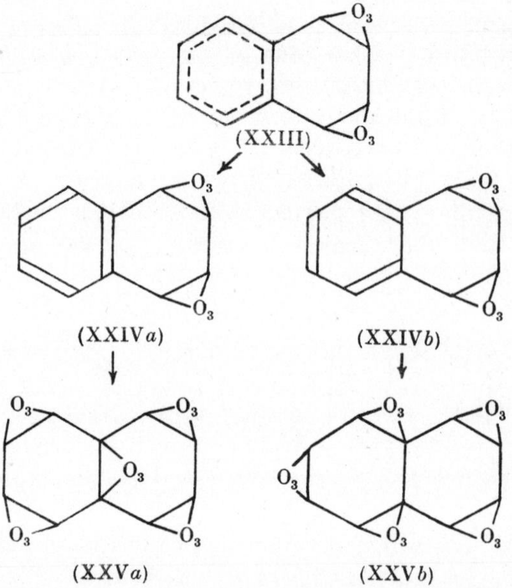

(XXIII)

(XXIV a)   (XXIV b)

(XXV a)   (XXV b)

(XXV b). This latter pentozonide is identical with the product which would be expected from the Erdmann naphthalene structure (XIX). The formation of methylglyoxal in the ozonization of 2:3-dimethylnaphthalene may therefore be explained by assuming that *part* of the material undergoes reaction according to the scheme (XXVI) → (XXVII) → (XXVIII).

THE AROMATIC 'DOUBLE' BOND

Similarly, the formation of methylglyoxal in the ozonization of 1:4-dimethylnaphthalene may be explained by assuming that *part* of the material undergoes reaction according to the scheme (XXIX) → (XXX) → (XXXI).

Quinoline might be expected to react with ozone in a manner very closely related to that of naphthalene, and this has proved to be the case.* Velocity measurements showed that the reaction with quinoline (XXXII) proceeds at a nearly constant rate until two moles of ozone have reacted with one mole of the aromatic compound. The reaction velocity then drops suddenly to a far lower value. It therefore seems that the primary product is a diozonide (XXXIII). This is confirmed by the formation of glyoxal and pyridinedialdehyde (XXXIV) by scission of the product. Incidentally, the formation of pyridinedialdehyde indicates that the primary addition of ozone is to the hydrocarbon ring, rather than to the nitrogen-containing ring.

* Wibaut and Boer, *Proc. K. Akad. Wet. Amst.* 1950, **53**, 19.

Analogous results were obtained by ozonolysis of 6:7-dimethylquinoline and of 5:8-dimethylquinoline. In the former case, diacetyl and pyridinedialdehyde were detected among the products of scission of the ozonide. Neither methylglyoxal nor pyridione was detected. With 5:8-dimethylquinoline, glyoxal was identified among the decomposition products, but no methylglyoxal. 2:3-Diacetylpyridine, the other product expected, was not detected, although its oxidation product, 2-acetylpyridine-3-carboxylic acid, was found. These results are in agreement with the idea that quinoline can react according to the structures (XXXV) and (XXXVI), but *not* (XXXVII).

(XXXV)   (XXXVI)   (XXXVII)

(XXXVIII)   (XXXIX)
(Mills-Nixon form)

Wibaut and de Jong* have also attempted to use this technique in the study of the Mills-Nixon effect. Thus the ozonolysis of 4:7-dimethylindane gave glyoxal and methylglyoxal. The former would not be expected at all if indane has the fixed structure postulated by Mills and Nixon. In point of fact, the quantity isolated (as the oxime) indicated that 4:7-dimethylindane reacts 75% according to the structure (XXXVIII) and only 25% according to the structure (XXXIX) favoured by Mills and Nixon. Similar results were also obtained using 5:6-dimethylindane. The molecular ratio of the oximes of diacetyl and methylglyoxal formed indicated that this compound reacts

---

* See Wibaut, *Comptes rendus de la Quinzième Conférence, Union Internationale de Chimie Pure et Appliquée*, 1949, p. 79.

THE AROMATIC 'DOUBLE' BOND

only 20% according to the Mills-Nixon form and 80% according to the alternative form.

It is interesting that these results are in agreement with the most recent theoretical studies of the Mills-Nixon effect. Thus Longuet-Higgins and Coulson[*] concluded that any differences in the bonds in indane are small but are in the opposite direction to those postulated by Mills and Nixon.[†]

THE ADDITION OF OSMIUM TETROXIDE

Osmium tetroxide also appears to be a double-bond reagent. It reacts readily with ethylenic double bonds (I) and, more slowly, with certain aromatic 'double' bonds, to form organo-osmium compounds having the probable structure (II). In the presence of a tertiary base, such as pyridine, the osmium is co-ordinated with two molecules of base, and the resulting complexes (III) are generally nicely crystalline brown compounds which may often be isolated in almost theoretical yield.

Hydrolysis of such complexes yields the corresponding *cis*-diols, and the reaction provides a very convenient, if costly, method for the preparation of these compounds. For example, in the presence of pyridine, osmium tetroxide adds to the reactive 9:10 bond of phenanthrene, and the complex, on hydrolysis, yields *cis*-9:10-dihydroxy-9:10-dihydrophenanthrene.[‡]

The addition to aromatic 'double' bonds takes place very much more slowly than to ethylenic bonds, and it is noteworthy

---
[*] *Trans. Faraday Soc.* 1946, **42**, 756.
[†] See also Berthier and Pullman, *Bull. Soc. chim. Fr.* 1950, **17**, 88.
[‡] Criegee, *Liebigs Ann.* 1936, **522**, 75; Criegee, Marchand and Wannowius, ibid. 1942, **550**, 99.

that only the most reactive aromatic bonds are attacked at all. Osmium tetroxide does not attack benzene under the usual conditions, and even with reactive polycyclic compounds only one bond is usually attacked, as is the case with phenanthrene.

(IV) → (V) → (VI)

Cook and Schoental[*] have examined the oxidation of a number of polycyclic compounds with osmium tetroxide, and in every case the reagent attacks the most reactive *bond* even when particularly reactive *centres* are present. With pyrene (VII), the addition is to the 1:2 bond, although ionic reagents such as Br^+, NO_2^+, etc., attack the reactive centres at positions 3, 5, 8 and 10. In this respect, therefore, osmium tetroxide closely resembles ozone.

In the same way, osmium tetroxide adds to the 6:7 bond of 3:4-benzopyrene (VIII), although ionic reagents normally attack the reactive centres at positions 5, 8 and 10.

Similarly, osmium tetroxide adds exclusively to the 3:4 bond of 1:2-benzanthracene (IX) and its derivatives,[*†] in spite of the fact that most reagents attack this hydrocarbon at the 9:10 positions.

In the case of anthracene (X) it has been shown that osmium tetroxide adds to the 1:2 and 3:4 bonds.[‡] The first molecule of osmium tetroxide evidently adds to the 1:2 bond, and the product (XI) may be considered as a β-vinylnaphthalene derivative. The

[*] *J. Chem. Soc.* 1948, p. 170.
[†] Badger, ibid. 1949, p. 2497.
[‡] Cook and Schoental, *Nature, Lond.*, 1948, **161**, 237.

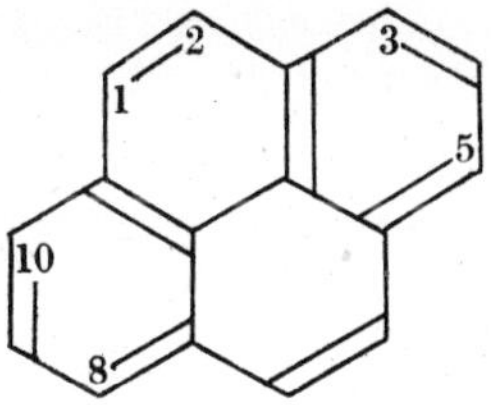

THE AROMATIC 'DOUBLE' BOND

(VII)

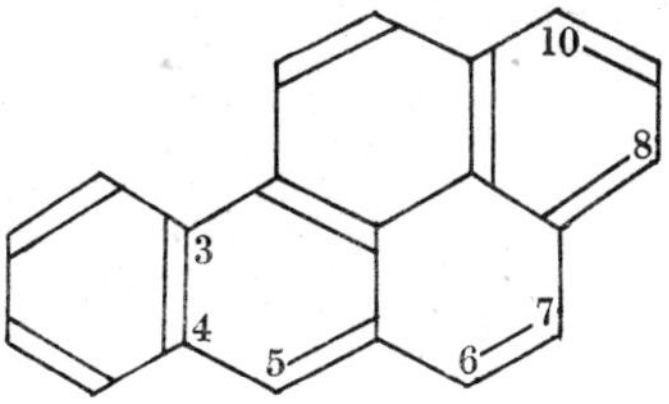

(VIII) 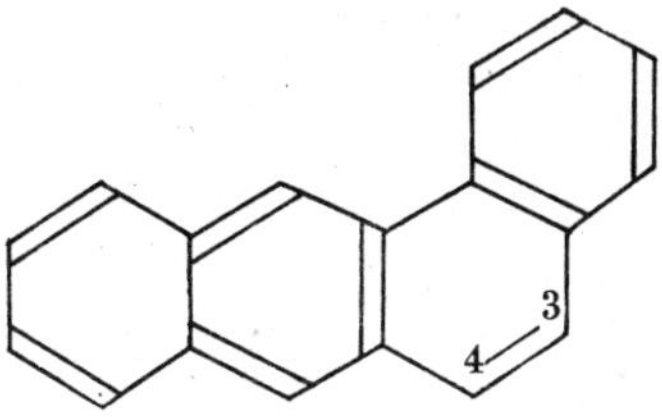 (IX)

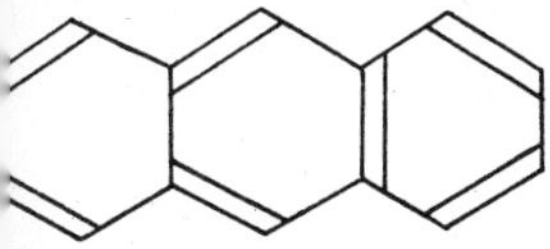

(X)

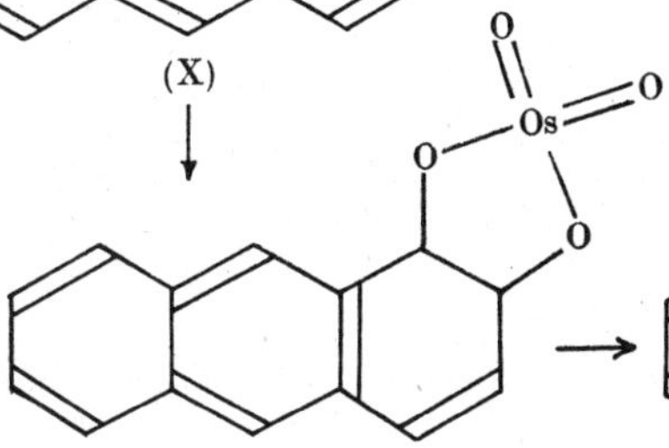

(XI)

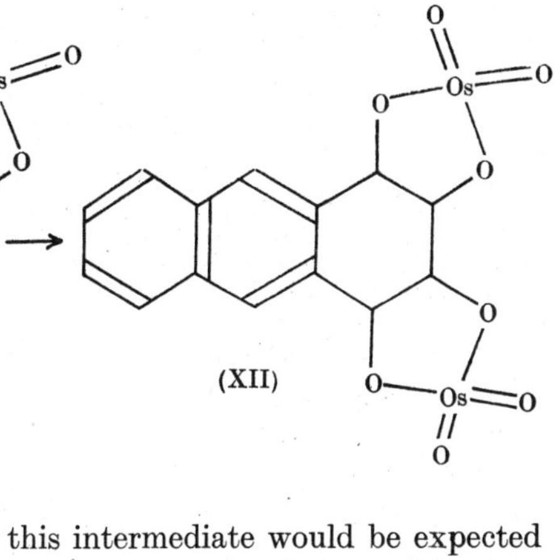

(XII)

ethylenic double bond in this intermediate would be expected to react very rapidly with an additional molecule of osmium tetroxide to give the complex (XII). Cook and Schoental confirmed the structure of this complex by hydrolysis to the tetrol, and subsequent oxidation to naphthalene-2:3-dialdehyde.

These experiments indicate that osmium tetroxide is a very useful reagent for the study of the bond structure of aromatic compounds, and particularly to probe for the most reactive bond in such substances. The reagent has also been used to measure the relative reactivity of aromatic bonds. It has been shown, for example, that the addition of osmium tetroxide to the 3:4

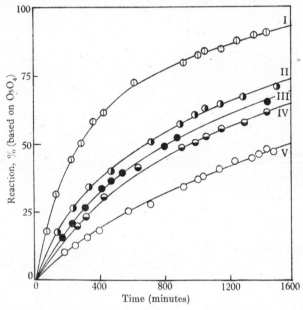

Fig. 5·1. Curves illustrating the effect of methyl and methylene groups on the rate of addition of osmium tetroxide to 1:2-benzanthracenes. I, 9:10-dimethyl-1:2-benzanthracene. II, methylcholanthrene (i.e. 6-methyl-5:10-dimethylene-1:2-benzanthracene). III, 10-methyl-1:2-benzanthracene. IV, 2′:7-dimethyl-1:2-benzanthracene. V, 1:2-benzanthracene. (From Badger, *J. Chem. Soc.* 1949, p. 456.)

bond of 1:2-benzanthracene proceeds about ten times more rapidly than to the 9:10 bond of phenanthrene.* In the same way it has been shown that the presence of alkyl substituents accelerates the addition to the 3:4 bond of substituted benzanthracenes, and that some other substituents, such as the cyano group, retard the rate of addition.†

* Badger and Reed, *Nature, Lond.*, 1948, **161**, 238.
† Badger, *J. Chem. Soc.* 1949, p. 456; 1950, p. 1809; Badger and Lynn, ibid. 1950, p. 1726.

The accompanying graph (fig. 5·1) shows how the rate of reaction is increased by methyl substituents, and it is noteworthy that the magnitude of the effect varies considerably with the position of the substituent.

BOND-LENGTH VARIATIONS IN AROMATIC SYSTEMS

Considering all the experimental work which has been carried out with the three double-bond reagents, diazoacetic ester, ozone and osmium tetroxide, it is impossible to escape the conclusion that some of the bonds in polycyclic compounds are more reactive than others. In other words, it seems that some aromatic bonds are more like ethylenic double bonds than others. The same must also apply in compounds such as thiophen, furan and pyrrole, the $\alpha\beta$ bonds of which seem to be almost as reactive as pure ethylenic bonds.

Moreover, this chemical evidence that aromatic bonds show a variation in character and reactivity is also supported by the physical measurements of the C—C bond lengths in such compounds.

It has long been known that the length of a C—C bond is intimately connected with its nature or character. Thus, a C—C single bond, such as occurs in diamond and in the paraffin hydrocarbons, has a length of 1·54 Å.; the length of a C—C double bond, such as occurs in ethylene, measures 1·34 Å.; and that of a C—C triple bond, such as occurs in acetylene, measures 1·20 Å. All aromatic bonds are, of course, hybrids (see Chapter 1), of length intermediate between that of a single and that of a double bond. In recent years, however, it has been shown that C—C bonds in aromatic compounds vary considerably in length. Some are much shorter, that is, nearer the length of a double bond, than others. It is significant that it is precisely those bonds which are shortest which the chemical evidence indicates to be the most reactive.

In the simple heterocyclic compounds there are marked differences in the C—C bond lengths, as the values attached to the accompanying formulae (I)–(III) indicate. The 'double bonds' of the classical structures are invariably much shorter (1·35 Å.) than the conventional single bonds (1·44–1·46 Å.).*

Bond-length variations have also been observed in polycyclic aromatic hydrocarbons. The molecular dimensions have been determined by X-ray crystallographic analysis, and although the observed variation in bond length is sometimes not much greater than the possible experimental error, in other cases the variations are considerable and of undoubted validity.†

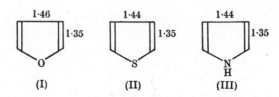

(I)   (II)   (III)

For both naphthalene (IV) and anthracene (V), the bond lengths have been determined with great accuracy by full three-dimensional analysis. The results show clearly that the $\alpha\beta$ bonds in both compounds are shorter (i.e. closer to double bonds) than the $\beta\beta$ bonds by approximately 0·03 Å., a result which confirms the chemical evidence that the former are more like double bonds than the latter.‡

(IV)   (V)

Bond-length variations for several other polycyclic aromatic compounds have been determined. Robertson and White§ found that although all the hexagon angles in pyrene (VI) are 120° as nearly as can be determined, there are very considerable

---

* Beach, J. Chem. Phys. 1941, 9, 54; Schomaker and Pauling, J. Amer. Chem. Soc. 1939, 61, 1769.
† For a review see Robertson, Acta Cryst. 1948, 1, 101.
‡ Abrahams, Robertson and White, ibid. 1949, 2, 233, 238; Sinclair, Robertson and Mathieson, ibid. 1950, 3, 251.
§ J. Chem. Soc. 1947, p. 358.

variations in bond length. The 1:2 and 3:4 bonds in this molecule, for example, were found to be 1·39 Å., as against 1·45 Å. for the central bond.

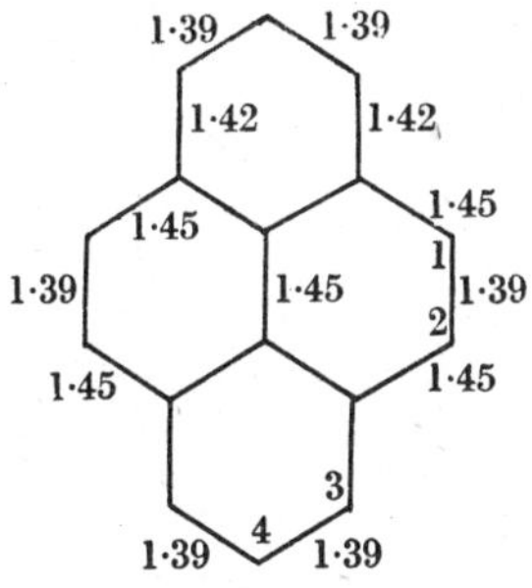

(VI)

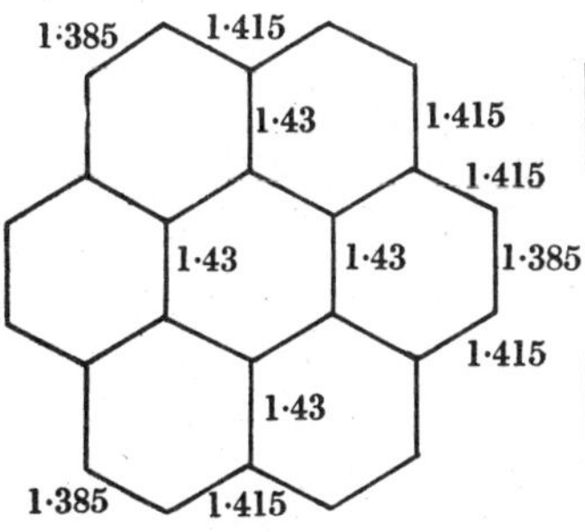

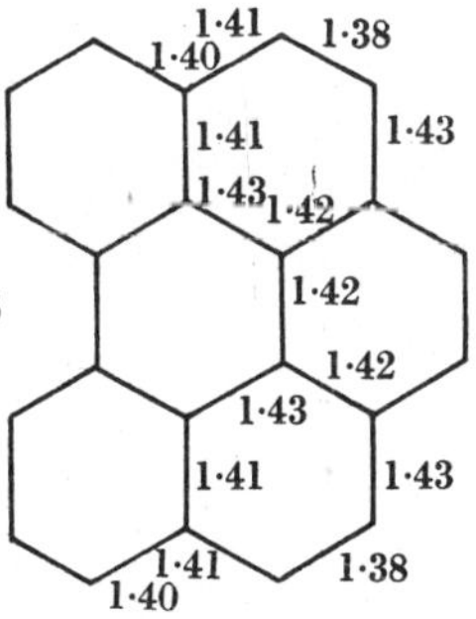

(VII) coronene

(VIII) 1:12-benzoperylene

The structure of coronene (VII) is particularly interesting on account of its high degree of symmetry. Robertson and White* found the lengths of the bonds of the central ring and of the 'spokes' to be 1·43 Å., and the bond lengths in the outer ring were found to be 1·385 and 1·415 Å. For the closely related hydrocarbon, 1:12-benzoperylene (VIII), the bond-length variations are very similar.†

The highly condensed hydrocarbon ovalene (IX) is also of interest in this connexion, for the C—C bond lengths have been obtained with great accuracy.‡

* Ibid. 1945, p. 607.   † White, ibid. 1948, p. 1398.
‡ Donaldson and Robertson, *Nature, Lond.*, 1949, **164**, 1002; Buzeman, *Proc. Phys. Soc.* 1950, **63**, 827.

The carcinogenic hydrocarbon 1:2:5:6-dibenzanthracene exists in both monoclinic (X) and orthorhombic (XI) crystal forms. Both forms have been examined, and the bond-length variations show reasonably satisfactory agreement.*

(IX)

Bond	Length	Bond	Length
A	1·404 Å	G	1·426 Å
B	1·441	H	1·435
C	1·345	J	1·416
D	1·433	K	1·461
E	1·403	L	1·442
F	1·428	M	1·383

(X)
Monoclinic form

(XI)
Orthorhombic form

* Robertson and White, *J. Chem. Soc.* 1947, p. 1001; Robertson, *Acta Cryst.* 1948, **1**, 101.

DOUBLE-BOND CHARACTER AND RELATED CONCEPTS

The observed variations in the reactivity, in the length, and in the character of aromatic bonds can be explained quite simply by application of the theory of resonance.

If the small contributions of the Dewar structures to the benzene hybrid (Chapter 2) are neglected, this molecule can be conceived as a resonance hybrid of the two Kekulé structures (I) and (II). Under these circumstances, as all the C—C bonds are entirely equivalent, each must have 50% double-bond character and 50% single-bond character.

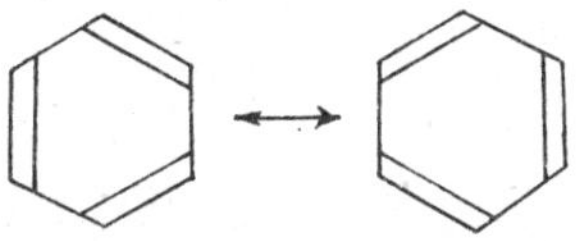

(I)   (II)

In graphite (III), each bond may be represented as a double bond in one-third of the structures, and as a single bond in two-thirds. The C—C bonds of graphite must therefore have $33\frac{1}{3}$% double-bond character.

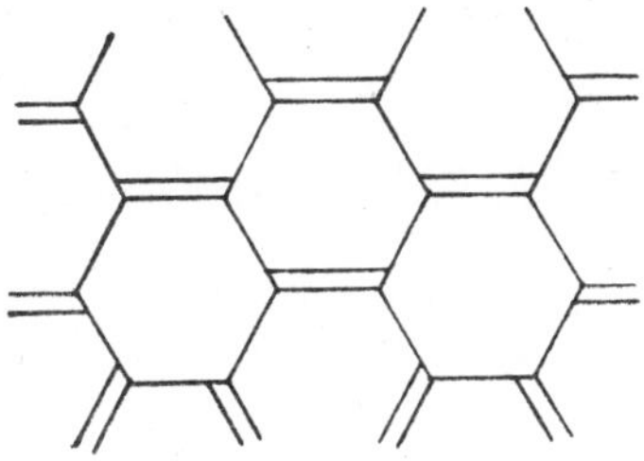

(III)

In naphthalene and other aromatic molecules the double-bond characters of the various bonds depend on the number and nature of the different contributing structures of the Kekulé type. For naphthalene, there are three structures of the Kekulé type, (IV), (V) and (VI).

The 1:2 bond is represented as a double bond in two of the three structures, (IV) and (V), and may therefore be said to have 66⅔% double-bond character. The 2:3 bond, on the other hand, is represented as a double bond in only one (VI) of the three structures, so it has 33⅓% double-bond character. By appropriate summation the double-bond characters of all the bonds may be evaluated, the complete figures for naphthalene being as in (VII).

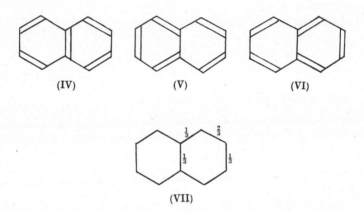

The double-bond characters of the bonds in other aromatic hydrocarbons can be evaluated in the same way. For anthracene there are four structures of the Kekulé type, (VIII)–(XI), and for phenanthrene there are five, (XII)–(XVI). These give the double-bond characters as represented in (XVII) and in (XVIII).

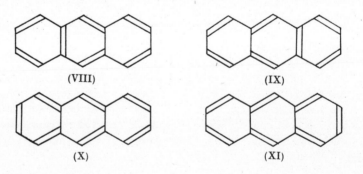

The relationship between bond length and double-bond character was first emphasized by Pauling, Brockway and Beach,* who showed that a smooth curve (fig. 5·2) can be drawn between the experimental points for diamond, graphite, benzene and ethylene.

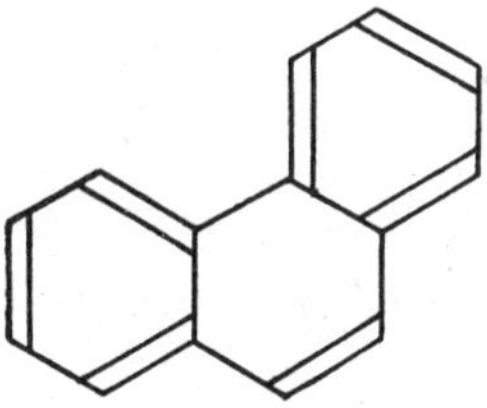

(XII) 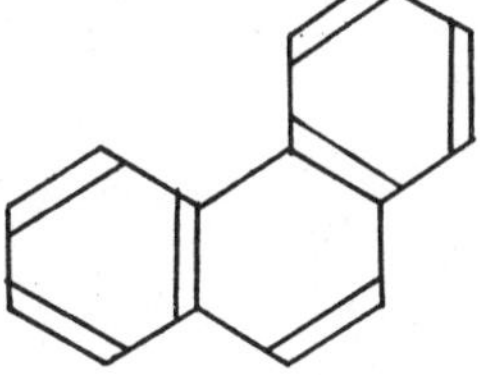   (XIII) 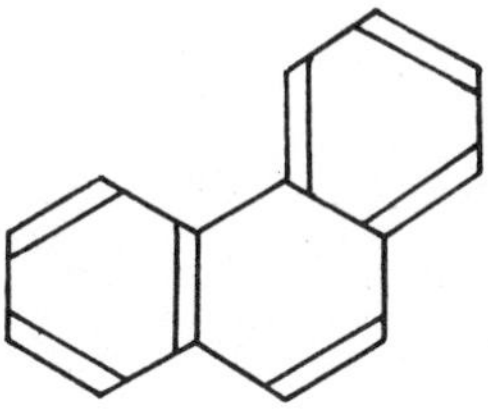   (XIV)

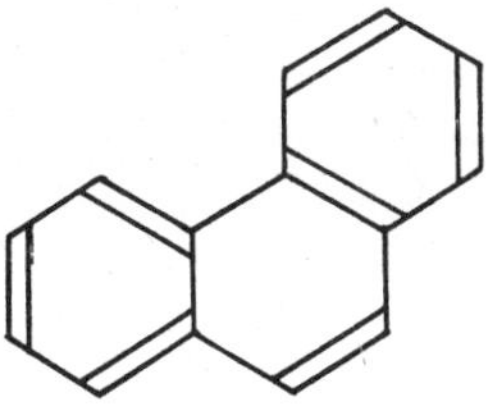

(XV) 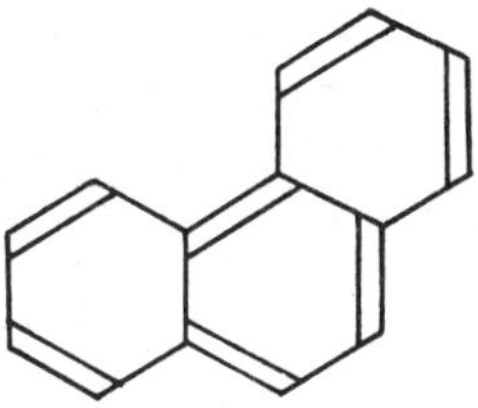   (XVI)

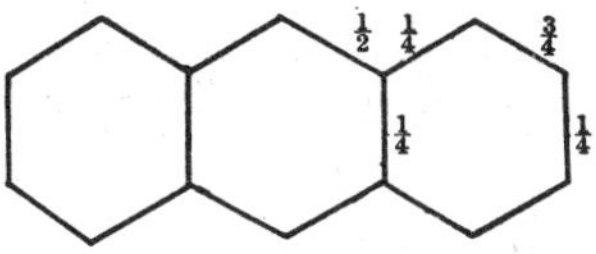

(XVII) 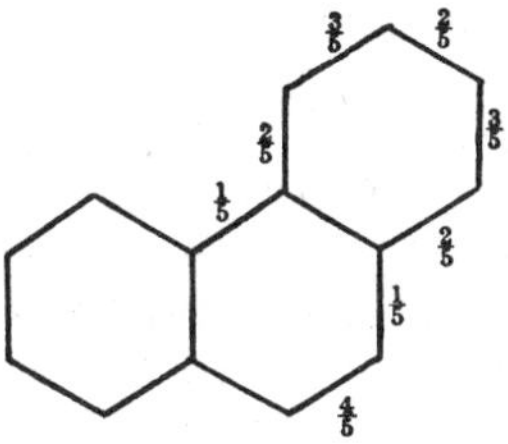   (XVIII)

The relationship between bond length and bond character has also been expressed by the equation

$$R = R_1 - (R_1 - R_2)\frac{3p}{2p+100}$$

or

$$p = \frac{100(R_1 - R)}{2R + R_1 - 3R_2},$$

* J. Amer. Chem. Soc. 1935, **57**, 2705; Pauling and Brockway, ibid. 1937, **59**, 1223.

where p is the percentage double-bond character, R is the interatomic distance, or bond length, R_1 is the length of a pure single bond, and R_2 is the length of a pure double bond. When $R = R_1$, p becomes zero, and when $R = R_2$, p becomes 100.

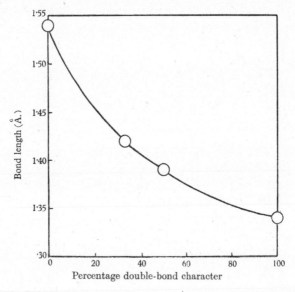

Fig. 5·2. Relationship between C—C bond lengths and percentage double-bond character. (After Pauling, Brockway and Beach, *J. Amer. Chem. Soc.* 1935, **57**, 2705.)

This relationship between bond length and double-bond character is of very considerable value in organic chemistry. For one thing, it is possible to predict the length of a bond from its double-bond character as calculated from the valence-bond structures of the molecule. On the other hand, using experimentally determined bond lengths, it is possible to construct a table of percentage double-bond characters (table 5·2). The observed variation in C—C bond length provides a convincing proof of the hybrid character of such bonds in conjugated molecules.

This description of a linkage in terms of percentage double-bond character, as suggested by Pauling, Brockway and Beach, was the first attempt to assign an 'order' to bonds. Although

TABLE 5·2. *Relationship between bond length and percentage double-bond character** *

Substance	C—C bond length	Percentage double-bond character
Diamond	1·54	0
Diphenyl	1·48†	12½
Glyoxal	1·47	15
cycloPentadiene	1·46	18
Butadiene	1·46‡	18
Furan	1·46‡	18
Thiophen	1·44‡	25
Graphite	1·42	33⅓
Benzene	1·39	50
Pyrazine	1·39	50
Resorcinol	1·39	50
Hexamethylbenzene	1·39	50
Diacetylene	1·36‡	75
Ethylene	1·34	100

very successful in explaining many of the properties of organic compounds, the method is somewhat over-simplified. Many attempts have therefore been made to describe the orders of bonds rather more rigorously. In all such methods the order is divided into two parts, that due to the σ-bonds, which is assumed to have the value unity, and that due to the π-bonds. The latter part, called the 'mobile order', p, has a value of 1 for ethylene and 2 for acetylene. The total bond orders of ethylene and acetylene are, therefore, 2·0 and 3·0, respectively, and ethane is given a bond order of 1·0, having no π-electrons.[§]

Penney[‖] showed that the method of Pauling is not completely satisfactory, as the mobile order is lowered by the addition of more excited canonical structures, although these structures increase the stability of the hybrid. He used the vector model of Dirac, and defined bond order in such a way that there is a

---

* Mainly after Wheland, *The Theory of Resonance*, see p. 108, John Wiley, New York, 1944.
† The bond joining the two benzene rings.
‡ This value applies to the length of the bond conventionally represented as a single bond.
§ These assumptions have, however, been criticized by Walsh (*Trans. Faraday Soc.* 1946, **42**, 779), who has developed a definition of bond order in terms of ionization potentials. In this method the bond order of diamond is 1·07, of ethylene 2·00, of acetylene 3·10, and of benzene 1·51.
‖ *Proc. Roy. Soc.* A, 1937, **158**, 306.

THE AROMATIC COMPOUNDS

direct relation between bond order and bond energy. According to this method the bond order in benzene is found to be 1·623, and that in graphite 1·45.

(XIX)

TABLE 5·3. *Relationship between the 'numbers' of a bond and its Penney type of bond order*

'Numbers' of bond	Penney type of bond order	'Numbers' of bond	Penney type of bond order
(2)	1·827	(2, 3, 3)	1·493
(3)	1·873	(2, 3, 4)	1·508
(1, 1)	1·437	(2, 4, 4)	1·537
(1, 2)	1·524	(3, 3, 3)	1·537
(1, 3)	1·566	(3, 3, 4)	1·556
(2, 2)	1·612	(3, 4, 4)	1·578
(2, 3)	1·652	(2, 2, 3, 3)	1·324
(2, 4)	1·676	(2, 2, 3, 4)	1·346
(3, 3)	1·692	(2, 2, 4, 4)	1·367
(3, 4)	1·714	(2, 3, 3, 3)	1·367
(4, 4)	1·735	(2, 3, 3, 4)	1·387
(1, 2, 3)	1·367	(2, 3, 4, 4)	1·407
(1, 2, 4)	1·387	(2, 4, 4, 4)	1·428
(1, 3, 3)	1·407	(3, 3, 3, 3)	1·407
(1, 3, 4)	1·428	(3, 3, 3, 4)	1·428
(1, 4, 4)	1·440	(3, 3, 4, 4)	1·440
(2, 2, 2)	1·412	(3, 4, 4, 4)	1·469
(2, 2, 3)	1·452	(4, 4, 4, 4)	1·489
(2, 2, 4)	1·472		

The method of calculation is certainly tedious, especially for large molecules. It is therefore of considerable interest that Vroelant and Daudel* have published a table which enables an approximate value of (a) the Penney type of bond order and (b) the interatomic distance, to be determined for aromatic and other unsaturated *hydrocarbons* without calculation. This table depends on the observation that the order of any bond is a function of the bonds adjacent to the one studied. Each bond is given a 'number', 1, 2, 3 or 4, depending on whether it is linked

* *C.R. Acad. Sci., Paris*, 1949, **228**, 399.

to one, two, three or four other carbon atoms. For example, the various bonds in naphthalene may be given the 'numbers' represented in (XIX).

To find the mobile order of any bond, the table is entered with the 'numbers' of the *adjacent* bonds, and the bond order is read off directly. The 'numbers' of the bonds adjacent to the αβ bond in naphthalene, for example, are 2 and 3. The total bond order for this bond is, therefore, 1·652. Similarly, the ββ bond has a bond order of 1·612.

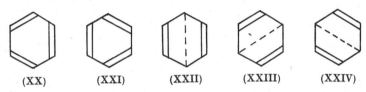

(XX)   (XXI)   (XXII)   (XXIII)   (XXIV)

Another approach has been made by Daudel and Pullman.* The complete canonical set for benzene includes three structures of Dewar type (XXII)–(XXIV) as well as the two structures of Kekulé type (XX) and (XXI); and each structure may be analysed in terms of the distribution of π-electrons, as follows:

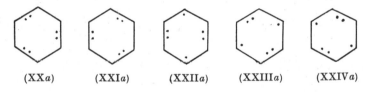

(XXa)   (XXIa)   (XXIIa)   (XXIIIa)   (XXIVa)

If the two Kekulé structures *alone* contribute to the benzene hybrid, then this hybrid can be represented by a regular hexagon having one π-electron along each side. If the Dewar structures *alone* contribute to the hybrid, then the structure can be represented by a regular hexagon having $\frac{1}{3}$ electron at each angle, and $\frac{2}{3}$ electron along each side. However, as Pauling and Wheland[†] have shown, all five structures contribute to the hybrid, and it is calculated that the Kekulé structures contribute 78 % to the hybrid, and the Dewar structures 22 %. Using these

---

* Ibid. 1945, **220**, 599; Daudel, Daudel, Jacques and Jean, *Rev. Sci.*, Paris, 1946, **84**, 489; Pullman, *Ann. Chim.* 1947, **2**, 5.
† *J. Chem. Phys.* 1933, **1**, 362.

figures, the distributions of π-electrons for the Kekulé structures and the Dewar structures may be represented as in (XXV) and (XXVI) respectively. The superposition of these two diagrams gives the 'molecular diagram' (XXVII) for benzene.

(XXV) 0·780

(XXVI) 0·073, 0·147, 0·073

(XXVII) 0·073, 0·927, 0·073

(XXVIII) 0·122, A 0·098, B
$A = 1·054$; $B = 0·752$

(XXIX) 0·250, 0·194, A 0·161, B
$A = 0·904$; $B = 0·774$

(XXX) 0·184, A, 0·172, B, 0·163, C, 0·197, D, 0·200
$A = 0·889$
$B = 0·767$
$C = 0·906$
$D = 0·891$

The figure $0·073e$, the 'concentration' of electrons at each carbon atom, represents what Daudel has called the *charge de sommet*, *indice de valence libre*, or free-valence number. The figure $0·927e$ represents the 'concentration' of electrons between each carbon atom, and has been called the *charge de liaison*. One-half of this value represents p, the mobile bond order, or *indice de liaison*, that is, the number of *pairs* of π-electrons between each two carbon atoms.*

The molecular diagrams for some other aromatic hydrocarbons are given in (XXVIII), (XXIX) and (XXX).

It will be noted that the most reactive positions have large

* Daudel and Daudel, *J. Chem. Phys.* 1948, 16, 639.

THE AROMATIC 'DOUBLE' BOND

free-valence numbers. In naphthalene (XXVIII) the 1-position has a larger free-valence number than the 2-position. In anthracene (XXIX) the 9-position has a larger free-valence number than either the 1- or 2-position; and in phenanthrene (XXX) the 9:10 positions are also characterized by large free-valence numbers. All these figures are in accord with the known reactivity of the 1-position in naphthalene, the 9:10 positions in anthracene, and 9:10 positions in phenanthrene.

These methods for the calculation of bond orders, based as they are on valence-bond structures, become increasingly tedious as the molecules become more complex. Coulson* has therefore proposed a new definition, based upon the method of molecular orbitals, which can be applied even in fairly complex cases without undue labour.

This method may briefly be described as follows. Each of the six mobile (π) electrons in benzene has a wave function of the form
$$\Psi = c_1\psi_1 + c_2\psi_2 + c_3\psi_3 + c_4\psi_4 + c_5\psi_5 + c_6\psi_6,$$
where all the carbon nuclei are numbered from 1 to 6, and ψ_1 is the dumb-bell orbital around nucleus 1, and so on. c_1^2 is the probability of finding the electron around nucleus 1, and c_2^2 is the probability of finding it around nucleus 2. Coulson defines the contribution of one π-electron to the mobile order, p, between neighbouring nuclei r and s, as $c_r c_s$. The total mobile order of the bond between r and s is therefore the sum of the contributions $c_r c_s$ from each of the six mobile electrons, i.e. $p = \Sigma c_r c_s$, and the total bond order is $1 + \Sigma c_r c_s$. The determination of the bond order therefore requires the evaluation of the coefficients in the wave function for each mobile electron. This evaluation has already been discussed in Chapter 2. In the ground state the three occupied molecular orbitals are:

$$\Psi_1 = \frac{1}{6^{\frac{1}{2}}}(\psi_1 + \psi_2 + \psi_3 + \psi_4 + \psi_5 + \psi_6),$$

$$\Psi_2 = \frac{1}{12^{\frac{1}{2}}}(2\psi_1 + \psi_2 - \psi_3 - 2\psi_4 - \psi_5 + \psi_6),$$

$$\Psi_3 = \frac{1}{4^{\frac{1}{2}}}(\psi_2 + \psi_3 - \psi_5 - \psi_6).$$

* *Proc. Roy. Soc.* A, 1939, **169**, 413.

We have, therefore,

Orbital	Population of orbital (n)	c_1	c_2	$n \times c_1 c_2$
1	2 electrons	$\dfrac{1}{\sqrt{6}}$	$\dfrac{1}{\sqrt{6}}$	$\dfrac{1}{3}$
2	2 electrons	$\dfrac{2}{\sqrt{12}}$	$\dfrac{1}{\sqrt{12}}$	$\dfrac{1}{3}$
3	2 electrons	0	$\dfrac{1}{\sqrt{4}}$	$\dfrac{0}{}$
			Mobile bond order	$\dfrac{2}{3}$

The mobile bond order is found to be $\tfrac{2}{3}$, whichever pair of carbon atoms is chosen.

The application of this method to benzene therefore gives a *total* bond order of 1·667, and also shows that all the C—C bonds

(XXXI) 1·725, 1·603, 1·725

(XXXII) 1·738, 1·586, 1·738

(XXXIII) 1·702, 1·623, 1·705, 1·775

are equivalent. Similar calculations for graphite give a bond order of 1·53; and the calculated bond orders for some aromatic hydrocarbons are attached to the formulae (XXXI)–(XXXVI).*

The physical reality of bond orders is confirmed by the fact that there is a strong smooth-curve relationship between bond order and experimentally determined bond lengths. This rela-

* Coulson, *Proc. Roy. Soc.* A, 1939, **169**, 413; Moffitt and Coulson, *Proc. Phys. Soc.* 1948, **60**, 309; Berthier, Coulson, Greenwood and Pullman, *C.R. Acad. Sci., Paris*, 1948, **226**, 1906; Coulson and Longuet-Higgins, *Rev. Sci., Paris*, 1947, **85**, 929.

THE AROMATIC 'DOUBLE' BOND

tionship is illustrated by the accompanying figure in which the points are those for ethane, graphite, benzene, ethylene and acetylene.

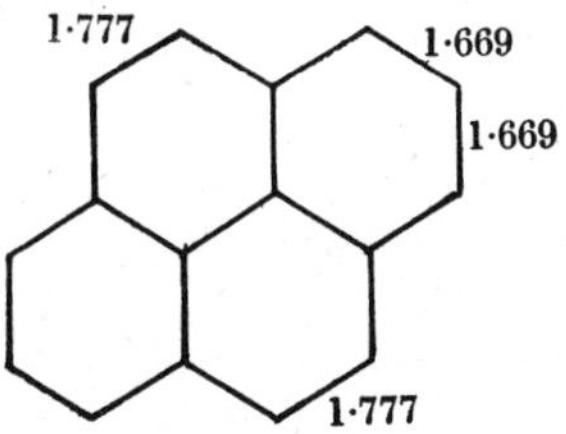

(XXXIV) with values 1·777, 1·669, 1·669, 1·777

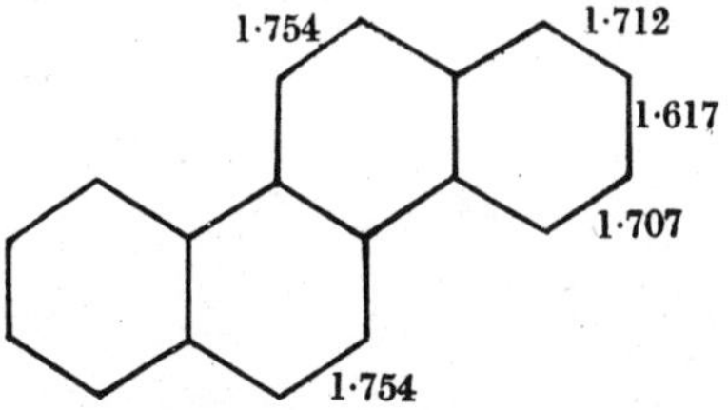

(XXXV) with values 1·754, 1·712, 1·617, 1·707, 1·754

(XXXVI) with values 1·695, 1·628, 1·700, 1·731, 1·593, 1·732, 1·783

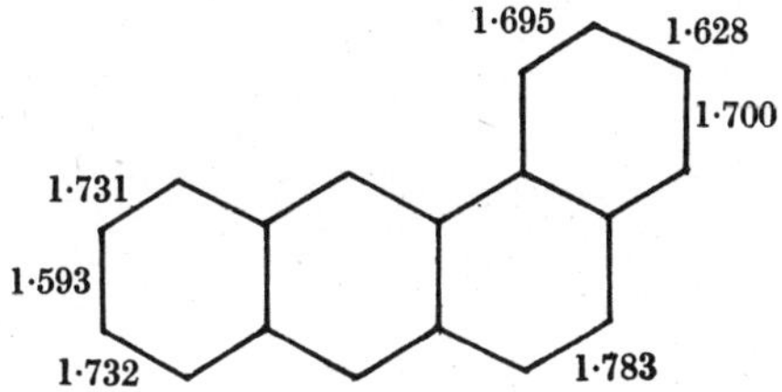

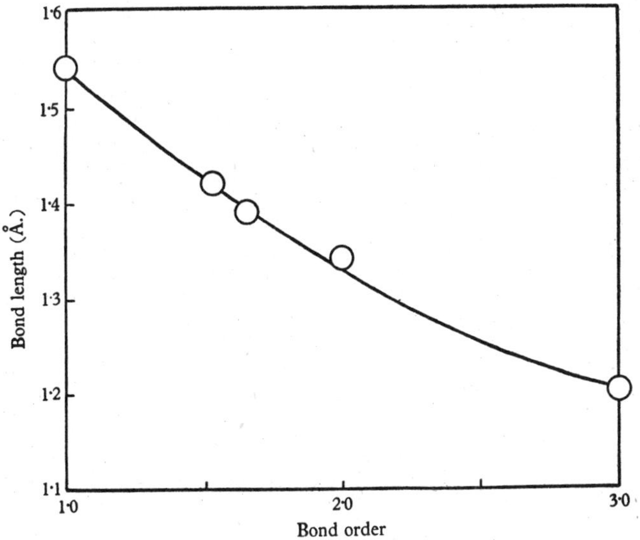

Fig. 5·3. Relationship between bond length and bond order (molecular orbital method).

THE AROMATIC COMPOUNDS

This relationship may also be expressed mathematically by the equation

$$x = s - \frac{s-d}{1 + 0.765\frac{(1-p)}{p}},$$

where x is the bond length, s is the length of a pure single bond, d the length of a pure double bond and p is the mobile bond order. In this way it is possible to calculate bond lengths from calculated bond orders, or, alternatively, to derive 'experimental' bond orders from measured or experimental bond lengths. Many interesting comparisons can be made in this way. For example, the measure of the agreement is illustrated by the remarkable correspondence between the observed and calculated bond lengths for the hydrocarbon ovalene (XXXVII) as enumerated in table 5·4.*

(XXXVII)

The concept of free valence has also been introduced into the molecular orbital treatment.† The free-valence number, F_r, at an atom r is defined as

$$F_r = N_{\text{max.}} - N_r,$$

where N_r is the sum of the orders of all bonds joining the atom r to the remainder of the system, and $N_{\text{max.}}$ is a constant depending

---

* Buzeman, *Proc. Phys. Soc.* 1950, **63**, 827.
† Coulson, *Disc. Faraday Soc.* 1947, **2**, 9; *J. Chim. Phys.* 1948, **45**, 243; Burkitt, Coulson and Longuet-Higgins, *Trans. Faraday Soc.* 1951, **47**, 553.

THE AROMATIC 'DOUBLE' BOND

TABLE 5·4. *Comparison between calculated and observed bond lengths for ovalene* (XXXVII)

Bond	Bond length	
	Calculated	Observed
A	1·410 Å.	1·404 Å.
B	1·427	1·441
C	1·380	1·345
D	1·425	1·433
E	1·418	1·403
F	1·421	1·428
G	1·424	1·426
H	1·428	1·435
J	1·422	1·416
K	1·430	1·461
L	1·425	1·442
M	1·387	1·383

on the chemical nature of the atom r and its state of hybridization. In aromatic ring systems the carbon atoms are all in the trigonal state of hybridization and N_{max} has a value of $3 + \sqrt{3}$, $3 + \sqrt{2}$ or $3 + \sqrt{1}$, according as the atom in question is bonded to three, two or one other trigonal carbon atoms. For primary, secondary and tertiary carbon atoms, therefore, F_r is related to the partial bond orders (p) by the three equations

$$F_r = 1 - \sum_s p_{rs},$$

$$F_r = 1 \cdot 414 - \sum_s p_{rs},$$

or

$$F_r = 1 \cdot 732 - \sum_s p_{rs}.$$

In benzene, for example, each C—C bond has a mobile order of $0 \cdot 66\dot{6}$ and each C—H bond a mobile order of 0. The free-valence number, F_r, for each carbon atom in benzene is, therefore,

$$1 \cdot 414 - (0 \cdot 66\dot{6} + 0 \cdot 66\dot{6} + 0) = 0 \cdot 081.$$

Using the concepts of bond order and free valence it is possible to interpret most of the reactions of aromatic hydrocarbons. In every case the most reactive bond, and the one preferentially attacked by double-bond reagents, is found to be the bond having the highest order. Moreover, the bonds having the highest orders are precisely those which, by virtue of certain reactions such as diazo coupling, the Claisen reaction, etc., were earlier supposed to be 'fixed' double bonds.

The 1:2 bond of naphthalene has a higher bond order than the 2:3 bond, and double-bond reagents preferentially attack the former. The 9:10 bond of phenanthrene has a higher order than any other bond in this molecule, and it is significant that osmium tetroxide and diazoacetic ester both attack this bond. The 1:2 bond of pyrene has the highest order and is also attacked preferentially by ozone and by osmium tetroxide. Similarly, osmium tetroxide attacks the bond of highest order in 1:2-benzanthracene (the 3:4 bond), in chrysene (the 1:2 bond), in 3:4-benzopyrene (the 6:7 bond), and so on.

Kooyman and Ketelaar* have suggested that the initial process in the addition of a double-bond reagent to an aromatic bond is the conversion of the aromatic bond into a 'pure' double bond. That is, two π-electrons have to be 'localized', the remaining π-electrons of the system rearranging to form a conjugated 'residual molecule' of minimum energy. The energy required to bring about this localization has been called the *bond-localization energy*.

The magnitude of the bond-localization energy for any particular bond can be determined without difficulty. In the simple treatment of Kooyman and Ketelaar,* who first introduced the quantity, the localization energy was estimated by subtracting the resonance energy of the 'residual' molecule (XXXIX) from that of the original molecule (XXXVIII). The 'residual' molecule formed when one double bond is localized in benzene simulates butadiene. Taking the resonance energy of benzene to be 39 kcal./mole, and that of butadiene to be 3·5 kcal./mole, the localization energy for one double bond in benzene is 35·5 kcal./mole. Similarly, the energy for the localization of a double bond in the 1:2 position of naphthalene is 29 kcal./mole, this being the difference between the resonance energies of naphthalene and benzene, minus the energy of conjugation of a double bond with an aromatic nucleus: $(75-39) - 7 = 29$ kcal./mole.

* *Rec. trav. chim. Pays-Bas*, 1946, **65**, 859.

THE AROMATIC 'DOUBLE' BOND

This treatment suffers from the disadvantage that very few experimental resonance energy values are known. For this reason Ketelaar and van Dranen* and Brown† have calculated the bond-localization energies for a number of compounds using the molecular orbital method.

(XXXVIII)   (XXXIX)

(XL)   (XLI)

According to Brown, the bond-localization energy for benzene is 30 kcal./mole; that for the 1:2 bond of naphthalene is 22 kcal./mole; and that for the 2:3 bond of naphthalene, 37 kcal./mole (see (XLII)). This large difference in the bond-localization energies for the 1:2 and 2:3 bonds is the reason naphthalene appears to react as if a double bond is available only at the 1:2, 3:4, 5:6 and 7:8 positions; the localization of a double bond at the 2:3 and 6:7 positions requires such a large energy that this process can never occur. Similarly, the enhanced reactivity and double-bond character of the 9:10 bond of phenanthrene is indicated by the small bond-localization energy of 16 kcal./mole (see (XLIII)).

Brown‡ has also shown that the bond-localization energies can be used to derive relative reaction rates for the addition of double-bond reagents, and a remarkable agreement between the predicted and observed relative rates is obtained. Somewhat similar calculations have also been made by Ketelaar and van

* Ibid. 1950, **69**, 477.
† Aust. J. Sci. Res. A, 1949, **2**, 564; J. Chem. Soc. 1950, p. 3249.
‡ Ibid.

Dranen* in their discussions on the rate of ozonization of benzene and naphthalene. It seems, therefore, that the entropy of activation remains nearly constant from compound to compound, and that the bond-localization energy is a good approximation to the variable portion of the heat of activation.

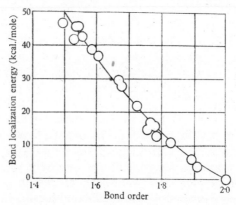

The significance of bond-localization energy is also illustrated by the fact that there is a smooth-curve relationship (fig. 5·4) between bond-localization energy and bond order.†

Fig. 5·4. Relationship between bond order and bond-localization energy. (From Badger, *Quart. Rev. Chem. Soc., Lond.*, 1951, 5, 168.)

Bond-localization energies are also very closely related to the oxidation-reduction potentials of the corresponding *ortho*-quinones, for such potentials are measures of the energy required to convert the quinols (XLIV) into the quinones (XLV).

* *Rec. trav. chim. Pays-Bas*, 1950, **69**, 477.
† *Aust. J. Sci. Res.* A, 1949, **2**. 564; Badger, *Quart. Rev. Chem. Soc., Lond.*, 1951, **5**, 147.

A linear relationship can be demonstrated between the two quantities so that the bond-localization energy can be evaluated provided the redox potential of the corresponding *ortho*-quinone has been determined.*

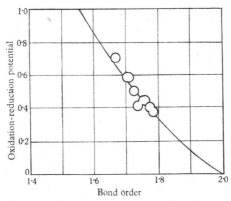

(XLIV)   (0·41 volt) ⇌   (XLV)

This linear relationship between redox potential and bond-localization energy, and the smooth-curve relationship between bond-localization energy and bond order, implies that there

Fig. 5·5. Relationship between bond order and the redox potential of the corresponding *ortho*-quinone. (From Badger, *Quart. Rev. Chem. Soc., Lond.*, 1951, **5**, 169.)

must be a relationship between bond order and bond-localization energy (fig. 5·5). It is possible to obtain a 'theoretical' curve, and the experimental values agree very closely.†

Summarizing, therefore, it seems that the bond corresponding to an *ortho*-quinone having a low redox potential not only has

---

* Compare Evans, Gergely and de Heer, *Trans. Faraday Soc.* 1949, **45**, 312.
† Badger, *J. Chem. Soc.* 1950, p. 1809; *Quart. Rev. Chem. Soc., Lond.*, 1951, **5**, 147.

a low bond-localization energy, but is also characterized by its high reactivity, its high bond order, and its short length.

These relationships are illustrated in table 5·5.

TABLE 5·5. *Relation between bond order, bond length, localization energy, and the redox potential of the corresponding* ortho-*quinone*

Compound	Bond	Bond order (calculated)	Bond length (experimental) (Å.)	Bond-localization energy (calculated) (kcal./mole)	'Corrected' redox potential of corresponding *ortho*-quinone (experimental) (volts)
Benzene	1:2	1·667	1·39	30	0·71
Naphthalene	1:2	1·725	1·36	22	0·51
	2:3	1·603	1·395	37	—
Anthracene	1:2	1·738	1·36	20	0·42
	2:3	1·586	1·39	38	—
Phenanthrene	1:2	1·705	—	24	0·59
	2:3	1·623	—	34	—
	3:4	1·702	—	24	0·57
	9:10	1·775	—	16	0·41
1:2-Benz-anthracene	3:4	1·783	—	15	0·38
Chrysene	1:2	1·754	—	18	0·45
Picene	5:6	1·758	—	20	0·45

Finally, the relationship just discussed between bond order and the redox potential of the corresponding *ortho*-quinone makes it possible to re-examine the use of substitution reagents in the early attempts to demonstrate bond fixation in naphthalene derivatives. There is little doubt that the transition state which is formed during such substitution reactions has a quasi-quinonoid character. In the case of 2-naphthol the transition state (XLVI) corresponding to substitution in the 1-position is related to 1:2-naphthaquinone; that corresponding to substitution in the 3-position (XLVII) is related to 2:3-naphthaquinone. The above relationships between redox potential, bond order, and bond-localization energy clearly indicate that the formation of the latter transition state would involve such a high energy that it never, or hardly ever, occurs.

It may be concluded, therefore, that in the substitution of *mono-substituted* aromatic compounds the reagent enters the

THE AROMATIC 'DOUBLE' BOND

ortho position which is linked to the substituent by the bond of the higher order. It never, or hardly ever, enters the alternative *ortho* position which is linked to the substituent by the bond of smaller order.*

(XLVI)   (XLVII)

* Badger, *Quart. Rev. Chem. Soc., Lond.*, 1951, **5**, 147; compare Waters, *J. Chem. Soc.* 1948, p. 727.

CHAPTER 6

THE EFFECTS OF SUBSTITUENTS

THE INDUCTIVE EFFECT

It has been known since the time of Faraday that molecules can be polarized by induction. When placed in an electrical field, the positive and negative electrical centres of each molecule become temporarily displaced, with the formation of a dipole. These dipoles are induced in such a way as partially to neutralize the applied field. The temporary polarization induced in this way should be independent of temperature, for the inertia is negligible and thermal motion is not a factor, but with some substances experiment has shown that the polarization *is* dependent on the temperature. In 1912 Debye suggested that these apparent anomalies can be readily explained if it is assumed that some molecules possess a *permanent* dipolar character, and that the total polarization is the sum of the induced polarization (polarizability) and the permanent polarization. The thermal motion would tend to prevent the permanent dipoles aligning themselves in the field, and as thermal motion is a function of temperature, this is, of course, the basis of the temperature dependence.

Further work has confirmed Debye's assumption, and methods have now been devised* for the quantitative evaluation of the permanent polarization in terms of *dipole moments*. A dipole moment (μ) is the product of the distance separating the 'electrical centres' of the molecule, and the magnitude of the charge at these points. It is generally measured in electrostatic units (e.s.u.), and 10^{-18} e.s.u. is called a Debye (D.).

These methods measure the *resultant* dipole moment of the *molecule*, but permanent polarization is really a property of *bonds*. Practically all covalent bonds are more or less polar. The only completely non-polar bonds are those which exist in diatomic molecules such as N_2, H_2, etc., which are composed of like atoms, and in completely symmetrical bonds, such as the C—C bond in CH_3—CH_3. The coordinate covalent bonds are,

* See Le Fèvre, *Dipole Moments*, 2nd ed., Methuen, London, 1948.

THE EFFECTS OF SUBSTITUENTS

of course, highly polar. The combination of the electrically neutral NMe_3 molecule and an uncharged oxygen atom, for example, gives a bond which can only be represented as Me_3N^+—O^-. Such coordinate covalent bonds represent the extreme case, it is true, but ordinary covalent bonds also have some polar character for the simple reason that some atoms are more electronegative than others. As Lewis wrote in 1923:*
'The pair of electrons which constitutes the bond may lie between two atomic centers in such a position that there is no electric polarization, or it may be shifted towards one or the other atom in order to give to that atom a negative, and consequently to the other atom a positive charge.'

Fig. 6·1. Diagrammatic representation of the molecular orbital for hydrogen chloride.

In terms of molecular orbital theory this can be expressed by saying that although the molecular orbital for a homonuclear diatomic molecule such as H_2 has a boundary surface of the symmetrical 'sausage type', that for a heteronuclear molecule is unsymmetrical. Hydrogen chloride, for example, has a permanent polarization. It has a dipole moment of 1·04 D., so that there must be an excess of negative charge on one or other of the atoms, and consequently an excess of positive charge on the other. A variety of evidence indicates that chlorine is more electronegative than hydrogen, and that it is the chlorine atom in hydrogen chloride that acquires the negative character. This means that the boundary surface of the molecular orbital cannot be symmetrically disposed about the plane which is perpendicular to the bond, but that there must be a greater electron-cloud density around the chlorine atom. The formation of the molecular orbital for hydrogen chloride may therefore be represented diagrammatically as in fig. 6·1.

* *Valence and the Structure of Atoms and Molecules*, p. 83, Chemical Catalog Co., New York, 1923.

THE AROMATIC COMPOUNDS

As already mentioned, the formation of a bond having some polar character is dependent on the relative electronegativity of the two atoms. Pauling* has devised a table of such electronegativities, and has assigned relative values of electronegativity to the elements.

TABLE 6·1. *Pauling's electronegativity values for some elements*

H 2·1							
	Li 1·0	Be 1·5	B 2·0	C 2·5	N 3·0	O 3·5	F 4·0
	Na 0·9	Mg 1·2	Al 1·5	Si 1·8	P 2·1	S 2·5	Cl 3·0
	K 0·8	Ca 1·0			As 2·0	Se 2·4	Br 2·8
	Rb 0·8	Sr 1·0					I 2·4

There seems to be an increase in electronegativity as one moves from left to right in each period in the Periodic Table, and a decrease in electronegativity with increasing atomic weight in each group. It seems likely, therefore, that electronegativity is associated with the magnitude of the positive charge on the nucleus of an element, and that as the atoms become larger the effect is weakened by the greater distance between the atoms in a molecule, and by the greater screening effect of the increased number of non-valency electrons.

In heteronuclear diatomic molecules such as hydrogen chloride, the magnitude of the dipole moment may be taken as a measure of the difference in electronegativity of the two atoms. The very great majority of molecules, however, have more than two atoms, and the situation becomes more complex. The experimental methods give only the *resultant* dipole moment of the molecule. In some cases two or more dipoles may neutralize one another and the molecule may have no resultant dipole moment, indicating that dipole moments are vector quantities. Carbon

* *The Nature of the Chemical Bond*, 2nd ed. p. 64, Cornell University Press, 1940.

THE EFFECTS OF SUBSTITUENTS

tetrachloride, for example, has no resultant dipole moment, as the four polar C—Cl bonds are symmetrically disposed towards the corners of a regular tetrahedron. This condition does not apply in chloroform, on the other hand, and this substance has a resultant dipole moment of 1·05 D. Similarly, carbon dioxide has no dipole moment, although carbon monoxide has a moment of 0·11 D. The former compound must be linear, O=C=O, so that the two dipoles are neutralized. For similar reasons, water, having a dipole moment of 1·84 D., must be triangular.

Very many dipole moments have now been determined. In the following table (table 6·2) a series of substituted methanes has been collected, the magnitude of the dipole moment in each case being the resultant of all the *bond* moments in the molecule.

TABLE 6·2. *Dipole moments of* CH_3—X *compounds**

Compound	Dipole moment (D.)	Compound	Dipole moment (D.)
CH_3—CH_3†	0·0	CH_3—Br	1·79
CH_3—CO_2H	0·8	CH_3—F	1·81
CH_3—NH_2	1·28	CH_3—Cl	1·86
CH_3—OCH_3	1·32	CH_3—CO_2Et	1·86
CH_3—SCH_3	1·40	CH_3—CHO	2·72
CH_3—OH	1·68	CH_3—NO_2	3·0
CH_3—I	1·62	CH_3—CN	3·2

The fact that dipole moments are vector quantities also makes it possible to determine the *direction* of polarization. Toluene (I) has a dipole moment of 0·4 D. Chlorobenzene (II) has a moment of 1·6 D. *para*-Xylene and *para*-dichlorobenzene have zero moments, and *para*-chlorotoluene (IV) has a moment of 1·9 D. As this latter figure is approximately the sum of the moments for chlorobenzene and for toluene, it must be concluded that the C—Cl and C—CH_3 moments act in such a way that they reinforce one another. That is, one substituent attracts electrons *from* the ring and the other donates electrons *to* the ring. Similarly, nitrobenzene (III) has a moment of 4·0, and as *para*-chloronitrobenzene (V) has a moment of 2·6 D., it seems that the C—NO_2 and C—Cl moments act in such a way as partially to neutralize

* Compiled from Le Fèvre, *Dipole Moments*, p. 110.
† *All* paraffin hydrocarbons have zero moments, indicating that a CH_3 group is neither electron-donating nor electron-attracting in *saturated* systems. In *unsaturated* systems it acts as an electron-donor.

one another. The nitro group may be represented as $-N^+{\overset{O^-}{\underset{O}{<}}}$, so there can be little doubt that the polar character of a C—NO_2 bond must be represented as C↦NO_2. The polar characters of the C—CH_3 and C—Cl bonds must, by the above reasoning, be C↤CH_3 and C↦Cl respectively.

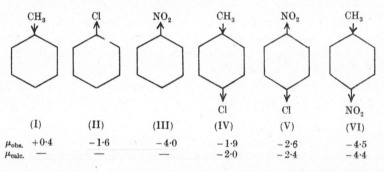

	(I)	(II)	(III)	(IV)	(V)	(VI)
$\mu_{obs.}$	+0·4	−1·6	−4·0	−1·9	−2·6	−4·5
$\mu_{calc.}$	—	—	—	−2·0	−2·4	−4·4

For *ortho-* and *meta-*disubstituted benzene compounds, the resultant dipole moment can also be calculated by vector addition. In all cases the resultant moment, μ_R, is given by

$$\mu_R = \sqrt{(\mu_1^2 + \mu_2^2 + 2\mu_1\mu_2 \cos \theta)},$$

where μ_1 and μ_2 are the moments of the mono-substituted benzenes, and θ is the angle between the axes of the two substituents. For *ortho-*disubstituted compounds (VII), $\theta = 60°$; for

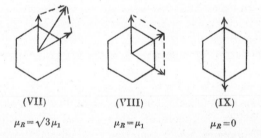

(VII)   (VIII)   (IX)
$\mu_R = \sqrt{3}\mu_1$ $\mu_R = \mu_1$ $\mu_R = 0$

meta compounds (VIII), $\theta = 120°$; and for *para* compounds (IX), $\theta = 180°$. If μ_1 and μ_2 are identical, the resultant moments are therefore $\sqrt{3}\,\mu_1$, μ_1 and zero. For the chlorobenzenes, therefore, the observed and calculated dipole moments can be compared, as follows:

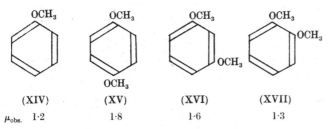

	(X)	(XI)	(XII)	(XIII)
$\mu_{obs.}$	1·6	0·0	1·5	2·25
$\mu_{calc.}$	—	0·0	1·5	2·8

The agreement between $\mu_{obs.}$ and $\mu_{calc.}$ is good, except in the case of the *ortho*-disubstituted compound. Interaction between the substituents probably occurs in this case, leading to a smaller observed dipole moment.*

Of course, this treatment assumes that the substituent is axially symmetrical, a condition which is fulfilled with some substituents, such as —NO_2, —Cl, etc., but not by others. The methoxy group, for example, is not axially symmetrical, and the methoxybenzenes (XIV)–(XVII) have the dipole moments indicated.

	(XIV)	(XV)	(XVI)	(XVII)
$\mu_{obs.}$	1·2	1·8	1·6	1·3

The study of the dipole moments of organic molecules therefore demonstrates conclusively that substituents do exert a permanent electrical effect, either attracting or releasing electrons. It is shown below that the general conclusions of this work are confirmed by a study of the effects of substituents on the

---

* With non-identical substituents the agreement is usually satisfactory only with *meta* derivatives. The substituents in *ortho* and in *para* derivatives usually interact with one another. See Syrkin and Dyatkina, *Structure of Molecules and the Chemical Bond* (translated by Partridge and Jordon), pp. 219 et seq., Butterworths Scientific Publications, London, 1950.

strengths of acids and bases. Moreover, the study of acid and base strengths provides valuable information on the way in which effects are transmitted from one part of the molecule to the other.

Organic acids in water undergo dissociation according to the scheme
$$R.CO_2H + H_2O \rightleftharpoons R.CO_2^- + H_3O^+,$$
and the dissociation constant, K_a, is given by
$$K_a = \frac{[H_3O^+][R.CO_2^-]}{[R.CO_2H]}.$$

The acidic strength is usually expressed as pK_a, which is defined by the expression $pK_a = -\log K_a$.

Evaluated in this way, the strengths of organic acids vary over a wide range, depending on the nature of the radical, R. Certain radicals favour the right-hand side of the equilibrium, while others favour the left-hand side. In other words, the nature of the radical, R, governs whether the removal of a hydrogen ion is facilitated or impeded, as compared with the case of $R = H$.

TABLE 6·3. *Strengths of substituted acetic acids**
$X.CH_2.CO_2H$

X	pK_a	X	pK_a
CO_2^-	5·7	I	3·0
CH_3	4·9	Br	2·9
H	4·7	Cl	2·8
Ph	4·3	CO_2H	2·8
OH	3·8	F	2·7
OCH_3	3·6	CN	2·4
OPh	3·1	NH_3^+	2·3

The effect of substituents on the strength of acids is most clearly shown in the series of substituted acetic acids having the general formula $X.CH_2.CO_2H$. The strengths of these acids are given in table 6·3.

In this series of acids, the only variable is the group X. If the substituent produces an increase in acid strength it means that the hydrogen of the carboxylic acid group has an increased

* Mainly after Branch and Calvin, *The Theory of Organic Chemistry*, p. 224, Prentice-Hall, 1941.

THE EFFECTS OF SUBSTITUENTS

tendency to separate as a proton, and X must be electron-attracting. Alternatively, if the substituent produces a decrease in acid strength, it must be electron-repelling. That is, the substituent makes it more difficult for the hydrogen to separate as a proton. Part of the effect may be an electrostatic field effect, but part is certainly transmitted by induction through the covalent bonds from atom to atom. It is not possible, however, to separate these two effects; both together form the *inductive effect*. The inductive effect is usually represented by an arrow on the valence bond in the direction of the permanent electron displacement.

When X is electronegative relative to hydrogen, and attracts electrons, the molecule may be represented as

$$X \mathbin{-\!\!\!\leftarrow} CH_2 \mathbin{-\!\!\!\leftarrow} CO_2 \mathbin{-\!\!\!\leftarrow} H.$$

In this case, X is said to exert a $-I$ effect. On the other hand, when X, relative to hydrogen, releases electrons, the molecule becomes $X \mathbin{\rightarrow\!\!\!-} CH_2 \mathbin{\rightarrow\!\!\!-} CO_2 \mathbin{\rightarrow\!\!\!-} H$, and X is said to exert a $+I$ effect. The resulting acid is weaker than acetic acid.

Naturally enough the NH_3^+ group is very strongly electronegative, and may be said to exert a large $-I$ effect. The positive charge exerts a profound pull on all the electrons in the molecule. In the same way, the CO_2^- group releases electrons to the molecule, and is said to exert a large $+I$ effect. It is sometimes referred to as an electron 'source', while $-I$ substituents are electron 'sinks'. Of the groups which have no formal charge, only the alkyl groups have $+I$ effects. All the other substituents have $-I$ effects and attract electrons, thereby increasing the strength of the acid. It is interesting to note that the $-I$ effects of the halogens are in the order $F > Cl > Br > I$, as expected from Pauling's electronegativity values.

The inductive effect of a substituent as evaluated in this way closely follows that estimated by a study of the dipole moments of compounds such as CH_3-X. A smooth curve is obtained by plotting the logarithms of the dissociation constants of a series of substituted acetic acids against the dipole moments of the corresponding substituted methanes.* The existence of this

---

* Nathan and Watson, *J. Chem. Soc.* 1933, p. 890; see also Dippy and Lewis, ibid. 1937, p. 1008.

relationship clearly shows that either set of data may be used as a measure of the inductive effect in such non-resonating molecules.

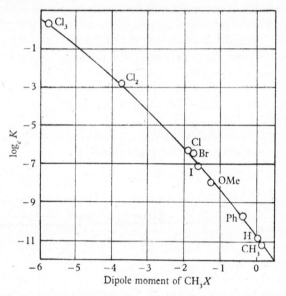

Fig. 6·2. Curve showing the relationship between the dissociation constants of acids and the dipole moments of substituted hydrocarbons. (From Nathan and Watson, *J. Chem. Soc.* 1933, p. 894.)

The inductive effect of certain substituents can also be evaluated by consideration of their effect on the strengths of bases. The strength of an aliphatic amine is lowered by the introduction of substituents such as Cl, NO_2, OH, CO_2Et, but is increased by alkyl groups. The basicity is dependent on the availability of the lone pair of electrons on the nitrogen. Substituents which exert a $-I$ effect (Cl, NO_2, OH, CO_2Et) evidently reduce the basicity by attracting these electrons, and substituents which exert a $+I$ effect (alkyl groups) increase the availability of these electrons. The greater the inductive effect, the greater the effect on the basicity.* These effects may be recognized in the following short table of basicities (table 6·4), in which the base strength is given by the pK_a value.

* See also Albert, *Chem. and Ind.* 1947, p. 51.

This table shows that the substituents NH_2, OH and $CO.CH_3$ have $-I$ effects, and that the ethyl group exerts a $+I$ effect.

TABLE 6·4. *Strengths of substituted amines*, NH_2—X

Base	pK_a
NH_3	9·2
NH_2—NH_2	8·2
NH_2—OH	6·0
NH_2—Et	10·7
NH_2—$CO.CH_3$	-0.5

These methods provide very useful information as to the sign and magnitude of the inductive effects of substituents. They may also be adapted to determine how far the effect of a substituent can be transmitted. It can be shown, for example, that the effect of a substituent on the strength of an aliphatic amine is the greater the nearer it is to the nitrogen atom. Similarly, the effect of a substituent on the strength of an aliphatic acid is also rapidly damped as the distance between the substituent and the carboxylic acid group increases. This may be appreciated by a study of the effects of halogenation on the homologous series of fatty acids (table 6·5).

TABLE 6·5. *Strengths of halogenated fatty acids*

Acid	pK_a	Δ	Acid	pK_a	Δ
$CH_3.CO_2H$	4·7		$CH_3.CO_2H$	4·7	
$CH_2Br.CO_2H$	2·9	1·8	$CH_2Cl.CO_2H$	2·8	1·9
$CH_3.CH_2.CO_2H$	4·9		$CH_3.CH_2.CO_2H$	4·9	
$CH_2Br.CH_2.CO_2H$	4·6	0·3	$CH_2Cl.CH_2.CO_2H$	4·0	0·9
$CH_3.CH_2.CH_2.CO_2H$	4·8		$CH_3.CH_2.CH_2.CO_2H$	4·8	
$CH_2Br.CH_2.CH_2.CO_2H$	4·6	0·2	$CH_2Cl.CH_2.CH_2.CO_2H$	4·5	0·3

In the homologous series of bromo acids, bromoacetic acid is much stronger than acetic acid; but as the distance between the bromine atom and the acid group increases, the acid strength becomes rapidly weaker, approaching that of the unsubstituted acid. γ-Bromobutyric acid is only very slightly stronger than butyric acid itself. Similarly, the chloro acids become rapidly weaker as the distance between the chlorine atom and acid group increases, and γ-chlorobutyric acid is only slightly stronger than butyric acid.

The same conclusion results from a study of the α-, β- and γ-chlorobutyric acids, the strengths of which are given in table 6·6.

TABLE 6·6. *Strengths of chlorobutyric acids*

Acid	pK_a
$CH_3.CH_2.CHCl.CO_2H$	2·85
$CH_3.CHCl.CH_2.CO_2H$	4·05
$CH_2Cl.CH_2.CH_2.CO_2H$	4·5
$CH_3.CH_2.CH_2.CO_2H$	4·8

All these observations indicate that the inductive effect of the halogens can only be transmitted through a saturated chain of about four carbon atoms, after which the effect is so small as to be negligible.

THE MESOMERIC EFFECT

Substituents have a profound effect on the strengths of aromatic as well as aliphatic acids, but the effects in the two series are by no means identical. A methoxy group, for example, increases the strength of acetic acid; but a *para*-methoxy group in benzoic acid is acid-weakening. A hydroxy group increases the strength of acetic acid, but a *para*-hydroxy group in benzoic acid is acid-weakening. On the other hand, a methyl group is acid-weakening in acetic acid and a *para*-methyl group also weakens the strength of benzoic acid. Similarly, *para*-chloro, bromo and nitro substituents in benzoic acid are acid-strengthening, as might be expected from a consideration of the inductive effects of these substituents. A different series of effects can be observed if the *meta*-substituted benzoic acids are considered. A *meta*-methyl group is acid-weakening, but all other substituents in this position (table 6·7) bring about an increase in acid strength. Moreover, nearly all substituents in the *ortho* position, regardless of the sign of their inductive effects, are acid-strengthening. It seems clear that the inductive effect is not the *only* effect operating in substituted aromatic compounds.

THE EFFECTS OF SUBSTITUENTS

TABLE 6·7. *Strengths of substituted benzoic acids**

Acid	pK_a	Acid	pK_a
Benzoic	4·2	o-Hydroxybenzoic	3·0
o-Chlorobenzoic	2·9	m-Hydroxybenzoic	4·1
m-Chlorobenzoic	3·8	p-Hydroxybenzoic	4·5
p-Chlorobenzoic	4·0	o-Methoxybenzoic	4·1
o-Bromobenzoic	2·8	m-Methoxybenzoic	4·1
m-Bromobenzoic	3·9	p-Methoxybenzoic	4·5
p-Bromobenzoic	3·0	o-Toluic	3·9
o-Nitrobenzoic	2·2	m-Toluic	4·3
m-Nitrobenzoic	3·4	p-Toluic	4·4
p-Nitrobenzoic	3·4		

Similar anomalies may be observed among the aromatic bases. All the alkylamines, whether primary, secondary or tertiary, are strong bases with pK_a of the order of 10·0–11·3. Ammonia itself, and *cyclo*hexylamine, are also strong bases. Aniline, however, is only a weak base, with pK_a 4·6, much weaker than would be expected from a consideration of the $-I$ effect of the phenyl radical, at least as determined by the acid strength of phenylacetic acid (table 6·3).[†] Furthermore, *para*-methoxy and *para*-ethoxy substituents increase the basic strength, while the same substituents in the *meta* position decrease it. Again, a *meta*- or *para*-amino substituent increases the basic strength of aniline in spite of the fact that this substituent has the opposite effect in aliphatic amines (and in compounds such as NH_2—NH_2). On the other hand, chloro and nitro substituents decrease the basicity in much the same way as might be expected (table 6·8).

A third type of anomalous result is observed with the aromatic hydroxy compounds, the phenols. The acid strength of phenol itself is very considerably greater than that of any aliphatic alcohol. This does not arise from the inductive effect of the phenyl group, for, as already mentioned, this effect is relatively small. Some other factor must therefore be involved. Furthermore, various anomalies are apparent in the effects of certain

---

* From *International Critical Tables* and *Landolt-Börnstein Tabellen*; see also Shorter and Stubbs, *J. Chem. Soc.* 1949, p. 1180; Dippy, *Chem. Rev.* 1939, 25, 151.

† Benzylamine has a pK_a of 9·4, and is therefore only slightly less basic than the simple alkylamines (ethylamine, pK_a 10·7). This relatively small reduction in basicity is due to the $-I$ effect of the phenyl group.

TABLE 6·8. *Strengths of substituted anilines**

Base	pK_a	Base	pK_a
Aniline	4·6	o-Phenetidine	4·5
o-Chloroaniline	2·8	m-Phenetidine	4·5
m-Chloroaniline	2·1	p-Phenetidine	5·3
p-Chloroaniline	3·5	o-Phenylene diamine	4·4
o-Nitroaniline	−0·2	m-Phenylene diamine	4·9
m-Nitroaniline	+2·6	p-Phenylene diamine	6·1
p-Nitroaniline	+1·1	o-Toluidine	4·4
o-Aminophenol	4·3	m-Toluidine	4·7
m-Aminophenol	4·8	p-Toluidine	5·1
p-Aminophenol	5·5		
o-Anisidine	4·5		
m-Anisidine	4·2		
p-Anisidine	5·3		

substituents on the acid strengths of phenols. The halogens increase the acid strength more or less as might be expected, but the effect of a nitro substituent is much greater in the *ortho* or *para* position than when in the *meta* position (table 6·9). This is clearly demonstrated by the fact that 2:4-dinitrophenol is a much stronger acid than 3:5-dinitrophenol. The substituents evidently act by withdrawing electrons from the O—H bond, allowing the hydrogen to separate more readily as a proton, but clearly the inductive effect is not the only process operating in these cases.

Further anomalies are revealed by a comparison of the dipole moments of substituted aromatic and aliphatic compounds, and this comparison reveals the nature of the additional effect which operates in aromatic (or unsaturated) systems.

The most striking example of the difference is to be found by comparison of the moments of aliphatic and aromatic amines. It has been shown, for example, that the directions of the dipoles in the two types of compound are reversed. In aliphatic compounds the —NR_2 group is electron-attracting (−I); in aromatic compounds, however, the —NR_2 group is electron-releasing.[†]

---

* After Albert, *Chem. and Ind.* 1947, p. 51; Hall and Sprinkle, *J. Amer. Chem. Soc.* 1932, **54**, 3469.

† Højendahl, *Studies of Dipole Moments*, Bianco Lunos Bogtrykkeri, Copenhagen, 1928; Ingold, *Ann. Rep. Chem. Soc.* 1926, **23**, 144.

THE EFFECTS OF SUBSTITUENTS

TABLE 6·9. *Strengths of substituted phenols**

Phenol	pK_a	Phenol	pK_a
Phenol	9·95	o-Iodophenol	9·0
o-Fluorophenol	8·8	m-Iodophenol	9·4
m-Fluorophenol	9·3	p-Iodophenol	9·7
p-Fluorophenol	9·95	o-Nitrophenol	7·2
o-Chlorophenol	8·5	m-Nitrophenol	8·35
m-Chlorophenol	9·0	p-Nitrophenol	7·1
p-Chlorophenol	9·4	2:4-Dinitrophenol	3·0
o-Bromophenol	9·0	3:5-Dinitrophenol	6·7
m-Bromophenol	9·4	2:6-Dinitrophenol	3·7
p-Bromophenol	9·8		
o-Cresol	10·2		
m-Cresol	10·0		
p-Cresol	10·1		

Other, less striking, differences have also been observed. Sutton[†] compared the dipole moments of a series of substituted aromatic compounds (Ph—X) with those of a series of similarly substituted aliphatic compounds of the general formula Me_3C—X ('Alphyl—X') in which the substituent is linked to a tertiary carbon atom (table 6·10). The moments in the two series were found to be very different. Using the conventional positive and negative signs to indicate the direction of the moments, it was found that the algebraic sum of the difference between the aromatic and aliphatic series is *positive* when the substituent is *ortho-para*-directing, and *negative* when the substituent is *meta*-directing.

Furthermore, after allowing for the moments induced in the non-polar parts of a molecule by the presence of a primary dipole, Groves and Sugden[‡] were able to obtain consistent 'bond moments' for various polar bonds from the data for saturated aliphatic compounds. The observed dipole moments for the phenyl halides were, however, found to be much less than those calculated from the 'bond moments' (table 6·11). That is, the halogens do not exert their full inductive effects when present

---

* Judson and Kilpatrick, *J. Amer. Chem. Soc.* 1949, **71**, 3110; Bennett, Brooks and Glasstone, *J. Chem. Soc.* 1935, p. 1821; Pauling, *The Nature of the Chemical Bond*, 2nd ed. p. 206, Cornell University Press, 1940.
† *Proc. Roy. Soc.* A, 1931, **133**, 668.
‡ *J. Chem. Soc.* 1937, p. 1992.

as substituents in an aromatic ring. This is equivalent to saying that an additional effect is operating which *releases* electrons to the ring. The sign and magnitude of this additional effect can be estimated by comparing the calculated and observed dipole moments for such substituted aromatic substances.

TABLE 6·10. *Dipole moments of aromatic and aliphatic compounds*

X	Aryl—X	Alphyl—X	(Aryl—X)–(Alphyl—X)	Directive influence
Me	+0·45	0	+0·45	ortho-para
Cl	−1·56	−2·15	+0·59	ortho-para
Br	−1·52	−2·21	+0·69	ortho-para
I	−1·27	−2·13	+0·86	ortho-para
CH_2Cl	−1·82	−2·03	+0·21	ortho-para
CN	−3·89	−3·46	−0·43	meta
NO_2	−3·93	−3·05	−0·88	meta

TABLE 6·11. *Dipole moments of halogenated compounds*[*]

Methyl halide	$\mu_{obs.}$[†]	Phenyl halide	$\mu_{calc.}$	$\mu_{obs.}$[†]	Additional moment
CH_3—F	−1·81	Ph—F	−2·61	−1·61	+1·00
CH_3—Cl	−1·87	Ph—Cl	−2·70	−1·73	+0·97
CH_3—Br	−1·80	Ph—Br	−2·60	−1·71	+0·89
CH_3—I	−1·64	Ph—I	−2·37	−1·50	+0·87

TABLE 6·12. *Moments of saturated and unsaturated compounds*[‡]

Compound	μ	Compound	μ
CH_3—F	1·81	CH_2=CH—F	—
CH_3—Cl	1·87	CH_2=CH—Cl	1·44
CH_3—Br	1·78	CH_2=CH—Br	1·41
CH_3—I	1·59	CH_2=CH—I	1·26

The same sort of effect can be observed in the moments of the vinyl halides (table 6·12), all of which are considerably less than those of their saturated analogues.

[*] After Groves and Sugden, *J. Chem. Soc.* 1937, p. 1992.
[†] The negative sign refers to the fact that the substituent attracts electrons from the hydrocarbon radical or ring system.
[‡] Wiswall and Smyth, *J. Chem. Phys.* 1941, **9**, 356; Hugill, Coop and Sutton, *Trans. Faraday Soc.* 1938, **34**, 1518.

It is also noteworthy that Brockway, Beach and Pauling[*] have found that the C—Cl bonds in chloroethylenes are shorter than such bonds in chloroparaffins. Similar results have also been observed in other vinyl compounds.[†] Moreover, it is well known that bonds such as C—OH, C—Cl, C—Br, C—NR_2, etc., are significantly shorter in aromatic compounds than in saturated aliphatic compounds.[‡]

In order to explain the effects of *meta* and *para* substituents[§] in aromatic compounds, it has been suggested that atoms or groups carrying unshared valency electrons, or a dipole, can become conjugated with the ring system, and that a displacement of electrons takes place by a tautomeric or resonance mechanism. This additional electronic effect which is thought to operate in aromatic compounds is known as the *mesomeric effect*.

The displacement of electrons by this mechanism is usually represented by means of curved arrows, such as ⌒. This conjugation of the substituent with the ring system, which also occurs in other unsaturated compounds such as the vinyl halides, often leads to a displacement of electrons in opposition to the inductive effect. This is evidently the reason for the fact that the C—NR_2 bond in saturated aliphatic compounds has a dipole moment in the sense C→NR_2, but that in aromatic compounds it is in the direction C←NR_2. In the latter compounds the nitrogen 'lone pair' enters into conjugation with the ring system in such a way that charge is transferred, by a tautomeric mechanism, to the aromatic ring. In the same way, the lone pair of oxygen, or of the halogens, enters into conjugation, and charge is transferred to the ring. On the other hand, the dipole present in the —NO_2 group induces an electron displacement, *from* the ring, which is in the *same* direction as the inductive effect. In this way substituents such as —NH_2, —OH, etc., lead to an enhancement of electron density at the *ortho* and *para* positions, but substituents such as —NO_2 have the opposite effect and preferentially remove electrons from the *ortho* and *para* positions. These displacements are represented in structures (I)–(III).

[*] *J. Amer. Chem. Soc.* 1935, **57**, 2693.
[†] Compare Hugill, Coop and Sutton, *Trans. Faraday Soc.* 1938, **34**, 1518.
[‡] Brockway and Palmer, *J. Amer. Chem. Soc.* 1937, **59**, 2181; Robertson, *Proc. Roy. Soc.* A, 1936, **157**, 79; Robertson and Ubbelohde, ibid. 1938, **167**, 122.
[§] The effects of *ortho* substituents are considered in a subsequent section.

200 THE AROMATIC COMPOUNDS

Alternatively, it is possible to represent tautomeric electronic displacements by ionic structures such as (IV), (V) and (VI), in which the substituent becomes linked to the ring system by a double bond. Such structures, of course, have no real and

independent existence. They may be regarded as contributing structures in the resonance hybrids of phenol, aniline and nitrobenzene respectively. In fact, the greater the contributions of such structures to the hybrids, the greater the conjugation of the substituent with the ring, and the greater the magnitude of the mesomeric effect.

On the basis of the inductive and mesomeric effects of substituents it is possible to account for the apparently anomalous properties of substituted aromatic compounds outlined above.

For example, the fact that the ionic structures (IV), (V) and (VI) contribute to the resonance hybrids means that the C—X bonds in such compounds must have some double-bond character, and this is evidently the reason why these bonds are significantly shorter in aromatic compounds (and in the vinyl compounds) than in paraffin derivatives. Indeed, the degree of shortening, and therefore the percentage double-bond character, is dependent on the degree of conjugation, and the contribution of ionic structures such as (IV), (V) and (VI) to the resonance hybrids.

The degree of shortening of the bond linking the hydroxy group to the nucleus in phenol[*] indicates that this bond has about 16 % double-bond character. The ionic structures (IVa), (IVb) and (IVc) therefore contribute 16 % to the phenol hybrid. With this information it is possible very approximately to estimate the increase in electronic charge at the *ortho* and *para* positions. For phenol, this increase is $0.053e$ (16 % of 0.333, the contribution from the three ionic structures), and the complete distribution of electronic charge is therefore:

HO 1·841

1·053   1·053
1·0   1·0
   1·053

(VII)

This technique has been utilized extensively by Daudel and Pullman.[†]

---

[*] Robertson and Ubbelohde, *Proc. Roy. Soc.* A, 1938, **167**, 122.
[†] See, for example, Pullman, *Bull. Soc. chim. Fr.* 1948, **15**, 533; Daudel and Martin, *Bull. Soc. chim. Fr.* 1948, **15**, 559.

The acid strength of phenol, as compared with aliphatic alcohols, is also explained by the mesomeric effect. This effect, transferring charge to the ring structure, makes it easier for the hydrogen to separate as a proton. In the same way, the powerful $+M$ effect of the $-NR_2$ group makes the nitrogen 'lone pair' of electrons less available for salt formation, and aniline is therefore a much weaker base than ammonia.

The direction of the mesomeric effect can easily be determined experimentally, especially by a comparison of the dipole moments of the compounds alphyl—X and aryl—X as was carried out by Sutton.* With the halogens, for example, the moments of the alkyl compounds are greater than those of the corresponding aromatic compounds, so the mesomeric effect must operate in opposition to the inductive effect, which for these atoms is $-I$. The magnitude of the mesomeric effects of the halogens was determined by Groves and Sugden,† for the 'additional moments' already discussed (p. 197) really represent the 'mesomeric moments'. In other words, the halogens have $+M$ effects in the order $F > Cl > Br > I$ (see table 6·11).

Similarly, it may be shown that the mesomeric effects of groups such as —CN, —NO_2 and —CO_2Et act in the same direction as the inductive effects, so that the resultant effect is reinforced. Alkyl groups represent a special case, which is discussed in detail later, but they are generally believed to have inductive and mesomeric effects which reinforce one another, transferring charge to the ring system.

The effect of some of the common substituents may therefore be summarized:

MeO	$-I, +M$	NO_2	$-I, -M$
HO	$-I, +M$	CN	$-I, -M$
F	$-I, +M$	CO_2H	$-I, -M$
Cl	$-I, +M$	CO_2Et	$-I, -M$
Br	$-I, +M$	Me	$+I, +M$
I	$-I, +M$		

When both the inductive and mesomeric effects of the substituents are considered, it also becomes possible to explain some

---

* *Proc. Roy. Soc.* A, 1931, **133**, 668.   † *J. Chem. Soc.* 1937, p. 1992.

of the apparently anomalous effects of substituents on the strengths of aromatic acids, phenols and bases.

In *para* methoxybenzoic acid (VIII), for example, the methoxy group is conjugated with the carboxy group, and the resulting electronic displacements ($+M$) make it more difficult for the hydrogen to ionize. The $+M$ effect is more powerful than the $-I$ effect of a *para*-methoxy group, and a *para*-methoxy group is therefore acid-weakening. On the other hand, in *meta*-methoxybenzoic acid (IX), the methoxy group is not conjugated with the carboxy group, so that the mesomeric effect is not operative and the inductive effect ($-I$) of the methoxy substituent alone acts, facilitating the removal of hydrogen as a proton. A *meta*-methoxy group is therefore acid-strengthening.

(VIII) Acid-weakening

(IX) Acid-strengthening

Similarly, substitution of a hydroxy group in the *para* position of benzoic acid is acid-weakening by a $+M$ mechanism, and *meta* substitution is acid-strengthening by virtue of the $-I$ effect. On the other hand, the halogen substituents also have $-I$ and $+M$ effects, and yet chlorine and bromine are acid-strengthening in both the *meta* and *para* positions of benzoic acid. It seems that the $-I$ effect of the halogens is always sufficiently powerful to overcome the $+M$ effect which evidently operates to a small extent when the halogen is in the *para* position. The halogens are also uniformly acid-strengthening when present as substituents in phenol. In the nitrophenols (table 6·9) it is noteworthy that the nitro group is significantly more acid-strengthening when in the *para* position than when present in the *meta* position. Here again the explanation evidently involves a combination of the mesomeric and inductive effects. In the *para* position a nitro group is conjugated with the phenolic hydroxy group, so that the resultant acid-strengthening effect is the sum of the $-M$ and $-I$ effects. In the *meta* position, however, only the $-I$ effect can operate, leading to a less marked acid-strengthening effect.

Very similar considerations apply to the substituted anilines A *meta*-methoxy group is base-weakening by virtue of the $-I$ effect, but a *para*-methoxy group is base-strengthening because the $+M$ effect is sufficiently powerful to overcome the weak $-I$ effect of a substituent in the *para* position. Again, the effects of nitro and methyl substituents are more marked when in the *para* position, because in this position the mesomeric and inductive effects reinforce one another.

HAMMETT'S σ CONSTANTS

In a comprehensive examination of the effects of substituents on numerous side-chain reactions of benzene derivatives, Hammett observed some interesting relationships.* He found that there is a linear relationship between the rate or equilibrium constants for one side-chain reaction which involves a series of *meta*- and *para*-substituted benzenes, and the corresponding constants for some *other* side-chain reaction. There is, for example, a linear relationship between the logarithms of the ionization constants of *meta*- and *para*-substituted benzoic acids and the logarithms of the rate constants for the hydrolysis of a similar series of substituted benzoic esters. This relationship is shown in fig. 6·3. Each point on this graph is determined by the nature of a particular substituent, and the straight line may be represented by the equation

$$\log k_h = \rho \log K_i + A,$$

where k_h is the rate constant for the hydrolysis, K_i is the ionization constant of the correspondingly substituted benzoic acid, ρ is the slope of the curve, and A is the intercept.

Similar linear relationships have been observed with many other side-chain reactions of benzene derivatives. Some examples of these side-chain reactions are: the esterification of benzoic

* Hammett, *Chem. Rev.* 1935, **17**, 125; Burkhardt, *Nature, Lond.*, 1935, **136**, 684; Burkhardt, Ford and Singleton, *J. Chem. Soc.* 1936, p. 17; Burkhardt, Evans and Warhurst, ibid. 1936, p. 25; Hammett, *J. Amer. Chem. Soc.* 1937, **59**, 96; Hammett, *Trans. Faraday Soc.* 1938, **34**, 156.

acids; the hydrolysis of benzamides; the reaction of anilines with benzoyl chloride; the ionization constants of phenylacetic acids; the hydrolysis of benzoyl chlorides; and so on. Hammett has investigated over fifty such side-chain reactions.* He chose the ionization constants of substituted benzoic acids as the

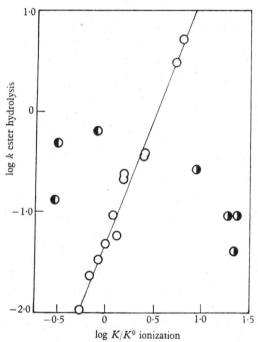

Fig. 6·3. Comparison of hydrolysis rates of esters with ionization constants of acids. Aliphatic acid derivatives ◐ ; *meta*- and *para*-substituted benzoic acid derivatives ○; *ortho*-substituted benzoic acid derivatives ◑. (From Hammett, *Physical Organic Chemistry*, p. 185, McGraw-Hill Book Co., 1940.)

standard of reference. All the other side-chain reactions of *meta*- and *para*-substituted benzenes show a linear relationship with the ionization constants of the correspondingly substituted benzoic acids. In general, these linear relationships all have the form

$$\log k = \rho \log K_i + A,$$

i.e. $\quad \log k = \rho (\log K_i - \log K_i^0) + (A + \rho \log K_i^0),$

* Hammett, *Physical Organic Chemistry*, McGraw-Hill Book Co., 1940. See pp. 184 et seq.

in which k is any rate or equilibrium constant, K_i is the ionization constant of the substituted benzoic acid, and K_i^0 is the ionization constant of benzoic acid itself. The quantity $(A + \rho \log K_i^0)$ can clearly be replaced by $\log k^0$, where k^0 is the rate or equilibrium constant of the unsubstituted reactant. The expression $(\log K_i - \log K_i^0)$ is dependent only on the nature and position of the substituent. Hammett called this expression 'σ', the *substituent constant*. It is a property of the substituent itself and is independent of the nature of the side-chain reaction investigated. The equation therefore becomes

$$\log k - \log k_0 = \rho \sigma.$$

The constant, ρ, is called the *reaction constant*. It determines the slope of the linear relationship and is a constant for all substituents. It depends only on the nature of the reaction.

The numerical values of the substituent constants can easily be determined if the ionization constant of the substituted benzoic acid is known. In other cases it is possible to determine ρ by plotting $\log k$ against σ, and to read off the value of σ for a given substituent. In this way the values of ρ for a large number of reactions, and the values of σ for a large number of different substituents, have been determined. The latter figures are reproduced in table 6·13.

The importance of this table lies in the fact that when the substituent constant σ has a negative value, it means that the substituent transfers charge to the ring, and when it is positive it means that it attracts electrons from the ring. The magnitude of σ is indicative of the magnitude of the charge transferred.

Examination of the relative values of the various constants provides much information of value. It will be noted, for example, that a *para*-amino group is a very strong electron-donor, and that even in the *meta* position an amino group is activating. Similarly, alkyl groups are also activating whether in the *meta* or *para* position, but it will be seen that the effect is greater when the group is present in the *para* position. This confirms the fact that alkyl groups act as electron-donors by a tautomeric mechanism as well as by an inductive mechanism. Alkoxy groups, on the other hand, are activating in the *para*

THE EFFECTS OF SUBSTITUENTS

position, but deactivating in the *meta* position. The $-I$ effect of such groups tends to decrease the π-electron density on all the annular carbon atoms, but the $+M$ effect tends to increase it specifically at the *ortho* and *para* positions. As indicated in the previous section, a *para*-alkoxy group is conjugated with the side chain, but a *meta*-alkoxy group is not. In this case, therefore, the opposing electronic effects lead to negative values of σ when the substituent is in the *para* position, and positive values when in the *meta* position.

TABLE 6·13. *Hammett's substituent constants*

Substituent	σ	Substituent	σ
p-NH$_2$	-0.660	p-C$_6$H$_5$	$+0.009$
p-CH$_3$O	-0.268	p-F	$+0.062$
p-C$_2$H$_5$O	-0.25	m-CH$_3$O	$+0.115$
m-Me$_2$N	-0.211	m-C$_2$H$_5$O	$+0.15$
p-Me$_2$N	-0.205	m-C$_6$H$_5$	$+0.218$
p-t.-C$_4$H$_9$	-0.197	p-Cl	$+0.227$
p-CH$_3$	-0.170	p-Br	$+0.232$
m-NH$_2$	-0.161	p-I	$+0.276$
p-C$_2$H$_5$	-0.151	m-CH$_3$CO	$+0.306$
p-i.-C$_3$H$_7$	-0.151	m-F	$+0.337$
m-CH$_3$	-0.069	m-I	$+0.352$
p-CH$_3$S	-0.047	m-CO$_2$H	$+0.355$
		m-Cl	$+0.373$
		m-CHO	$+0.381$
		m-Br	$+0.391$
		m-CN	$+0.608$*
		p-CN	$+0.656$*
		m-NO$_2$	$+0.710$
		p-CO$_2$H	$+0.728$
		p-NO$_2$†	$+0.778$
		p-CH$_3$CO	$+0.874$
		p-CHO	$+1.126$
		p-NO$_2$‡	$+1.27$

With the halogens, the $-I$ and $+M$ effects are also opposed. In this case the σ constants indicate that the halogens produce a decrease in π-electron density all over the ring, but a smaller decrease in the *para* position, for in the *para* position the halogen is conjugated with the side chain.

* Given by Roberts and McElhill, *J. Amer. Chem. Soc.* 1950, 72, 628.
† To be used for all reactions of benzene derivatives except those of aniline and phenol.
‡ To be used for the reactions of derivatives of aniline and phenol.

In the case of —CHO, —CN, —NO$_2$ and —CO$_2$H substituents, the $-I$ and $-M$ effects reinforce one another when the substituent is in the *para* position, and for this reason the σ constants have greater values for the *para* than for the *meta* positions.

THE INDUCTOMERIC AND ELECTROMERIC EFFECTS

Physical methods such as the measurement of acid and base strengths, the measurement of dipole moments, and the determination of the lengths of C—X bonds, etc., all give an indication of the magnitude of the inductive and mesomeric effects in the *static* molecule. During a reaction, however, the electronic distribution of a molecule may be profoundly perturbed by the presence of the attacking ions, by polar molecules, and even by the proximity of the molecules of polar solvents. It is customary to suppose that perturbations which are brought about at the demand of the attacking reagent also take place by mechanisms of the same type as the inductive and mesomeric effects. Four effects are therefore probably operating in many cases: the static inductive effect (I_s), and the dynamic inductive effect (I_d), which is usually called the inductomeric effect; the static tautomeric effect, otherwise known as the mesomeric effect (M), and the dynamic tautomeric effect, or electromeric effect (E). The effects of substituents on the electron distribution of aromatic compounds may therefore be summarized as in table 6·14.

TABLE 6·14. *Effect of substituents on aromatic compounds**

Electronic mechanism	Electrical classification	
	Static (polarization)	Dynamic (polarizability)
Inductive (→—) (I)	Inductive (I_s)	Inductomeric (I_d)
Tautomeric (⌢) (T)	Mesomeric (M)	Electromeric (E)

There is only slender independent *physical* evidence that the inductomeric effect is a real one. The electromeric effect is,

* Ingold, *Chem. Rev.* 1934, **15**, 225.

however, a very powerful one, and the evidence for its reality obtained by refraction studies is considerable.*

When electromagnetic waves, such as light, pass through a substance, the electric field brings about a temporary polarization, by causing slight movements of the electrons, and the ultimate result, in general, is that the velocity of the waves is reduced. The refractive index, n, of a substance is the ratio of the velocity of light in a vacuum to its velocity in that substance. This property, therefore, is a measure of the polarizability of the molecules forming that substance. As proposed by Lorenz and Lorentz, the specific refractivity, r, of a substance is given by

$$r = \frac{n^2 - 1}{n^2 + 2} \frac{1}{d},$$

where d is the density. The molecular refractivity is given by the product of r and the molecular weight.

TABLE 6·15. *Atomic refractivities for calculating molecular refractivities*[†]

Atom or group	$[r_M]_D$	Atom or group	$[r_M]_D$
—CH$_2$—	4·618	I	13·900
C	2·418	N (primary amine)	2·322
H	1·100	N (secondary amine)	2·499
O (carboxyl)	2·211	N (tertiary amine)	2·840
O (ether)	1·643	S (mercaptan)	7·69
O (hydroxyl)	1·525	CN	5·459
Cl	5·967	C=C (double bond)	1·733
Br	8·865	C≡C (triple bond)	2·398

Experiment has shown that molecular refractivities for any particular wave-length of light (such as the D line of sodium) are fairly additive in character, and several workers[‡] have therefore assigned values for 'atomic refractivities' to a number of atoms and linkages (table 6·15), such that the molecular refractivity of any organic substance can be calculated by appropriate summation of the individual contributing atoms and groups.

* The chief evidence for these effects is provided by a study of the rates of substitution reactions.
[†] After Eisenlohr, Z. Phys. Chem. 1911, **75**, 585.
[‡] Brühl, Z. Phys. Chem. 1891, **7**, 140; Eisenlohr, ibid. 1911, **75**, **585**; **Vogel**, *Practical Organic Chemistry*, p. 898, Longmans, Green and Co., 1948.

Although these atomic refractivities very frequently give excellent values for the molecular refractivities of molecules, they do not always do so. It has been found that the additive nature of the constants fails in *conjugated* substances. The difference between the measured molecular refractivity and that calculated from the atomic refractivities is called the molecular exaltation. The specific exaltation, E_r, is given by

$$E_r = \frac{\text{molecular exaltation}}{\text{molecular weight}} \times 100.$$

The specific exaltation, as a polarizability phenomenon, may be taken as a measure of the extent of conjugation by the electromeric effect. Large exaltations are shown by precisely those substances which might be expected to show large electromeric effects. Among the substituted benzenes, for example, as given in table 6·16, it would be expected that styrene and phenylbutadiene would show very large electromeric effects, and the observed exaltations confirm this. Similarly, the substituents —CHO, —CO$_2$Et, —NH$_2$, —OMe, etc., show large exaltations, confirming the existence of significant electromeric effects in these cases.

TABLE 6·16. *Specific exaltation of the refractivities**

Compound	$[E_r]_D$	Compound	$[E_r]_D$
Benzene	−0·22	Ethoxybenzene	0·39
Chlorobenzene	−0·09	Ethyl benzoate	0·49
Bromobenzene	−0·03	Ethyl *p*-chlorobenzoate	0·60
Iodobenzene	+0·03	Ethyl *p*-bromobenzoate	0·57
Phenol	+0·13	Aniline	0·92
Toluene	0·15	Benzaldehyde	1·02
p-Chlorotoluene	0·22	Styrene	1·13
p-Bromotoluene	0·08	1-Phenylbutadiene	3·69
Anisole	0·31		

THE EFFECTS OF HALOGEN SUBSTITUENTS

All the available evidence indicates that the halogens exert $-I_s$ and $+M$ effects, and it may be assumed that the polarizability factors, the inductomeric and electromeric effects, have

* Cornubert, *Rev. gén. sci.* 1922, **33**, 433, 471; Auwers and Frühling, *Liebigs Ann.* 1921, **422**, 160; Auwers, *Ber. dtsch. chem. Ges.* 1912, **45**, 2781.

THE EFFECTS OF SUBSTITUENTS

similar signs, that is, $-I_d$ and $+E$. It is not possible to predict whether the *resultant* effect will be either positive or negative for any given reaction, for the polarizability effects vary with the conditions; experimentally, however, it is found that the inductive effects of the halogens are greater than the opposing tautomeric effects.

The relationship of the halogens among themselves is of special interest and there is, as yet, no measure of agreement on the subject. There is little doubt that the static inductive effect (I_s) decreases in the order $F > Cl > Br > I$.* This is the order expected from Pauling's electronegativity values, and it is confirmed by comparison of the strengths of the halogenated acetic acids and by other evidence. The mesomeric effect also seems to decrease in the same order, $F > Cl > Br > I$.† It does not follow, however, that the polarizability factors will decrease in the same order. Indeed, it has been suggested that the polarizability of the halogen-nucleus bond decreases in the reverse order, $I > Br > Cl > F$, and that this may be either inductomeric or electromeric in character.‡

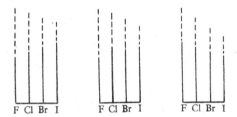

Fig. 6·4. Relative effects of halogen substituents.
(After Dippy and Lewis, *J. Chem. Soc.* 1936, p. 644.)

Be this as it may, Dippy and Lewis§ have pointed out that as the total inductive effects and total tautomeric effects of the halogens are opposed, their relative magnitude is of great importance. In the adjacent diagrams (fig. 6·4), for example, the total line in each case represents the magnitude of the $-I$ effect. The dotted part represents the opposing $+T$ effect, and the solid line represents the resultant electron-attracting effect.

* Ingold, *Chem. Rev.* 1934, 15, 225.
† Audsley and Goss, *J. Chem. Soc.* 1942, p. 497; Groves and Sugden, ibid. 1937, p. 1992; Baker and Hopkins, ibid. 1949, p. 1089.
‡ de la Mare and Robertson, ibid. 1948, p. 100.   § Ibid. 1936, p. 644.

From these diagrams it may be appreciated that although the total $-I$ and $+T$ effects are always represented in the order $F > Cl > Br > I$, the relative magnitudes may cause the resultant effect, represented by the solid line, to be in *any* order. In the first case the resultant is $I > Br > Cl > F$; in the second, $Br > Cl = I > F$; in the third, $F > Cl > Br > I$. Experimentally, it is found that any order can be observed.*

THE EFFECTS OF ALKYL SUBSTITUENTS

The electrical effects of alkyl groups in aromatic compounds have been described above as $+I$ and $+M$. Some discussion of this conclusion is, however, necessary. For one thing, *all* paraffin hydrocarbons have zero dipole moments, so that alkyl groups do not always act as electron-donors. Furthermore, it is not obvious why an alkyl group *should* have a mesomeric effect, for it has no unshared valency electrons, and is completely saturated.

There can be little doubt that alkyl groups *do* act as electron-donors under certain conditions, and it seems reasonably certain that part of this effect is transmitted by an inductive mechanism. The dissociation constants of the fatty acids clearly show that the introduction of alkyl substituents results in acid-weakening, and this may be regarded as satisfactory evidence for the $+I_s$ effect (see table 6·17).

TABLE 6·17. *Dissociation constants of fatty acids*

Acid	pK_a
$H.CO_2H$	3·7
$CH_3.CO_2H$	4·7
$CH_3.CH_2.CO_2H$	4·9
$(CH_3)_3C.CO_2H$	5·0

Such an inductive effect has been confirmed by a study of the dipole moments of alkyl halides, alkyl cyanides and the nitro-paraffins.† On the other hand, as mentioned above, the methyl group does not confer polar characteristics on saturated aliphatic

---

* de la Mare and Robertson, *J. Chem. Soc.* 1948, p. 104.
† Groves and Sugden, ibid. 1937, p. 158.

compounds, and this suggests that alkyl groups exert only those polar effects which are impressed upon them by other groups present in the molecule.

As the larger members of an homologous series are all derived from the preceding members by successive replacement of hydrogen by a methylene group, the inductive effects of the alkyl groups must be in the order: $Me < Et < Pr^\alpha < Bu^\alpha$ and $Me < Et < Pr^\beta < Bu^\gamma$. This series has actually been observed in a number of instances.* On the other hand, it is not always observed. In some cases the effects of the alkyl groups are found to be in the order: $Me > Et > Pr^\beta > Bu^\gamma$.

To explain such an order, Baker and Nathan[†] suggested that the two electrons forming the C—H bond of an alkyl group are less localized than in C—C bonds, and that, in consequence, alkyl groups can enter into conjugation with the ring system, or with any unsaturated system. In other words, Baker and Nathan postulated an electron release by a tautomeric mechanism, which is superimposed on the inductive effect:

$$H—C—C=C$$

This electron release has been called the Baker-Nathan effect, but it is also known as *hyperconjugation* because it involves conjugation of an alkyl group with other groups containing multiple bonds.[‡] Assuming that only C—H bonds are effective, the expected order of electron release by this mechanism is: $Me > Et > Pr^\beta > Bu^\gamma = 0$. This order has been substantially confirmed. Hughes, Ingold and Taher[§] showed conclusively that the electromeric effects of the alkyl groups are in the order $Me > Et > Pr^\beta > Bu^\gamma$; and Baker and Hemming[||] later showed that the mesomeric effects are in the same order.

Hyperconjugation is necessary to account for the relative rates of bromination of various alkyl benzenes, and it has even

---

* Ayling, *J. Chem. Soc.* 1938, p. 1014; Evans, Gordon and Watson, ibid. 1938, p. 1439.
† Ibid. 1935, p. 1844.
‡ Crawford, *Quart. Rev. Chem. Soc., Lond.*, 1949, 3, 226; Mulliken, Rieke and Brown, *J. Amer. Chem. Soc.* 1941, 63, 41.
§ *J. Chem. Soc.* 1940, p. 949. || Ibid. 1942, p. 191.

been postulated that the β- and γ-hydrogen atoms in certain alkyl groups also enter into hyperconjugation.*

There seems to be no reason why hyperconjugation should be confined exclusively to C—H bonds, and hyperconjugation with σ C—C bonds has indeed been suggested, but the evidence is still somewhat inconclusive.

THE '*ORTHO*' EFFECT

The great majority of the effects of *meta* and *para* substituents in benzene derivatives can be explained in terms of their inductive and mesomeric effects. This is not so, however, when substituents are in the *ortho* position. In this case, the results are often anomalous. *Ortho*-substituted benzoic acids are stronger than benzoic acid, regardless of the electronic influence of the substituent in question, and the *ortho* acids are also generally stronger than the *meta* and *para* isomerides. Similarly, substituents in the *ortho* positions decrease the basic strength of aniline, regardless of the nature of their inductive and mesomeric effects. Clearly an additional factor is involved, and this may be called the '*ortho*' effect.

The nature of the '*ortho*' effect is still somewhat obscure, but it is probable that a variety of effects can operate.

In many cases the effect can be traced to the formation of hydrogen bonds. This is exemplified in the strengths of the hydroxybenzoic acids. Although the ionization constants of *meta*- and- *para*-hydroxybenzoic acids are of the same order as that of benzoic acid itself, the *ortho*-hydroxy derivative (salicylic acid) has an ionization constant 17 times as great. Moreover, 2:6-dihydroxybenzoic acid, in which both substituents are in *ortho* positions, has an ionization constant some 800 times that of benzoic acid. In this case, and in the case of salicylic acid, it seems reasonable to suppose that the effect is due to hydrogen bonding between the *ortho*-hydroxy groups and the carboxylic acid oxygen atoms. Such hydrogen bonding would be expected to stabilize the anions (I and II), thereby increasing the acid

* Berliner and Bondhus, *J. Amer. Chem. Soc.* 1946, **68**, 2355; ibid. 1948, **70**, 854; Berliner and Berliner, ibid. 1949, **71**, 1195; ibid. 1950, **72**, 222.

strength of the hydroxybenzoic acid.* A variety of evidence, both chemical and physical, indicates that hydrogen bond formation does often occur in these cases.† It is known, for example, that hydrogen bond formation occurs in *ortho*-nitroacetanilide in 2-nitroresorcinol, in 3-nitrocatechol, in salicylaldehyde (III), in *ortho*-nitrophenol (IV), in ethyl salicylate (V), and in other similar compounds.‡

(I) (II) (III) (IV) (V)

Many of the anomalous properties of *ortho*-substituted compounds can be explained by considerations of this nature; but simple steric factors are also involved in very many cases. The fact that some *ortho*-substituted compounds can be separated into optically active forms certainly confirms the existence of steric hindrance (Chapter 11). The effect has also been confirmed by examination of the ultra-violet absorption spectra of such compounds (Chapter 10). Further data are provided by a study of the dipole moments of suitable *ortho*-substituted compounds.

The dipole moments of substituted aromatic compounds are derived from a summation of the inductive and mesomeric

* Baker, *Nature, Lond.*, 1936, **137**, 236; Branch and Yabroff, *J. Amer. Chem. Soc.* 1934, **56**, 2568.
† For example, Martin, *Nature, Lond.*, 1950, **166**, 474.
‡ See Hunter, *Symposium on the Hydrogen Bond*, Royal Institute of Chemistry, 1950.

effects of the substituents. The dipole moment of Me_3C-NO_2 is 3·30 D.; that of nitrobenzene is larger, namely 3·95 D., because in aromatic compounds the mesomeric effect also operates and, in this case, both effects reinforce one another. Now durene, which is a symmetrical molecule, has no dipole moment, so that, in the absence of any special factor, the moment of nitrodurene should be the same as that of nitrobenzene. In fact, the dipole moment of nitrodurene is 3·62 D., which is much smaller than that for nitrobenzene.* The explanation seems to be as follows: If a substituent exerts a mesomeric effect it means that certain ionic structures, in which the substituent is linked to the ring by a double bond, contribute to the resonance hybrid. This presupposes that the substituent and the ring system are coplanar. In the presence of *ortho* substituents, however, the original substituent may be prevented from assuming a coplanar configuration, and thereby prevented from exerting its mesomeric effect. This seems to be the case with nitrodurene. The *ortho*-methyl groups interfere sterically with the nitro group, and prevent that group from taking up a coplanar configuration with the ring. In other words, the mesomeric effect of the nitro group is inhibited, and the dipole moment of nitrodurene is almost entirely that due to the inductive effect of the nitro group. It is therefore smaller than the moment of nitrobenzene, in which both the inductive and mesomeric effects are involved.

For the same reason, the moment of dimethylaniline (1·58 D.) is reduced to 1·03 D. by the introduction of two *ortho*-methyl groups, as in 2:4:6-trimethyl-*N*-dimethylaniline.†

It is interesting that *ortho*-methyl groups inhibit the mesomeric effect of the carbonyl group in acetophenone, but not in benzaldehyde. Evidently the steric effect is negligible in benzaldehyde, but not in acetophenone.‡

Of course, the inhibition of the mesomeric effect can occur only if the steric hindrance to coplanarity is sufficiently marked. *Ortho*-methyl groups, for example, are not sufficiently bulky to interfere with a hydroxy group, for hydroxydurene has almost the same moment as phenol. Moreover, *ortho* substituents can

---

* Birtles and Hampson, *J. Chem. Soc.* 1937, p. 10; Kofod, Sutton, De Jong, Verkade and Wepster, *Rec. trav. chim. Pays-Bas*, 1952, **71**, 521.
† Ingham and Hampson, *J. Chem. Soc.* 1939, p. 981.
‡ Kadesch and Weller, *J. Amer. Chem. Soc.* 1941, **63**, 1310.

have little effect on spherically symmetrical substituents (such as Br) and the dipole moments of bromodurene and of bromobenzene are almost identical. These and other dipole moments are collected in table 6·18.

TABLE 6·18. *Influence of* ortho *substituents on dipole moments**

Substituent		Dipole moment (μ)		
A	B	A—⬡—B	H_3C CH_3 / A—⬡—B / H_3C CH_3	H_3C / A—⬡—CH_3 / H_3C
NO_2	H	3·95 D.	3·62 D.	3·64 D.
NMe_2	H	1·58	—	1·03
NH_2	H	1·53	1·39	1·40
OH	H	1·61	1·68	—
F	H	1·46	—	1·36
Cl	H	1·56	—	1·55
Br	H	1·52	1·55	1·52
I	H	1·27	—	1·42
$CH_3.CO$	H	2·88	2·68	2·71
H.CO	H	2·92	—	2·96
Cl.CO	H	3·32	—	2·95
NO_2	NMe_2	6·87	4·11	—
NO_2	NH_2	6·10	4·98	—
NO_2	OH	5·04	4·08	—
NO_2	Br	2·65	2·36	—

When two substituents are present, and *both* are hindered, the reduction in the dipole moment becomes very considerable. The dipole moment of *p*-nitro-*N*-dimethylaniline (6·87) is reduced to 4·11 when both substituents are hindered by *ortho*-methyl groups, as in the corresponding derivative of durene.

Steric inhibition of the conjugation of a substituent with an aromatic ring system can also be proved by studies of molecular refractivity. This method, of course, determines whether or not there is a reduction in the *electromeric* (*E*) effect of a substituent, as refraction is a polarizability phenomenon.

* After Birtles and Hampson, *J. Chem. Soc.* 1937, p. 10; Ingham and Hampson, ibid. 1939, p. 982; Kadesch and Weller, *J. Amer. Chem. Soc.* 1941, **63**, 1310; Dewar, *Electronic Theory of Organic Chemistry*, Oxford University Press, 1949; Kofod, Sutton, De Jong, Verkade and Wepster, *Rec. trav. chim. Pays-Bas*, 1952, **71**, 521.

The molecular refractivity of N-dimethyl-m-5-xylidine is normal as compared with that of N-dimethylaniline (VI), N-dimethyl-m-toluidine and dimethyl-p-toluidine. In each case the molecular refractivity of the amine is about 14·65 c.c. greater than that of the corresponding hydrocarbon (table 6·19).

    N(CH$_3$)$_2$ N(CH$_3$)$_2$ N(CH$_3$)$_2$
                         CH$_3$ H$_3$C   CH$_3$

    (VI)   (VII)   (VIII)

Thomson[*] has found, however, that whenever a methyl or a chloro substituent is introduced *ortho* to the NMe$_2$ group, there is a reduction of approximately 1 c.c. in the molecular refractivity, and that when both *ortho* positions are occupied by methyl groups the molecular refractivity is lowered by about 1·6 c.c. The molecular refractivity of N-dimethyl-o-toluidine (VII), for example, is only 13·56 c.c. greater than that of toluene, and that of N-dimethyl-m-2-xylidine (VIII) is only 13·06 c.c. greater than that of *meta*-xylene.

From these results it may be shown that the atomic refractivity of the nitrogen in N-dimethyl-m-2-xylidine is lowered almost to the value which is observed for aliphatic tertiary amines, indicating that the electromeric effect is entirely inhibited. In N-dimethyl-m-5-xylidine, however, in which the methyl groups are *not* in the *ortho* positions, the atomic refractivity of the nitrogen is the normal value for aromatic tertiary amines, and the electromeric effect has its normal value. Finally, the atomic refractivity of the nitrogen in N-dimethyl-m-4-xylidine, in which only one methyl group occupies a position *ortho* to the amine grouping, has a value intermediate between that for normal aromatic tertiary amines and that for aliphatic tertiary amines, indicating that the electromeric effect has only half its normal value.

There can be little doubt, therefore, that the presence of methyl groups *ortho* to a dimethylamino group has a pronounced effect,

[*] *J. Chem. Soc.* 1944, p. 408.

both on the mesomeric and electromeric effects of the substituent.

It might be expected that the steric inhibition of conjugation (resonance) of this nature would have an important effect on the acid strength of phenols, and on the basic strength of anilines, and this has also been found to be the case.

TABLE 6·19. *Molecular refractivities of dimethylanilines*[*]

Compound	$[R_L]_D$	Δ
Dimethylaniline Benzene	40·85 26·17	14·68
Dimethyl-p-toluidine Toluene	45·68 31·06	14·62
Dimethyl-m-toluidine Toluene	45·68 31·06	14·62
Dimethyl-m-5-xylidine m-Xylene	50·58 35·93	14·65
Dimethyl-o-toluidine Toluene	44·62 31·06	13·56
Dimethyl-o-chloroaniline Chlorobenzene	44·95 31·38	13·57
Dimethyl-m-4-xylidine m-Xylene	49·55 35·93	13·62
Dimethyl-m-2-xylidine m-Xylene	48·99 35·93	13·06

The acid strength of phenol is dependent on the degree of conjugation of the hydroxy group with the ring system. The greater the conjugation, the greater the transfer of charge to the ring, and the greater the acid strength. Phenol (XI) itself has a pK_a value of about 10·0, and the introduction of alkyl substituents weakens the acid strength. m-2-Xylenol (IX), with a pK_a of 10·6, is slightly weaker than m-5-xylenol (X), with a pK_a of 10·1. Very little if any '*ortho*' effect is probably involved here, for dipole moment studies do not show any steric interference between an hydroxy group and two *ortho* substituents. On the other hand, anomalous acid strengths are observed in the nitro derivatives of the above compounds. 5-Nitro-m-2-xylenol (XII) is actually a stronger acid than 2-nitro-m-5-xylenol (XIII), although the opposite might have been predicted from

[*] After Thomson, *J. Chem. Soc.* 1944, p. 408.

a study of the xylenols themselves. The explanation seems to be that, in 2-nitro-*m*-5-xylenol (XIII), the methyl groups offer steric hindrance to the nitro group, and prevent it from taking up a coplanar configuration. The mesomeric effect of the nitro group is therefore inhibited, and the effect on the acidity of the phenol is due (*a*) to the small inductive effect of the methyl groups, and (*b*) to the inductive effect of the nitro group. In 5-nitro-*m*-2-xylenol (XII), however, there is no steric hindrance of the nitro group, which exerts both its mesomeric and inductive effects. This compound therefore has an acid strength of the same order as that of *para*-nitrophenol (XIV).*

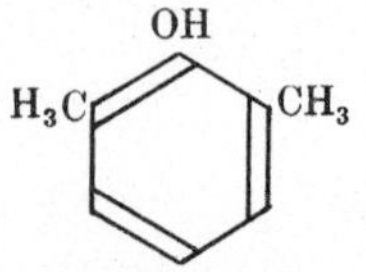

pK_a, 10·6 (IX)

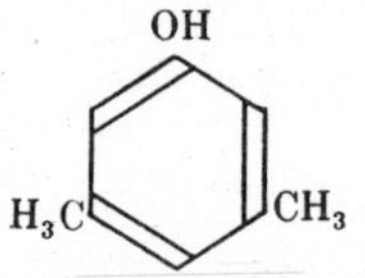

pK_a, 10·1 (X)

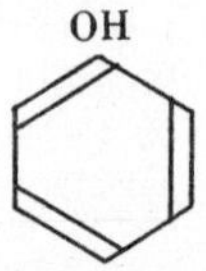

pK_a, 10·0 (XI)

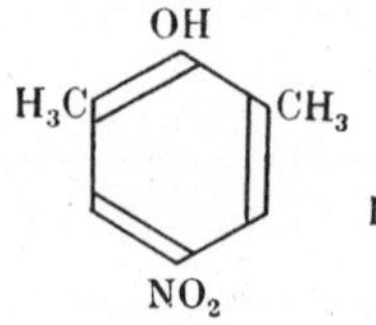

pK_a, 7·2 (XII)

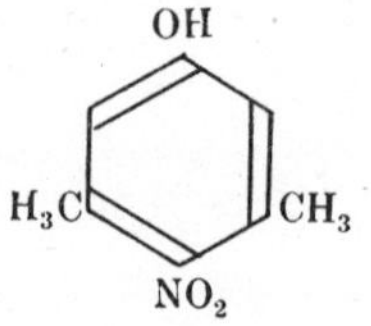

pK_a, 8·2 (XIII)

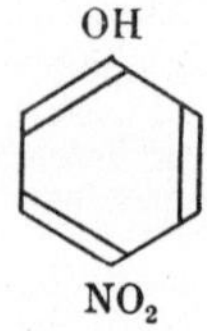

pK_a, 7·2 (XIV)

Similar results may be observed among the derivatives of *N*-dimethylaniline (XV). The presence of a single *ortho*-methyl group (XVI) significantly increases the basicity, presumably by steric inhibition of resonance with the ionic structures which demand a coplanar configuration. It might be expected that the presence of a second *ortho*-methyl group would increase the base strength still further, but this is not the case. *N*-Dimethyl-*m*-2-xylidine (XVII) in fact has a pK_a of 4·7, intermediate between that for *N*-dimethylaniline (4·3) and that for *N*-

* Wheland, *The Theory of Resonance*, see p. 185, John Wiley, New York, 1944; Wheland, Brownell and Mayo, *J. Amer. Chem. Soc.* 1948, **70**, 2492.

THE EFFECTS OF SUBSTITUENTS

dimethyl-*o*-toluidine (5·1). Moreover, *N*-dimethyl-*m*-2-xylidine (XVII) is found to be only very slightly stronger than *N*-dimethyl-*m*-5-xylidine (XVIII), which has *no ortho* substituent.*

$N(CH_3)_2$	$N(CH_3)_2$	$N(CH_3)_2$	$N(CH_3)_2$
	CH_3	H_3C CH_3	H_3C CH_3
pK_a, 4·3	pK_a, 5·1	pK_a, 4·7	pK_a, 4·5
(XV)	(XVI)	(XVII)	(XVIII)

Here again, however, an explanation can be offered in terms of steric hindrance.† The conversion of an amine into its ion must involve a change from a trivalent nitrogen to a tetravalent state, and the —NH_3^+ group has dimensions roughly comparable with those of a methyl group. Large *ortho* substituents might therefore be expected to offer steric resistance to the formation of the ion, which would *weaken* the base. This does seem to be the case with aniline itself and its *ortho* substituents. The introduction of a single *ortho*-methyl group, as in *ortho*-toluidine, weakens the base strength from 4·6 to 4·4; but a second *ortho*-methyl group reduces the pK_a value to 3·4. Steric hindrance therefore seems to favour the left-hand side of the equilibrium:

H_3C—[ring with NH_2 and CH_3] ⇌ H_3C—[ring with NH_3^+ and CH_3]

(XIX)   (XX)

and the net effect is to decrease the apparent strength of the base.

With *substituted* amino groups (e.g. —NR_2) the problem is more complex, because the *ortho* groups interfere with the nitrogen not only in its tetravalent state but also in the tertiary state. *N*-Dimethylaniline (XV) has a pK_a of 4·3. The introduction of a single *ortho*-methyl group, as in *N*-dimethyl-*o*-

---

* Thomson, *J. Chem. Soc.* 1946, p. 1113.
† Brown and Cahn, *J. Amer. Chem. Soc.* 1950, **72**, 2939.

toluidine (XVI), increases the pK_a value to 5·1 because the methyl group offers steric hindrance to the resonance of the dimethylamino group with the ring. The net effect is that the base strength is increased, but not to the extent that would have occurred in the absence of the *second* steric effect, namely, the steric hindrance offered to the formation of the quadrivalent ionic state. In N-dimethyl-m-2-xylidine (XVII), resonance is still further reduced by the presence of two *ortho* substituents, but the second steric factor, the inhibition of the formation of the ion, becomes even more important, and actually becomes the major factor in controlling the base strength. This additional steric factor has been called 'steric strain'.

TABLE 6·20. *Effect of* ortho *substituents on base strength of aromatic amines**

Substance	pK_a
Aniline	4·25
o-Toluidine	4·0
m-2-Xylidine (XIX)	3·4
m-4-Xylidine	4·6
m-5-Xylidine	4·5
p-Xylidine	4·2
N-Dimethylaniline (XV)	4·3
N-Dimethyl-o-toluidine (XVI)	5·1
N-Dimethyl-m-2-xylidine (XVII)	4·7
o-t.-Butyl-N-dimethylaniline	4·3

Summarizing, therefore, it seems that two steric effects control the strength of *ortho*-substituted anilines, (i) a steric inhibition of resonance, increasing the base strength, and (ii) a steric strain factor, a steric hindrance to the formation of the ion, decreasing the base strength. It is obvious that the *resultant* effect will depend on the size of the groups involved both as *ortho* substituents and as N- substituents.

The opposition of the two effects can also be observed by observing the base strengths of dimethylanilines having a single, but variable, *ortho* substituent. As the size of the substituent increases, the base strength is first *increased*, because the

* In 50% ethanol, at 25°. Quoted by Brown and Cahn, *J. Amer. Chem. Soc.* 1950, 72, 2939.

resonance effect becomes increasingly damped; but as the size of the substituent is still further increased the base strength *decreases*, because the steric strain factor becomes increasingly important. Thus *N*-dimethyl-*o*-toluidine (XVI) is a stronger base than *N*-dimethylaniline (XV), but *o-t.*-butyldimethylaniline has about the same basicity as the parent amine (see table 6·20).

THE EFFECTS OF SUBSTITUENTS IN POLYCYCLIC COMPOUNDS

In static aromatic compounds substituents exert three important effects: (i) an inductive effect, which may be either positive or negative; (ii) a mesomeric effect which may either reinforce or oppose the inductive effect; and, in suitable cases, (iii) an '*ortho*', or steric, effect. In benzene itself all the annular carbon atoms are, of course, entirely equivalent, and a substituent will exert the same effects no matter to which carbon atom it is linked. In the heterocyclic analogues of benzene, however, and in polycyclic compounds, both homocyclic and heterocyclic, this condition does not apply, and an additional factor is therefore involved.

In naphthalene there are two possible positions for monosubstitution; in anthracene there are three; in phenanthrene there are five; and in some other polycyclic compounds there may be many more. Substituents in these cases need not always exert the same effect. In some cases a steric factor may be involved, but in any case substituents in different positions of such compounds are not, in general, conjugated with the ring system to the same extent.

In naphthalene, the experimental evidence indicates that a given substituent is conjugated with the ring system to a greater extent when it is present in the 1-position than when it is present in the 2-position. After analysis of the dipole moments of the halogenated naphthalenes, for example, Ketelaar and Oosterhout[*] concluded that the C—Cl bond in 1-chloronaphthalene has about 14·2 % double-bond character, as against 13·0 % for the isomeric 2-chloronaphthalene. As the double-bond character

[*] *Rec. trav. chim. Pays-Bas*, 1946, **65**, 448.

of this linkage is governed by the extent of conjugation of the substituent with the ring system, this means that a 1-chloro substituent must be conjugated with the ring system to a greater degree than a 2-chloro substituent in naphthalene. It is also known* that β-naphthylamine is a stronger base (pK_a 3·9) than α-naphthylamine (pK_a 3·7). The extent of conjugation governs the availability of the nitrogen 'lone pair' for salt formation, and the greater the conjugation the weaker the base.

Other experimental evidence of the same nature has been obtained by a study of the rate of oxidation of the azonaphthalenes (to the azoxy compounds) with perbenzoic acid. Badger and Lewis† have found that $\beta\beta'$-azonaphthalene (I) is oxidized much more rapidly than $\alpha\alpha'$-azonaphthalene. This reaction evidently involves the addition of the oxygen to the nitrogen 'lone pair', and the greater the conjugation the less 'available' this 'lone pair' becomes. It seems, therefore, that the conjugation is the greater when the substituent is present in the α-position.

A somewhat similar case was observed by Badger.‡ It was found that osmium tetroxide adds to the ethylenic double bond

* See Albert, *Chem. and Ind.* 1947, p. 51. † *Nature, Lond.*, 1951, **167**, 403.
‡ Ibid. 1950, **165**, 647.

of the three *trans*-dinaphthylethylenes at a rate $\beta\beta > \alpha\beta > \alpha\alpha$. In this case the greater the conjugation with the ring system, the smaller the double-bond character of the classical double bond. As osmium tetroxide adds most rapidly to those bonds which have the greatest double-bond character, it is apparent that this work confirms the fact that substituents are conjugated with the ring system to a greater extent when present in the α-position. It is interesting in this connexion that Coulson* has calculated that the bond orders of the bond to which the addition occurs are in the order $\beta\beta > \alpha\beta > \alpha\alpha$, as expected from qualitative considerations (table 6·21).

TABLE 6·21. *Addition of* OsO_4 *to* trans-*dinaphthylethylenes*

Compound	Time for 50% reaction	Bond order
$\beta\beta$-Dinaphthylethylene	2 min.	1·814
$\alpha\beta$-Dinaphthylethylene	3 min.	1·803
$\alpha\alpha$-Dinaphthylethylene	5 min.	1·792

There is also considerable spectrographic evidence that the extent of conjugation of a substituent varies with the position of substitution. As explained in Chapter 10, substituents produce a bathochromic shift in the positions of the absorption or fluorescence bands of a compound, and the magnitude of the bathochromic shift is determined by the extent of conjugation. The shift in the wave-length of the absorption bands for α-substituted naphthalenes is always greater than that for isomeric β-substituted derivatives. In the anthracene series, a substituent in the *meso* position produces a greater shift than when present in the 1-position, and this, in turn, is greater than that produced when the substituent is in the 2-position. It seems therefore that a *meso* substituent in anthracene is conjugated with the ring system to a greater extent than when in either the 1- or 2-position. There is independent chemical evidence which supports this observation. 9:10-Dimethylanthracene is very readily brominated or oxidized in the methyl groups[†] but methyl groups at other positions in the molecule are not readily attacked.

* *J. Chem. Soc.* 1950, p. 2252.
† Barnett and Matthews, *Ber. dtsch. chem. Ges.* 1926, **59**, 1429; Badger and Cook, *J. Chem. Soc.* 1939, p. 802.

That the different positions in polycyclic compounds should have different conjugating abilities is perfectly reasonable. The magnitude of the conjugation is represented by the contributions which the ionic structures make to the resonance hybrid. The double-bond character of the C—X bond is a measure of this conjugation, and as has already been mentioned the double-bond character of this linkage is greater in the case of α-substituted naphthalenes than in β-substituted derivatives. According to the valence-bond treatment this may be explained by the fact that for α-substituted naphthalenes there are seven such contributing structures ((III)–(IX)):

(III)   (IV)   (V)

(VI)   (VII)   (VIII)

(I)

but for β-substituted naphthalenes there are only six ((X)–(XV)).

The number of such polar structures is related to the free-valence number, as determined by the valence-bond method, for the positions of substitution. It therefore follows that the extent of conjugation parallels the free-valence numbers at the positions of substitution.* In support of this generalization it may be stated that the extent of the bathochromic shift produced by methyl substitution in 1.2-benzanthracene roughly parallels the free-valence numbers of the positions of substitution.†

(X)   (XI)

(XII)

(XIII)   (XIV)

(XV)

According to the calculations of Daudel‡ there is a linear relationship between the free-valence number at the point of substitution in an aromatic compound, and the bond order of the linkage joining the substituent to that position.

By the molecular orbital method conjugating ability is determined by the self-polarizabilities (χ) of the various positions.§

* Pullman, C.R. Acad. Sci., Paris, 1946, **222**, 1396.
† Ibid. 1947, **224**, 1354; Badger, Pearce and Pettit, J. Chem. Soc. 1952, p. 1112.
‡ C.R. Acad. Sci., Paris, 1950, **230**, 99.
§ Wheland and Pauling, J. Amer. Chem. Soc. 1935, **57**, 2086; Coulson and Longuet-Higgins, Proc. Roy. Soc. A, 1947, **191**, 39; **192**, 16; 1948, **195**, 188; Sci. J. R. Coll. Sci. 1948, **18**, 13; J. Chem. Soc. 1949, p. 971.

THE AROMATIC COMPOUNDS

The problem of steric hindrance is also a very real one in many polycyclic compounds. Even in naphthalene, many substituents, in the α-position, are sufficiently bulky to interfere with the *peri* hydrogen atoms. A methoxy group can interfere in this way, and free rotation must be inhibited, even if not entirely prevented. It is probable, therefore, that the configuration of 1-methoxynaphthalene is such that the —O—CH₃ group is disposed away from the *peri* hydrogen. Evidence exists that this is the case. Everard and Sutton* have shown that the dipole moments of 1:5- and 1:4-dimethoxynaphthalene are 0·67 and 2·09 D. respectively, while that of the analogous quinol dimethyl ether is 1·73 D. These differences can only be accounted for if it is assumed that the configurations of the first two compounds are more or less fixed, as in (XVI) and (XVII).

(μ = 0·67 D.) (μ = 2·09 D.)
(XVI)   (XVII)

A 9-methoxy group in anthracene is sterically hindered on both sides. Jones† has calculated that a methoxy group must be strained even when rotated at 90° to the plane of the anthracene ring. Under these circumstances the methoxy group in the 9-position of anthracene cannot be conjugated with the ring system to anything like the extent that occurs in benzene.

Many other substituents are hindered in this way, in polycyclic compounds. The available evidence indicates that, whenever it is possible to relieve the strain by rotation, this does occur, with resultant reduction in the conjugation. A discussion of the spectrographic evidence on this point is given in a later chapter.

* J. Chem. Soc. 1949, p. 2312.   † J. Amer. Chem. Soc. 1945, **67**, 2127.

THE EFFECTS OF SUBSTITUENTS

Considerable evidence has also accumulated in recent years that steric hindrance sometimes forces a substituent group *as a whole* either above or below the plane of the ring system. This is the basis of optical isomerism of the 4:5-phenanthrene type (Chapter 11). Other evidence also exists. The dipole moment of 4:8-dichloro-1:5-dimethoxynaphthalene (0·95 D.) is found to be greater than that of the parent ether. The explanation seems to be that the *peri* substituents force one another out of the plane of the ring system, giving rise to three different configurations, two of which are not centrosymmetrical.*

THE EFFECTS OF HETERO ATOMS

The effects of hetero atoms are very similar to those of substituents in aromatic compounds. Nitrogen is more electronegative than carbon, so that pyridine does not exhibit the degree of symmetry shown by benzene. The electrons tend to crowd round the nitrogen atom.

TABLE 6·22. *Dipole moments of heterocyclic compounds*†

Compound	Dipole moment	
	Benzene solution	Dioxan solution
Pyridine	2·3 D.	—
Pyridazine	—	3·9 D.
Pyrimidine	2·0	2·4
Pyrazine	0·0	0·6

The extent to which this crowding of electrons takes place is measured by the dipole moments of such molecules. The observed values of the dipole moments of the more important nitrogen-containing analogues of benzene are given in table 6·22.

If one assumes a constant value of 1·9 D. for the moment of the C—N linkage, and 0·4 D. for the C—H bond, the moments for a series of heterocyclic compounds may be calculated by vector addition. In such series, Schneider found the observed

* Everard and Sutton, *J. Chem. Soc.* 1949, p. 2312.
† Schneider, *J. Amer. Chem. Soc.* 1948, **70**, 627; Hückel, *Ber. dtsch. chem. Ges.* 1944, **77**, 810.

and calculated values to be in good agreement, so that the C—N bond does seem to have a constant value in such heterocyclic compounds.

The electron displacements in pyridine may be represented by the structures (I), (II) and (III).

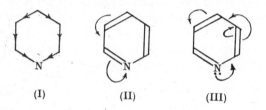

(I)   (II)   (III)

The effect of the electronegativity of nitrogen is to deactivate the whole ring system (structure (I)) and preferentially to remove electrons from the α- and γ-positions (structure (II)). On the other hand, it is possible that the lone pair of the nitrogen also interacts with the ring in such a way as to increase the charge at the β-positions, as in structure (III).

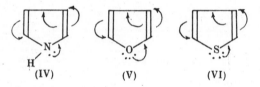

(IV)   (V)   (VI)

With the five-membered heterocyclic molecules such as pyrrole (IV), furan (V) and thiophen (VI), each hetero atom contributes two π-electrons to the aromatic sextet, so that complete electron symmetry is hardly to be expected even if the hetero atom in question has the same electronegativity as carbon. These compounds closely resemble aniline, phenol and thiophenol, and the electron displacements may be represented by tautomeric shifts similar to those postulated for these substituted benzenes.*

In other words, a pronounced increase of charge at both the α- and β-positions is to be expected in each case, and this is

* Remick, *Electronic Interpretations of Organic Chemistry*, 2nd ed., see p. 371, John Wiley, New York.

THE EFFECTS OF SUBSTITUENTS

confirmed by the fact that these compounds are all extremely reactive chemically.

Pullman* has calculated the π-electron densities at the various positions by an approximate valence-bond method. In pyridine (VII), the 2- and 4-positions were found to have electron densities of less than unity; in pyrrole (VIII), furan (IX) and in thiophen (X), all the positions were found to have electron densities greater than unity, and the α-positions to have very high values.

(VII) pyridine: 0·867 (2-position), 1·0 (4-position), 0·867

(VIII) pyrrole: 1·047, 1·073

(IX) furan: 1·020, 1·030

(X) thiophen: 1·040, 1·060

It is possible to calculate somewhat similar π-electron density diagrams (e.g. (XI) and (XII)) by the molecular orbital method, and these are found to be in substantial agreement with those obtained by the valence-bond approximation.†

(XI): 0·822, 0·947, 0·849

(XII): 1·058, 1·096

(XIII): 0·136, 0·072, 0·177

Bond orders and free-valence numbers have also been obtained by the molecular orbital method. For pyridine it is noteworthy that although the greatest density of π-electrons is calculated to be at the β-positions, the α-positions have the greatest free-valence numbers (XIII).‡

---

* *Bull. Soc. chim. Fr.* 1948, **15**, 533.
† Coulson and Longuet-Higgins, *Rev. Sci., Paris*, 1947, **85**, 929; Longuet-Higgins and Coulson, *J. Chem. Soc.* 1949, p. 971.
‡ Coulson and Longuet-Higgins, *Rev. Sci., Paris*, 1947, **85**, 935.

By the same technique it is possible to calculate π-electron densities, bond orders, and free-valence numbers, etc., for polycyclic heterocyclic compounds. The π-electron densities for some compounds of this type are given in formulae (XIV), (XV) and (XVI).*

(XIV): 0·958, 0·772, 0·989, 0·978, 0·947, 0·789, 1·003, 1·633 (N)

(XV): 0·996, 0·938, 0·940, 0·942, 0·984, N 1·594, 0·948, 0·767

(XVI): 0·695, 0·937, 0·999, 0·925, 1·706 (N), 1·014

* Coulson, Daudel and Daudel, *Bull. Soc. chim. Fr.* 1948, **15**, 1181; Coulson and Longuet-Higgins, *Proc. Roy. Soc.* A, 1948, **192**, 16.

CHAPTER 7

AROMATIC SUBSTITUTION REACTIONS

ELECTROPHILIC SUBSTITUTION

In benzene itself all the carbon atoms are equivalent; but, as has been shown in the previous chapter, the presence of substituents, or of hetero atoms, profoundly affects the distribution of π-electrons. Certain groups have an electron-releasing effect and increase the π-electron density all round the aromatic ring and especially at the *ortho-para* positions (as in (I)). Other groups (—OMe, etc.) have two opposing effects and increase the π-electron density at the *ortho-para* positions, but reduce it at the *meta* positions. Nitro, carbonyl and carbethoxy, and certain other groups, on the other hand, have an electron-attracting effect and decrease the charge on the aromatic ring, especially at the *ortho-para* positions (as in (II)). Under these circumstances, the *meta* positions have a greater π-electron density than the *ortho-para* positions. The halogens also seem, in general, to have a resultant electron-attracting effect, but they are exceptional in that they appear to leave the *ortho-para* positions more charged than the *meta* positions.

Substituents are therefore of two main types: (i) those which cause the *ortho-para* positions to have a greater negative charge than the *meta* positions; and (ii) those which cause the *meta* positions to have a greater negative charge than the *ortho-para* positions.

This fact is of very considerable value in the study of aromatic substitution reactions, because in reactions such as nitration, halogenation, etc., it has long been known that a substituent already present controls the orientation of the product. Certain substituents direct the entering group mainly into the *ortho-para* positions, but others direct it mainly into the *meta* positions. The classification is not a rigid one, for there is a gradual transition from the very strongly *meta*-directing substituents, such as the —NMe_3^+ group, to the very strongly *ortho-para*-directing substituents, such as the —OMe group. Nevertheless, the classification is of very great utility, because the *ortho-para*-directing groups are now known to be those which cause the *ortho-para* positions to acquire a greater negative charge than the *meta* positions, and the *meta*-directing groups are now known to be those which cause the *meta* positions to have a greater negative charge than the *ortho-para* positions.

The situation with the heterocyclic analogues of benzene is rather similar. As indicated in the previous chapter, the 3-position in pyridine has a greater π-electron density than either the 2- or 4-position, and in ordinary substitution reactions the 3-position is attacked. Again, the 2-positions in pyrrole, furan and in thiophen have higher π-electron densities than the 3-positions, and the usual substitution reagents attack the former preferentially.

Again, in non-benzenoid aromatic hydrocarbons such as azulene, substitution appears to take place at the position which is calculated to have the greatest π-electron density.*

It seems, therefore, that in substitution reactions such as nitration, halogenation, etc., the entering group attacks those centres which have the greatest π-electron densities.[†] In the presence of an electron-releasing substituent the *ortho-para* positions have the greatest negative charge, and the entering group attacks these positions. In the presence of an electron-attracting substituent, however, the *meta* positions have the greatest negative charge, and this position is attacked. In other words, most of the common substitution reactions are *electrophilic* in character. In this connexion it is also important, as Holleman[‡] has shown, that predominant *ortho-para* substitution

---

* Anderson and Nelson, *J. Amer. Chem. Soc.* 1950, **72**, 3824; Brown, *Trans. Faraday Soc.* 1948, **44**, 984.
† Additional factors, such as steric hindrance, may affect the observed result.
‡ *Die direkte Einführung von Substituenten in den Benzolkern*, Leipzig, 1910.

usually occurs much more rapidly, and under milder conditions, than substitution which occurs mainly in the *meta* positions. Halogenated benzenes, however, are attacked in the *ortho-para* positions, but at a greatly reduced rate.

It would be expected that a positively charged substituent which is directly attached to the ring would be very strongly *meta*-directing, in view of its very powerful $-I$ effect, and this has been observed.* In nitration, the group —NMe_3^+ directs exclusively into the *meta* position so far as can be ascertained. It is interesting, however, to compare this *meta*-directing force with that of the analogous substituents, —$CH_2.NMe_3^+$, —$CH_2.CH_2.NMe_3^+$, etc., in which the positive charge is separated from the nucleus by one or more methylene groups. It is found that as the charge is removed farther and farther from the ring, the amount of *meta* substitution product which is formed rapidly diminishes. Thus:

$Ph.NMe_3^+$	$Ph.CH_2.NMe_3^+$	$Ph.CH_2.CH_2.NMe_3^+$
100%	88%	19%

$$Ph.CH_2.CH_2.CH_2.NMe_3^+$$
$$5\%$$

Similarly, the *meta*-directing force of a nitro group is rapidly damped when it is separated from the ring:

$Ph.NO_2$	$Ph.CH_2.NO_2$	$Ph.CH_2.CH_2.NO_2$
93%	67%	13%

Ingold and his collaborators[†] have determined the relative reactivities of the various *ortho*, *meta* and *para* positions in several substituted benzenes by competitive nitration. Relative speeds of nitration were determined in experiments in which benzene and a substituted benzene derivative were nitrated together, an insufficient quantity of nitric acid being used. The relative quantities of the products, $Ph.NO_2$ and $X.C_6H_4.NO_2$, gave the relative speeds of nitration of benzene and of $Ph.X$. When these results were combined with the relative quantities of *ortho*, *meta* and *para* isomers formed, it was possible to arrive

---

* Compare Roberts, Clement and Drysdale, *J. Amer. Chem. Soc.* 1951, **73**, 2181.
† *J. Chem. Soc.* 1927, p. 2918; 1931, p. 1959; 1938, pp. 905, 918.

at figures representing the relative reactivities of each of the possible substitution positions, as in table 7·1.

TABLE 7·1. *Relative reactivities towards nitration*

Compound	Relative reactivities		
	ortho	meta	para
Ph.H	1	1	1
Ph.CH$_3$	43	3	55
Ph.CO$_2$Et	0·0026	0·0079	0·0009
Ph.Cl	0·030	0·000	0·139
Ph.Br	0·037	0·000	0·106

The results show that a methyl group is a typical *ortho-para*-directing group, the relative reactivities of all three positions being greater than that of benzene. This is in accordance with the known electron-releasing effect of this substituent. On the other hand, the ester grouping lowers the reactivity of all three positions, but especially of the *ortho* and *para* positions, in accordance with the known electron-attracting effect of the group. It is reasonable to conclude, therefore, that nitration is not only an electrophilic reaction, but that the orientation of the product is controlled by the selective activation and de-activation of the various positions, especially the *ortho-para* positions.

Until very recently the mechanism of electrophilic substitution was only imperfectly understood. Electrophilic reagents were known to *add* to ethylenic double bonds under suitable conditions, and similar addition compounds have sometimes been obtained with certain polycyclic aromatic hydrocarbons. For example, addition products have sometimes been obtained following treatment of ethylenic compounds with nitric acid,[*] and Meisenheimer and Connerade[†] found that HO—NO$_2$ adds to the 9:10 positions of anthracene and that subsequent elimination of H—OH (with the formation of 9-nitroanthracene) occurs on treatment with mineral acid. For this reason, it was suggested[‡]

---

[*] Michael and Carlson, *J. Amer. Chem. Soc.* 1935, **57**, 1268; Wieland and Sakellarios, *Ber. dtsch. chem. Ges.* 1920, **53**, 201; Wieland and Rahn, ibid. 1921, **54**, 1770.
[†] *Liebigs Ann.* 1904, **330**, 133.
[‡] See, for example, Wieland and Sakellarios, *Ber. dtsch. chem. Ges.* 1920, **53**, 201.

AROMATIC SUBSTITUTION REACTIONS

that electrophilic substitution takes place by addition of the reagent $(X-Y)$, and subsequent elimination of $H-Y$, to give the substitution product, $Ar-X$.* Nitration, for example, was supposed to occur by addition of $HO-NO_2$, followed by elimination of $H-OH$, as in the sequence (III), (IV) and (V), the substituent, X, being assumed to be an electron-donating group.

(III)   (IV)   (V)

In the same way, bromination was supposed to occur by addition of bromine to an aromatic 'double' bond (or to the 1:4 positions) followed by elimination of hydrogen bromide. Hydrogen bromide is, of course, evolved during reactions of this type, and addition products of the kind visualized have actually been isolated from many aromatic hydrocarbons, including benzene, naphthalene, anthracene and phenanthrene. These addition products are always relatively unstable, and give rise to substitution products, with elimination of hydrogen bromide, on heating. This process has already been discussed in an earlier chapter (p. 97 et seq.).

The addition-elimination theory of aromatic substitution received considerable attention from organic chemists for many years, but evidence has now accumulated that the substitution of aromatic compounds proceeds by a true substitution mechanism, and that an addition product (of type (IV)) is not normally formed under substitution conditions. It has been shown, for example, that the *addition* of bromine to the 9:10 bond of phenanthrene is largely photochemical,† but that in the presence of certain catalysts, such as aluminium chloride, stannic

---

* For a review see Fieser in Gilman's *Organic Chemistry*, vol. 1, John Wiley, New York, 1943.
† Kharasch, White and Mayo, *J. Org. Chem.* 1938, **2**, 574.

chloride and iodine, etc., hydrogen bromide is evolved and the substitution product is formed. According to the addition-elimination theory, these catalysts would have to act by promoting the elimination of hydrogen bromide from the addition product, but Price* was able to show that the addition of iodine to a solution of pure phenanthrene dibromide produced no change, and did not result in the elimination of hydrogen bromide. These catalysts must therefore promote the elimination of hydrogen bromide in some other way, and the addition-elimination theory is definitely excluded.

A variety of evidence indicates that electrophilic substitution must proceed by an ionic mechanism. It is known, for example, that the rate of bromination of an aromatic compound is largely dependent on the polar character of the solvent; that non-catalytic brominations in the vapour phase (at least in the lower temperature ranges) take place exclusively on the (polar) walls of the reaction vessel; and that halogenations are catalysed by such substances as aluminium chloride, ferric bromide, iodine, etc., which form complexes as follows:

$$\text{Br---Br} + \text{AlCl}_3 \rightleftharpoons \text{Br}^+ \ldots \text{Br}^-\text{AlCl}_3.$$

Evidence for the presence of a bromine cation, Br^+, has also been obtained in some cases.[†] It is also known that nitric and sulphuric acids interact to give the nitronium ion, NO_2^+;[‡] that aluminium chloride interacts with alkyl halides and acyl halides to give carbonium ions, R^+ and RCO^+, which are probably involved in Friedel-Crafts reactions; that sulphuric acid interacts with certain compounds to give carbonium ions which are also probably involved in electrophilic substitutions;[§] and so on.

On general grounds it would seem that the reagent in electrophilic substitutions is probably either a positive ion, or else a neutral *polarized* molecule so oriented that its positive end is directed towards the point of attack.

---

* *J. Amer. Chem. Soc.* 1936, **58**, 1834, 2101.

† Derbyshire and Waters, *Nature, Lond.*, 1949, **164**, 446; *J. Chem. Soc.* 1950, p. 564.

‡ For a review see Gillespie and Millen, *Quart. Rev. Chem. Soc., Lond.*, 1948, **2**, 277.

§ For example, Bogert and Davidson, *J. Amer. Chem. Soc.* 1934, **56**, 185; Price, Davidson and Bogert, *J. Org. Chem.* 1938, **2**, 540.

AROMATIC SUBSTITUTION REACTIONS

It is also reasonable to suppose that the reagent must approach the aromatic molecule (VI), (IX) from above the plane of the ring, displacing the hydrogen atom downwards.* The transition state (VII), (X) must be of a quasi-quinonoid character, and may be represented approximately as a resonance hybrid of all structures of this type. Finally, a proton must be ejected, and the substitution product (VIII), (XI) formed.†

The above account of the nature of electrophilic substitution places the emphasis on the π-electron density at the point of attack in the aromatic ring system, and this treatment seems to give a satisfactory explanation of the experimental facts with most substituted benzenoid compounds, or with heterocyclic substances.

On the other hand, Wheland‡ has shown that it is possible to consider the problem in a rather different manner by placing the emphasis on the transition state. He has pointed out that the reagent in electrophilic substitution must have a 'gap' in its octet of valency electrons (i.e. $\ddot{X}$:), and that, before the attack can occur, two π-electrons (to fit into the open sextet of the reagent) must be 'localized' on one or other of the annular

---

* Cowdrey, Hughes, Ingold, Masterman and Scott, *J. Chem. Soc.* 1937, p. 1252; Hughes and Ingold, ibid. 1941, p. 608.
† The great majority of electrophilic substitutions do involve the displacement of hydrogen; but this is not invariably the case. Under suitable conditions certain substituents such as $-CO_2H$, $-SO_3H$, $-NO_2$, etc., can also be displaced.
‡ *J. Amer. Chem. Soc.* 1942, **64**, 900.

THE AROMATIC COMPOUNDS

carbon atoms. The process is best illustrated by considering the substitution of benzene itself. On the approach of the reagent towards the aromatic ring system (XII) the conjugated system of π-electrons is perturbed in such a way that two π-electrons become 'localized' or fixed, the remaining four π-electrons being distributed over the five carbon atoms not involved in the attack, as in (XIII).

(XII)

(XIII)

Resonance energy must be lost in this process, and the energy required to bring about this localization clearly equals this loss of resonance energy. In the case of benzene, therefore, the energy required to bring about the localization is the difference between the resonance energy of benzene and that of the 'residual conjugated system', in this case the pentadiene cation.

Wheland points out further that electrophilic substitution should occur at those positions in an aromatic compound at which an unshared pair of electrons, to fit into the open sextet of the reagent, can be provided most readily. Substituents or hetero atoms may either increase or decrease the energy required to bring about the localization, depending on whether they have a resultant electron-attracting or electron-releasing effect, and therefore either hinder or assist the reaction. A methoxy group, for example, greatly reduces the energy which is required to localize two π-electrons at the *ortho* and *para* positions, so that electrophilic attack occurs predominantly at these positions and at a rate much greater than that in, say, benzene itself. A nitro group, on the other hand, increases the energy necessary to bring about the required localization, especially at the *ortho* and *para* positions. The rate of reaction, and therefore the orientation of the product (which is a function of this and other rates), therefore depends very largely on the magnitude of this energy value,* which can often be calculated with reasonable accuracy.

(XIV)

(XV)

This treatment is of special value in considering the substitution of di- and polycyclic aromatic hydrocarbons. In these substances (naphthalene, anthracene, etc.) the various carbon atoms all have the same π-electron density, but they are by no means equivalent. For example, in the case of naphthalene, calculation shows that it is easier to localize two π-electrons on the 1-position (XIV) than to bring about a similar localization on the 2-position (XV). In fact, Sixma† has calculated that the difference in the localization energies required is of the order of about 3 kcal./mole. It is to be expected therefore that the electrophilic substitution

---

* Compare Hughes and Ingold, *J. Chem. Soc.* 1941, p. 608.
† *Rev. trav. chim. Pays-Bas*, 1949, **68**, 915.

of naphthalene will take place predominantly at the α-position, and this is, of course, the observed experimental result.

Empirically, it is also found that in *unsubstituted polycyclic aromatic hydrocarbons* electrophilic attack occurs at the positions of greatest free-valence number. In naphthalene (XVI) attack occurs predominantly at the 1-position; in anthracene (XVII) at the 9- and 10-positions; in pyrene (XVIII) at the 3-, 5-, 8- and 10-positions; and in chrysene (XIX) at the 2-position. In each case these are the positions of greatest free valence as determined either by the method of molecular orbitals or by the valence-bond method. The former values are given in the formulae.

(XVI) 0·134, 0·086

(XVII) 0·202, 0·141, 0·090

(XVIII) 0·134, 0·151, 0·076, 0·090, 0·085

(XIX) 0·124, 0·122, 0·134, 0·139

Summarizing, therefore, it seems that electrophilic substitution very probably occurs predominantly at the positions at which two π-electrons can be provided (to fit into the open sextet of the reagent) most readily. In substituted (and heterocyclic) compounds it is observed empirically that the attack occurs predominantly at the positions of greatest π-electron density. This is reasonable enough if two π-electrons can be provided most easily at those positions which already have a high π-electron density. (In the majority of cases this does seem to be approximately true.) Finally, in unsubstituted polycyclic compounds it is also observed empirically that electrophilic attack occurs predominantly at the positions of greatest free valence. This implies that (in such compounds) two π-electrons can be provided or localized most easily at positions of greatest free valence.*

NUCLEOPHILIC SUBSTITUTION[†]

All substitution reactions do not proceed by an electrophilic mechanism. Nitrobenzene, for example, reacts with potassium diphenylamide to give N-(*para*-nitrophenyl)diphenylamide[‡] and with piperidine in the presence of sodamide to give N-(*para*-nitrophenyl)piperidine.[§] Similarly, nitrobenzene is converted into *ortho*-nitrophenol (and a trace of the *para* isomer) with fused potassium hydroxide and an oxidizing agent;[‖] *meta*-dinitrobenzene is converted into 2:6-dinitrophenol by the same treatment; and pyridine is converted into α-pyridone.[¶] The commercial manufacture of the dyestuff alizarin by fusing anthraquinone-β-sulphonic acid with caustic alkali and an oxidizing agent is another example of this type of reaction. Especially noteworthy is the fact that pyridine and related heterocyclic compounds react with sodium or potassium amide to give amino compounds.** Pyridine gives mainly 2-aminopyridine together with some of the 4-amino derivative; and quinoline gives a mixture of 2- and 4-aminoquinolines. In none of these examples can the reagent be electrophilic in character.

Reactions of this nature have not been observed with benzene itself. They occur only with compounds which are strongly deactivated, either by the presence of an electron-attracting

---

* In substituted (and heterocyclic) compounds, however, electrophilic substitution does *not* always occur at the positions of greatest free valence. The rule applies only to unsubstituted aromatic *hydrocarbons* of the benzenoid type. See Daudel, Daudel, Buu-Hoï and Martin, *Bull. Soc. chim. Fr.* 1948, **15**, 1202; Sandorfy, Vroelant, Yvan, Chalvet and Daudel, ibid. 1950, **17**, 304.

† For recent reviews see: Burnett and Zahler, *Chem. Rev.* 1951, **49**, 273; Miller, *Rev. Pure Appl. Chem.* 1951, **1**, 171.

‡ Bergstrom, Granara and Erickson, *J. Org. Chem.* 1942, **7**, 98.

§ Bradley and Robinson, *J. Chem. Soc.* 1932, p. 1254.

‖ Wohl, *Ber. dtsch. chem. Ges.* 1899, **32**, 3486.

¶ Tschitschibabin, ibid. 1923, **56**, 1879.

** Leffler, *Organic Reactions*, **1**, 91, John Wiley, New York, 1942.

atom (e.g. N) in the ring system, or by the presence of one or more electron-attracting substituents. Furthermore, as can be seen from the few examples cited, the reagent in this type of reaction seems preferentially to attack the positions which are *ortho* and *para* to the deactivating substituents. In other words, the reagent appears to seek out the most *positive* centres for attack.

(I) (II)

(III) (IV)

This conclusion has been further confirmed by a study of the action of sodium or potassium amide on various azahydrocarbons. As mentioned in a previous chapter, π-electron density diagrams for the azahydrocarbons can be calculated without difficulty, and it therefore becomes possible to correlate the position of attack with the π-electron density. Thus acridine (I) gives 5-aminoacridine;[*] *iso*quinoline (II) gives 1-amino*iso*quinoline;[†] quinoline (III) gives both 2-amino- and some 4-aminoquinoline;[‡] and phenanthridine (IV) gives 6-aminophenanthridine in 90% yield.[§] In each case the position of attack coincides with the position of lowest π-electron density (p. 232).

[*] Bergstrom and Fernelius, *Chem. Rev.* 1933, **12**, 163.
[†] Bergstrom, *Liebigs Ann.* 1934, **515**, 34; Tschitschibabin and Oparina, *J. Russ. Phys. Chem. Soc.* 1920, **50**, 543.
[‡] Bergstrom, *J. Org. Chem.* 1937, **2**, 411; Shreve, Riechers, Rubenkoenig and Goodman, *Industr. Engng Chem.* 1940, **32**, 173; Tschitschibabin and Zataepina, *J. Russ. Phys. Chem. Soc.* 1920, **50**, 553.
[§] Bergstrom and Fernelius, *Chem. Rev.* 1937, **20**, 472.

AROMATIC SUBSTITUTION REACTIONS

The mechanism of such reactions therefore appears to be entirely the reverse of that which occurs in electrophilic substitution. In this case the reagent is *nucleophilic* in character, and it seems likely that it is either a negative ion, or else a neutral polarized molecule so oriented that its negative end is directed towards the point of attack.* By analogy with electrophilic attack, nucleophilic substitution can be assumed to occur according to the mechanism (V)–(VII), the transition state (VI) being a resonance hybrid of all such quasi-quinonoid structures.

$$\underset{(V)}{\underset{\bar{Y}}{\bigcirc}\overset{X}{\longrightarrow}} \quad \underset{(VI)}{\underset{H}{\bigcirc}\overset{X}{\underset{Y}{\longrightarrow}}} \quad \underset{(VII)}{\bigcirc Y} + \bar{X}$$

This conception of the mechanism is supported by a number of facts. First, all the reactions mentioned apparently involve the displacement of a hydrogen atom as a negative hydride ion ($X = H$), and it is therefore significant that substances which can transform hydride ions into molecular hydrogen, into protons, or into water, facilitate substitutions of this type. This is evidently the function of the oxidizing agents (including air) in reactions such as the hydroxylation of nitrobenzenes with alkali.

Secondly, it might be expected that substitutions of this type would occur more readily when they involve the displacement, or substitution, of a group (X) which is of such a nature that it can form a stable anion (e.g. $X = Cl$, NO_2, OR, etc.). This has also been confirmed experimentally. Chlorobenzene, for example, can be converted into phenol by treatment with caustic alkali, according to the scheme (VIII)–(X).

Such reactions generally take place only very slowly unless very vigorous conditions are used. The conversion of chlorobenzene into phenol is typical and only proceeds at a useful rate under conditions of high temperature and pressure.

---

* It must be admitted, however, that some of the substitutions mentioned in this section have not yet been investigated in any detail. It is possible that some will prove to be radical reactions.

When deactivating substituents are present in the *ortho-para* positions, however, reactions of this nature occur very much more readily, and under much less vigorous conditions. As is well known, *para*-nitrochlorobenzene is very readily converted into *para*-nitrophenol with caustic alkali; 2:4-dinitrochlorobenzene is transformed into 2:4-dinitrophenol by brief boiling with dilute alkali; and 2:4:6-trinitrochlorobenzene is converted into picric acid under even milder conditions. It is also well known that *ortho*- and *para*-nitro groups facilitate the transformation of chlorobenzenes into amines or substituted amines. Chlorobenzene can be converted into aniline only by heating to a very high temperature with ammonia in an autoclave, but 2:4-dinitrochlorobenzene reacts with ammonia even at moderate temperatures.

(VIII)   (IX)   (X)

Again, both *para*-nitrochlorobenzene and *para*-nitroanisole are converted into *para*-nitrophenetole by boiling with ethanol and caustic alkali, and this substitution also occurs even more readily with the 2:4-dinitro derivatives.* In this connexion it is noteworthy that Meisenheimer† has obtained the same intermediate addition product (XII) from trinitroanisole (XI) and potassium ethoxide, and from trinitrophenetole (XIII) and potassium methoxide.

Ortho- and *para-* standing nitro groups can also 'activate' another nitro group, and advantage is taken of this fact in the purification of crude trinitrotoluene. The nitration of toluene gives about 4 % of the *meta*-nitro derivative, and this on further nitration is converted into unsymmetrical trinitrotoluenes such as (XIV). Such unsymmetrical trinitrotoluenes are much less stable than 2:4:6-trinitrotoluene (which is the major product of the nitration of toluene), and their removal is therefore desirable.

* Ogata and Okano, *J. Amer. Chem. Soc.* 1949, **71**, 3211.
† *Liebigs Ann.* 1902, **323**, 242.

AROMATIC SUBSTITUTION REACTIONS

The method usually used involves warming the crude product with aqueous sodium sulphite. The 'activated' nitro group is substituted by a sulphonic acid group, and the product (XV) is removed from the mixture by washing with water.

This treatment of nucleophilic substitution lays the emphasis on the π-electron density at the point of attack, and assumes that the reagent (which may be either a negative ion or else a neutral polarized molecule so oriented that its negative end is directed towards the point of attack) seeks out the position with the lowest density of π-electrons. As with electrophilic substitution, however, it is possible to consider the mechanism from the point of view of the transition state.

According to Wheland,[*] nucleophilic substitution should occur most readily at the positions at which an open sextet of electrons, to accommodate the unshared pair in the reagent, can be provided most easily. This requires that the aromatic ring system must be polarized in such a way that the six π-electrons are distributed over the five annular carbon atoms which are not involved in the attack. The activated complex can therefore

[*] J. Amer. Chem. Soc. 1942, **64**, 900.

be considered as a resonance hybrid of the structures (XVI), (XVII) and (XVIII).

The energy required to polarize the molecule in this way can be calculated, and Wheland assumes that the less the energy required, the easier the substitution with nucleophilic reagents. Electron-attracting substituents reduce the energy required to effect this polarization, especially at the *ortho* and *para* positions; electron-releasing substituents, on the other hand, increase the energy required and make it more difficult for the reaction to proceed.

(XVI) (XVII) (XVIII)

Nucleophilic substitution in di- and polycyclic aromatic hydrocarbons can also be considered in the same way. Calculation shows, for example, that the energy required to provide an open sextet (i.e. a positive charge) on the 1-position of naphthalene is about 3 kcal./mole less than that required to bring about a similar polarization on the 2-position.*†

It is to be expected, therefore, that nucleophilic substitution of naphthalene will occur predominantly at the 1-position. Experimentally, it is found that naphthalene undergoes nucleophilic substitution at the 1-position very much more readily than benzene. With sodium amide, for example, amination occurs at the 1-position.‡

Summarizing, it seems probable that nucleophilic attack occurs at those positions at which an open sextet of electrons (i.e. a positive charge) can be provided most readily. It is

---

* *Rec. trav. chim. Pays-Bas*, 1949, **68**, 915.
† In other words, the resonance energy of the transition state is higher for the α- than for the β-position (Dewar, *The Electronic Theory of Organic Chemistry*, p. 176, Oxford University Press, 1949). Less energy is required to convert naphthalene into the quasi-quinonoid transition state for α- substitution (which corresponds to α-naphthaquinone) than is necessary for the formation of the transition state for β-substitution (which corresponds to β-naphthaquinone).
‡ Sachs, *Ber. dtsch. chem. Ges.* 1906, **39**, 3006.

AROMATIC SUBSTITUTION REACTIONS

reasonable to suppose that in substituted and in heterocyclic compounds this will occur at those positions which already have the greatest deficiency of π-electrons, and this seems to be the case.

RADICAL SUBSTITUTION

In the reactions discussed above the reagents are always either ionic in character, or else neutral *polarized* molecules. There are some aromatic substitution reactions, however, which occur in a completely non-polar environment and which cannot, therefore, proceed by either an electrophilic or nucleophilic mechanism. Such reactions often occur in the vapour phase at high temperatures* or under irradiation, but several reagents are also known which can bring about non-ionic substitutions in solution. In all these cases the actual reagent seems to be an uncharged atom or free radical, formed by homolytic fission of a covalent bond in the reagent (equation (1)), as opposed to heterolytic fission (equations (2) and (3)):

$$A : B \to A\cdot + \cdot B \qquad (1)$$
$$A : B \to A{:}^{\ominus} + B^{\oplus} \qquad (2)$$
$$A : B \to A^{\oplus} + {:}B^{\ominus} \qquad (3)$$

Aromatic substitution with free radicals has been demonstrated with aryl radicals ($Ar\cdot$), with methyl radicals ($CH_3\cdot$), with halogen atoms ($Br\cdot$), with hydroxyl radicals ($HO\cdot$), and with several other free radicals.[†]

With reagents of this nature the normal orientation rules are not observed. It is usually found that *all* the available positions are attacked, but especially the *ortho* and *para* positions, whether the substituent already present is electron-releasing or electron-attracting. Nitrobenzene, for example, is readily hydroxylated in all three positions (but mainly in the *ortho* and *para* positions) by treatment with ferrous ion and hydrogen peroxide.[‡] Similarly, it is phenylated in all three positions, but especially in the

---

[*] At high temperatures adsorption of the reactants on the polar walls of the reaction vessel becomes negligible.
[†] Augood, Hey, Nechvatal and Williams, *Nature, Lond.*, 1951, **167**, 725; see also Hey, Nechvatal and Robinson, *J. Chem. Soc.* 1951, p. 2892.
[‡] Loebl, Stein and Weiss, *J. Chem. Soc.* 1949, p. 2074.

ortho and *para* positions.* In the same way it has been shown that phenylation takes place at all three positions with respect to a carbethoxy group, and to chloro and bromo substituents.† It is also known that the phenylation of pyridine gives a mixture of α-, β- and γ-phenylpyridines, the α-isomer predominating.‡

Not only are the normal rules of orientation not obeyed in radical substitution, but it seems that *all* substituents activate the nucleus‡ and facilitate reactions of this nature.

A nitro group is particularly effective in promoting radical reactions. Nitrobenzene and *meta*-dinitrobenzene, for example, are methylated with lead tetraacetate, the products being nitrotoluenes and nitroxylenes. Benzene, chlorobenzene, toluene and naphthalene, on the other hand, are not methylated under the same conditions but give the normal direct oxidation product.∥

The mechanism of radical substitution reactions is almost entirely unknown, and at present it may best be formulated by analogy with electrophilic and nucleophilic substitutions. The approach of the radical reagent $(X\cdot)$ towards the aromatic ring system (I) may be assumed to displace the hydrogen atom below the plane of the ring. The transition state would then be a resonance hybrid of all quasi-quinonoid structures of type (II), and subsequent ejection of the hydrogen as a hydrogen atom would complete the substitution and give (III).

If this mechanism is true, then the fate of the hydrogen is of some importance. Some may combine to give hydrogen molecules; some may be used in reducing one or other of the reactants, or the solvent; but further investigation is very necessary.

* DeTar and Scheifele, *J. Amer. Chem. Soc.* 1951, **73**, 1442.
† Hey, *J. Chem. Soc.* 1934, p. 1966; France, Heilbron and Hey, ibid. 1938, p. 1364.
‡ Haworth, Heilbron and Hey, ibid. 1940, p. 349.
§ Augood, Hey, Nechvatal and Williams, *Nature, Lond.*, 1951, **167**, 725; see also Hey, Nechvatal and Robinson, *J. Chem. Soc.* 1951, p. 2892.
∥ Fieser, Clapp and Daudt, *J. Amer. Chem. Soc.* 1942, **64**, 2052.

An alternative mechanism has been considered by some workers,* and may be illustrated by the following equations, representing the bromination of an aromatic substance:

$$Br_2 \rightleftharpoons Br\cdot + \cdot Br,$$
$$Br\cdot + ArH \rightarrow Ar\cdot \text{ (isomers)} + HBr,$$
$$Ar\cdot + Br_2 \rightarrow ArBr + Br\cdot.$$

Such a mechanism would not, however, offer a satisfactory explanation of the Pschorr synthesis (p. 316); nevertheless, it is conceivable that it may operate in some cases.

As the normal rules of orientation are not applicable in radical substitution, it is apparent that the orientation is not controlled by the π-electron densities at the possible positions of substitution. According to Wheland,† radical substitution should occur most readily at the positions at which an unshared electron (to pair with the unshared electron of the reagent) can be provided most easily. This requires that the aromatic ring system must be polarized in such a way that one π-electron becomes localized on the annular carbon atom attacked, and that the remaining five mobile electrons become distributed among the five annular carbon atoms not involved in the attack. The activated complex must therefore be considered as a resonance hybrid of the structures (IV), (V) and (VI).

The energy required to bring about this localization (which is equal to the loss of resonance energy) can be calculated, and the smaller this energy value, the easier the substitution with radical reagents. The presence of substituents affects the magnitude of this polarization energy, and, in general, a single unpaired electron

* For example, Sixma and Wibaut, *Rec. trav. chim. Pays-Bas*, 1950, **69**, 577.
† *J. Amer. Chem. Soc.* 1942, **64**, 900.

is more 'available' in the *ortho* and *para* positions, whatever the nature of the substituent. Calculation also shows that an unpaired electron is more available in the α-position than in the β-position of naphthalene, so that (other things being equal) radical reagents would be expected to attack the former preferentially. The energy required for localization in the β-position is greater than that for the α-position by about 3 kcal./mole, which also happens to be the difference in localization energy for both the electrophilic and nucleophilic substitution of naphthalene.[*]

Empirically, it is also found, both for heterocyclic compounds and for homocyclic polycyclic compounds, that radical substitution occurs preferentially at the positions of greatest free valence. It follows, therefore, that a single unpaired electron must be most available at the positions of greatest free valence.[†]

NITRATION

Nitration is a reaction of very considerable importance in aromatic chemistry. It is, of course, a valuable synthetic method, for the nitro group can be readily converted into a wide range of different substituent groups. Moreover, the reaction has been very extensively used in studying the effect of substituents on orientation in substitution reactions. For this purpose the nitration reaction is most convenient as it is irreversible, and under suitable conditions introduces only one nitro group.

In the early work it was shown that while certain substituents direct the entering group mainly into the *ortho-para* positions, others direct it mainly into the *meta* positions. The *ortho-para*-directing groups were also found to facilitate the reaction and to be electron-donating. Similarly, the *meta*-directing groups were found to be electron-attracting, and to retard the reaction. For example, toluene is attacked 24 times faster than benzene, but ethyl benzoate is attacked 300 times more slowly than benzene; and benzenesulphonic acid and nitrobenzene were

[*] *Rec. trav. chim. Pays-Bas*, 1949, **68**, 915.
[†] See Daudel, Daudel, Buu-Hoï and Martin, *Bull. Soc. chim. Fr.* 1948, **15**, 1202; Sandorfy, Vroelant, Yvan, Chalvet and Daudel, ibid. 1950, **17**, 304; Kooyman and Farenhorst, *Nature, Lond.*, 1952, **169**, 153.

found to be even less reactive.* This and similar evidence indicates that the reaction is electrophilic.

In spite of its great utility, the nitration reaction has only recently been carefully studied. Nitration can be effected with a wide variety of reagents, and under many different conditions. Nitric acid, fuming nitric acid, ethyl nitrate, acetyl nitrate, and other similar reagents have been used. Particularly effective are mixtures of these reagents with very strong acids.

Probably the most common nitrating agent in organic chemistry is 'mixed acid', a mixture of nitric and sulphuric acids. Several workers have supposed that the sulphuric acid acts as a dehydrating agent, facilitating the removal of water according to the equation

$$ArH + HNO_3 \rightarrow ArNO_2 + H_2O,$$

but this view was disproved by Spindler,[†] who showed that the reaction is irreversible.

As early as 1903, Euler[‡] suggested that the actual nitrating agent may be the nitronium ion, NO_2^+. This suggestion was supported in many quarters,[§] but satisfactory experimental confirmation has only recently been provided by several groups of workers, almost simultaneously.[||]

Several methods have been used in these investigations, among which the following may be mentioned.

Freezing-point depressions of solutions of nitric acid in sulphuric acid indicate a van't Hoff factor, i, of 4. This means that each molecule of nitric acid added to sulphuric acid solvent produces *four* solute particles. Only one explanation seems to be possible, namely, that the nitric acid must react with sulphuric acid according to the equation[¶]

$$HNO_3 + 2H_2SO_4 \rightleftharpoons NO_2^+ + H_3O^+ + 2HSO_4^-.$$

This is supported by the fact that solutions of nitric acid in sulphuric acid have a high conductivity.

* Hughes, Ingold and Reed, *J. Chem. Soc.* 1950, p. 2403.
† *Ber. dtsch. chem. Ges.* 1883, **16**, 1253.
‡ *Liebigs Ann.* 1903, **330**, 280; *Z. angew. Chem.* 1922, **35**, 580.
§ Walden, ibid. 1924, **37**, 390; Ri and Eyring, *J. Chem. Phys.* 1940, **8**, 433; Price, *Chem. Rev.* 1941, **29**, 51.
|| See Braude, *Ann. Rep. Chem. Soc.* 1949, **46**, 131.
¶ Gillespie, Graham, Hughes, Ingold and Peeling, *Nature, Lond.*, 1946, **158**, 480; *J. Chem. Soc.* 1950, p. 2504.

Similarly, solutions of dinitrogen pentoxide in sulphuric acid showed a van't Hoff factor of 6, indicating the formation of ions according to the equation

$$N_2O_5 + 3H_2SO_4 \rightleftharpoons 2NO_2^+ + H_3O^+ + 3HSO_4^-.$$

The existence of nitronium ions has also been confirmed by examination of the Raman spectra of these and similar solutions. Chédin* found that solutions of nitric acid in concentrated sulphuric acid exhibit two lines strongly, at 1050 and at 1400 cm.$^{-1}$, neither of which can be attributed to the acid molecules. Both lines are present in the spectrum of nitric acid itself, but only weakly. They become strong on the addition of sulphuric acid, and also of dinitrogen pentoxide. It seemed likely that the frequency 1400 cm.$^{-1}$ should be assigned to the nitronium ion (cf. CO_2), and that the frequency 1050 cm.$^{-1}$ should be assigned to the nitrate ion and/or to the bisulphate anion,[†] and this has now been satisfactorily confirmed. With perchloric acid, for example, nitronium perchlorate, $(NO_2^+)(ClO_4^-)$, is formed, and this showed only the 1400 cm.$^{-1}$ line.[‡]

That the nitronium ion is the nitrating agent under certain conditions follows from the above and from the kinetic data.

Martinsen[§] established that the speed of nitration in sulphuric acid is proportional to the concentration both of nitric acid and of the aromatic compound.

In H_2SO_4, rate $= k_2$ [ArH] [HNO_3].

In organic solvents, however, with the more reactive aromatic compounds zero-order kinetics have been established.[||] That is, the rate-determining stage does not involve the aromatic compound. This indicates that the measured rate must be that of a change in the nitric acid, and as the reagent is known to be electrophilic this can be interpreted as the rate of formation of nitronium ion. The rate of nitration under these conditions, with

---

* *Ann. chim.* 1937, **8**, 243.
† Bennett, Brand and Williams, *J. Chem. Soc.* 1946, p. 869.
‡ Ingold, Millen and Poole, *Nature, Lond.*, 1946, **158**, 480; Goddard, Hughes and Ingold, ibid. 1946, **158**, 480; Goddard, Hughes and Ingold, *J. Chem. Soc.* 1950, p. 2559; Millen, ibid. 1950, pp. 2589, 2600.
§ *Z. phys. Chem.* 1905, **50**, 385; 1907, **59**, 605.
|| Benford and Ingold, *J. Chem. Soc.* 1938, p. 929.

very reactive aromatic compounds, is therefore the rate of formation of nitronium ion, the aromatic compound consuming it very quickly:

$$ArH + NO_2^+ \to ArHNO_2^+ \to ArNO_2 + H^+.$$

That the ejection of the proton from the transition state is kinetically unimportant, and proceeds very rapidly, has been demonstrated by experiments using benzene derivatives labelled with tritium.[*]

It seems that the formation of the nitronium ions follows the equations

$$2HNO_3 \rightleftharpoons H_2NO_3^+ + NO_3^- \quad \text{(fast)}$$

and

$$H_2NO_3^+ \to NO_2^+ + H_2O \quad \text{(slow)}.$$

This follows from the fact that zero-order nitration is accelerated by strong acids and retarded by nitrate ion without change of order, and that it is retarded by water with change to first order.

Nitronium ions are probably produced by a wide variety of nitrating agents. For example, it is known that selenic acid and perchloric acid, as well as sulphuric acid, give nitronium ions. Furthermore, hydrogen fluoride and boron trifluoride both catalyse nitrations,[†] and the most reasonable explanation is that they promote the formation of nitronium ions:

$$BF_3 + HNO_3 = NO_2^+ + BF_3OH^-,$$

$$2HF + HNO_3 = NO_2^+ + H_3O^+ + 2F^-.$$

Sulphuric acid must likewise promote the formation of nitronium ions from reagents such as ethyl nitrate, dinitrogen tetroxide and dinitrogen pentoxide. Ethyl nitrate in sulphuric acid is a good nitrating agent,[‡] so that the reaction is probably

$$EtO \cdot NO_2 + 3H_2SO_4 = NO_2^+ + H_3O^+ + 2HSO_4^- + EtHSO_4.$$

Similarly, with dinitrogen tetroxide in sulphuric acid[§] the reaction is probably

$$N_2O_4 + 3H_2SO_4 = NO_2^+ + NO^+ + H_3O^+ + 3HSO_4^-.$$

---

[*] Melander, *Nature, Lond.*, 1949, **163**, 599; Gillespie *et al.* ibid. p. 599.
[†] Simons, Passino and Archer, *J. Amer. Chem. Soc.* 1941, **63**, 608; Thomas, Anzilotti and Hennion, *Industr. Engng Chem.* 1940, **32**, 408.
[‡] Raudnitz, *Ber. dtsch. chem. Ges.* 1927, **60**, 738.
[§] Pinck, *J. Amer. Chem. Soc.* 1927, **49**, 2536.

The case of dinitrogen pentoxide is interesting. Klemenc and Schöller* found that a solution of this reagent in sulphuric acid nitrated twice as fast as an equal concentration of nitric acid in sulphuric acid. Such a result is consistent with the view that the nitrating agent in each case is the nitronium ion, but that while HNO_3 gives rise to one NO_2^+, N_2O_5 gives two NO_2^+:

$$N_2O_5 + 3H_2SO_4 = 2NO_2^+ + H_3O^+ + 3HSO_4^-,$$
$$HNO_3 + 2H_2SO_4 = NO_2^+ + H_3O^+ + 2HSO_4^-.$$

It does not, of course, follow that the nitronium ion is the only effective nitrating agent. Indeed, in the presence of a large concentration of water the nitronium ion would be unlikely to exist, and nitration under these conditions may involve the ion $H_2NO_3^+$.† Nevertheless, with 75–85 % sulphuric acid as solvent, the reagent still seems to be the nitronium ion.‡ Other nitrating agents of the general formula $X-NO_2$ are theoretically possible, the efficiency of the reagent depending on the electron affinity of the group X. If X has only a weak electron affinity, the reagent can only be a weak nitrating agent. A list of actual or potential nitrating agents can be built up in order of increasing nitrating power, this order being also that of increasing strength of the acid HX:§

$EtO.NO_2$	ethyl nitrate
$HO.NO_2$	nitric acid
$AcO.NO_2$	acetyl nitrate
$NO_3.NO_2$	dinitrogen pentoxide
$Cl.NO_2$	nitryl chloride
$H_2O.NO_2^+$	nitracidium ion
NO_2^+	nitronium ion

Nitration with several of these molecular entities has been realized.∥

---

* *Z. anorg. Chem.* 1924, **141**, 231.
† Halberstadt, Hughes and Ingold, *Nature, Lond.*, 1946, **158**, 514.
‡ Williams and Lowen, *J. Chem. Soc.* 1950, p. 3312; Lowen, Murray and Williams, ibid. p. 3318.
§ Benford and Ingold, ibid. 1938, p. 955; Gillespie and Millen, *Quart. Rev. Chem. Soc., Lond.*, 1948, **2**, 277.
∥ Hughes, Ingold and Reed, *J. Chem. Soc.* 1950, p. 2400; Halberstadt, Hughes and Ingold, ibid. p. 2441; Gold, Hughes, Ingold and Williams, ibid. p. 2452; Gold, Hughes and Ingold, ibid. p. 2467.

HALOGENATION

The halogenation of aromatic compounds is certainly a complicated reaction. Under certain conditions, as already discussed, *addition* occurs. This is particularly the case if the reaction is carried out in the absence of catalysts other than ultra-violet light. Under irradiation, benzene readily furnishes a mixture of stereoisomeric hexachlorohexahydrobenzenes or hexabromohexahydrobenzenes as the case may be. The more reactive hydrocarbons, such as anthracene and phenanthrene, form addition compounds even in the ordinary light of the laboratory.

In the presence of suitable 'halogen carriers' or catalysts, however, *substitution* can be achieved; and substitution products can also be formed at high temperatures even in the absence of such catalysts.

Numerous catalysts have been used to effect halogenation. As early as 1862, Müller[*] found that the chlorination of benzene, to give chlorobenzene, is promoted by the addition of iodine and of antimony pentachloride. Other catalysts which have been used include iron filings, anhydrous ferric chloride, aluminium foil, aluminium chloride, aluminium-mercury couple, molybdenum pentachloride, and so on.

These catalysts all promote chlorination or bromination using molecular chlorine or bromine. Other halogenating agents are also effective. For example, brominating agents are of the general formula Br—X, the efficiency of the reagent depending on the affinity of X for electrons.[†] Br—Cl is a better brominating agent than Br—Br for this reason. As a matter of fact, nearly all brominating agents are of the general type Br—X, for the usual catalysts, $FeBr_3$, $AlBr_3$, $SbCl_5$, $SnCl_4$, I_2, etc., are all substances which are capable of forming complexes with bromine, and these complexes are probably of the type $Br^{\delta+}$—$FeBr_4^{\delta-}$, etc. For the same reason, iodine chloride is a much stronger iodinating agent than iodine,[‡] and iodine bromide also acts as an iodinating agent.[§]

[*] Ibid. 1862, **15**, 41.
[†] Ingold, Smith and Vass, ibid. 1927, p. 1245; Benford and Ingold, ibid. 1938, p. 955.
[‡] Ingold, *Rec. trav. chim. Pays-Bas*, 1929, **48**, 809.
[§] Finkelstein, *Z. phys. Chem.* 1926, **124**, 285.

Iodination is commonly brought about with iodine and an oxidizing agent such as iodic acid, sodium persulphate, nitric acid, or mercuric oxide.*

In recent years halogen cations have been shown to be involved in halogenation experiments *under certain conditions*.

In water, molecular bromine is a more powerful brominating agent than hypobromous acid. This has been demonstrated by Francis,[†] and it also follows from the fact that —OH is less electronegative than —Br. Thus, the brominating agents Br—Cl, Br—Br, Br—OH are in decreasing order of effectiveness.[‡] On the other hand, a solution of hypobromous acid in mineral acid is a very effective brominating agent, much more effective than a comparable solution of bromine in mineral acid. For example, Derbyshire and Waters[§] found that bromide-free solutions of hypobromous acid in sulphuric acid rapidly brominated benzene and benzoic acid. Similarly, Shilov and Kanyaev[‖] found that the reagent readily brominates anisole-*meta*-sulphonic acid at a rate which is nearly proportional to the hydrogen-ion concentration of the medium. Other experiments of the same nature have been recorded by Derbyshire and Waters.[¶]

This enhanced reactivity of hypobromous acid in mineral acid media can be explained if it is supposed that hydrogen ions interact with hypobromous acid to give free, or hydrated, bromine cations, and that these cations are the true brominating agents:

$$H^+ + HOBr \rightleftharpoons H_2O + Br^+$$

or $$H^+ + HOBr \rightleftharpoons (H_2OBr)^+.$$

Conclusive evidence that the chlorine cation sometimes participates in aromatic chlorine substitution has been obtained by de la Mare, Hughes and Vernon.** Using an acidified aqueous solution of hypochlorous acid, the rate of chlorination of reactive

---

* Klages and Liecke, *J. prakt. Chem.* 1900, **61**, 311; Elbs and Jaroslavzev, ibid. 1913, **88**, 92; Datta and Chatterjee, *J. Amer. Chem. Soc.* 1917, **39**, 437; *Organic Syntheses*, Coll. vol. 1, 323.

† *J. Amer. Chem. Soc.* 1925, **47**, 2340.

‡ Ingold, Smith and Vass, *J. Chem. Soc.* 1927, p. 1245; Ingold, *Chem. Rev.* 1934, **15**, 271.

§ *Nature, Lond.*, 1949, **164**, 446.

‖ *C.R. Acad. Sci. U.R.S.S.* 1939, **24**, 890; *Chem. Abstr.* 1940, **34**, 4062.

¶ *J. Chem. Soc.* 1950, pp. 564, 573.

** *Research*, 1950, **3**, 192.

aromatic compounds was found to be independent of the concentration and nature of the aromatic compound. Under the same experimental conditions, quinol dimethylether and phenol gave the same rate coefficients as anisole. For less reactive aromatic compounds, the rate was found to depend on the concentration of aromatic substance. That is, for the reactive substances, $v = k$ [HOCl], and for the less reactive substances $v = k$ [HOCl] [ArH]. This state of affairs is analogous to that found with nitration, and the fundamental process is clearly the fission of HOCl to give the chlorine cation

$$HOCl + H^+ \rightleftharpoons H_2O + Cl^+.$$

It is interesting, however, that the formation of chlorine cations from hypochlorous acid becomes significant only at much higher acidities than those needed for the production of bromine cations from hypobromous acid.*

It must not be supposed that halogen cations are *invariably* involved in aromatic halogenations. Indeed, the evidence indicates that halogenation, using molecular halogen, normally involves a reagent of the type $Br^{\delta+}\!\!-\!\!Br^{\delta-}$ or $Br^{\delta+}\!\!-\!\!BrX^{\delta-}$. The ionization of the halogen-halogen bond seems to occur *during* the reaction with the aromatic compound,† and the function of the halogenation catalysts seems to be to facilitate the initial polarization of the reagent and the subsequent fission of the halogen-halogen bond.

The halogenation reagents mentioned so far are all electrophilic or cationoid in character; but there is now ample evidence that halogenation can also occur by a radical mechanism. Conditions of high temperature are known to favour radical reactions, and with certain halogenations a gradual transition from an electrophilic mechanism to a radical mechanism with increasing temperature has been satisfactorily demonstrated.

Perhaps the most interesting of these experiments have been carried out using naphthalene as the aromatic component. With this substance the results are not complicated by the various

---

* Derbyshire and Waters, *J. Chem. Soc.* 1951, p. 73.
† Robertson *et al.* ibid. 1943, pp. 276, 279; 1947, p. 1167; 1948, p. 100; 1949, pp. 294, 933; Bradfield, Davies and Long, ibid. 1949, p. 1389; Wilson and Soper, ibid. 1949, p. 3376.

electrical and steric effects of substituents, and only two isomers are formed.

The non-catalytic bromination of liquid naphthalene in the temperature range 85–215°, and of gaseous naphthalene in the range 250–300°, has been found to yield mainly α-bromonaphthalene, together with small quantities of β-bromonaphthalene, depending on the temperature. At 85°, for example, 3% of the total quantity of monobromonaphthalene formed was β-bromonaphthalene; and at 215° the mixture contained 6·8% of the β-isomer.*

At temperatures above 300°, however, the quantity of the β-isomer formed was found to increase with temperature until, in the range 500–650°, equal amounts of the α- and β-bromonaphthalenes were formed.

There seems little reason to doubt that the reaction in the lower temperature ranges is electrophilic in character; and in the vapour phase, between 250° and 300°, adsorption on the wall or on the packing of the reaction vessel would facilitate polarization and lead to an electrophilic reaction.

It is also noteworthy that the experimental results are in good agreement with the equation of Scheffer:[†]

$$\ln \frac{C_\alpha}{C_\beta} = \frac{E_\beta - E_\alpha}{RT},$$

where C_α and C_β represent the quantities of α- and β-bromonaphthalenes respectively, and E_α and E_β represent the energies of activation for bromination in the α- and β-positions. This agreement indicates that the ratio of the two isomers formed is determined by the difference in the energy of activation required for substitution in the two positions. As a matter of fact, the experimental values for $(E_\beta - E_\alpha)$, namely, 2·5 kcal./mole for reaction in the liquid phase and 4·2 kcal./mole for reaction in the gaseous phase below 300°, are in good agreement with the theoretical value as calculated by Sixma.[‡] Sixma assumed the difference in activation energies to be identical with the difference

---
* Suyver and Wibaut, *Rec. trav. chim. Pays-Bas*, 1945, **64**, 65.
† *Proc. K. Akad. Wet. Amst.* 1913, **15**, 1109; Scheffer and Brandsma, *Rec. trav. chim. Pays-Bas*, 1926, **45**, 522.
‡ *Ibid.* 1949, **68**, 915.

in the energies required to localize two π-electrons at either the α- or the β-position, and obtained a value of 3 kcal./mole.

In the higher temperature ranges, however, adsorption of the reactants on the wall or on the packing of the reaction vessel would be negligible, and as equal quantities of the two isomers are formed, the ratio cannot be determined by the difference in the energies of activation. It seems reasonably certain, therefore, that a gas reaction (as distinct from a wall reaction) predominates in the high-temperature range, and that the reagent is the bromine radical. The ratio of isomers formed must depend solely on the probability of collision of a bromine atom with an α- or β-position, and as there are an equal number of such positions the two isomers are formed in equal amount.

Assuming this change in mechanism with increasing temperature to be true, Sixma and Wibaut[*] have derived a formula which accounts for the influence of temperature on the ratio of α- to β-bromonaphthalene formed in the whole temperature range of 215–650°, and the agreement between the observed and calculated ratios is astonishingly good. The calculated curve is given in fig. 7·1.

The *catalytic* bromination of naphthalene is even more complicated. Under the influence of catalysts such as ferric bromide, the reaction is reversible:

$$C_{10}H_8 + Br_2 \underset{FeBr_3}{\rightleftharpoons} C_{10}H_7Br + HBr.$$

The bromination equilibrium lies well over on the side of bromonaphthalene, but the reversibility of the reaction results in the indirect conversion of α-bromonaphthalene into β-bromonaphthalene, and vice versa, as in (I), (II) and (III).

At 150° the equilibrium mixture contains 62·3 % β-bromonaphthalene, so that the catalytic bromination of naphthalene can yield varying quantities of the two isomers depending on whether equilibrium is established or not. Considerable quantities of β-bromonaphthalene are also formed in the temperature range 250–400° owing to this reversible bromination-debromination, but the quantity diminishes somewhat with temperature. Above 450°, however, the adsorption on the catalyst is almost

[*] Ibid. 1950, **69**, 577.

negligible. Under these conditions the catalyst is found to exert practically no influence on the reaction, which, as the temperature is increased, closely resembles the non-catalytic reaction in

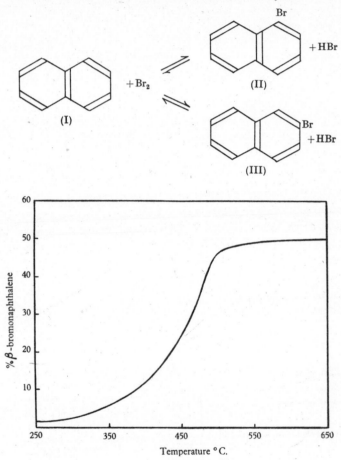

Fig. 7·1. Non-catalytic bromination of gaseous naphthalene. (After Sixma and Wibaut, *Rec. trav. chim. Pays-Bas*, 1950, **69**, 577.)

the high-temperature range. This is expressed in the curve (fig. 7·2) showing the percentage β-bromonaphthalene against the temperature.*

* Wibaut, Sixma and Suyver, *Rec. trav. chim. Pays-Bas*, 1949, **68**, 525.

AROMATIC SUBSTITUTION REACTIONS

The mechanism of the chlorination of naphthalene also appears to change with increasing temperature;* and changes in the mechanism of halogenation with temperature have also been recognized with bromobenzene and with pyridine.

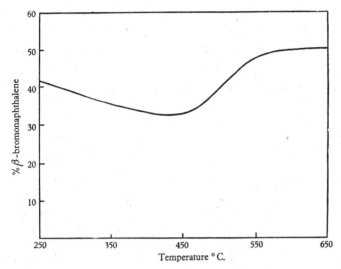

Fig. 7·2. Catalytic bromination of naphthalene. At low temperatures the reversible nature of the reaction gives rise to a mixture containing a high percentage of β-bromonaphthalene. At high temperatures the catalyst becomes ineffective; the reaction is presumably of the radical type, and equal quantities of α- and β-bromonaphthalene are formed. (After Wibaut, Sixma and Suyver, *Rec. trav. chim. Pays-Bas*, 1949, **68**, 525.)

The non-catalytic bromination of bromobenzene appears to be a wall reaction in the lower temperature ranges (below 410°) and to give mainly *ortho*- and *para*-dibromobenzenes. In the range 410–600°, however, adsorption of the reactants on the walls decreases, and a radical reaction predominates. The orientation of the product becomes increasingly dependent on the probability of collision at the various positions of substitution, and the high-temperature bromination of bromobenzene is therefore characterized by the formation of a relatively high percentage of the *meta* isomer.†

---

* Wibaut and Bloem, ibid. 1950, **69**, 586.

† Wibaut, Sixma and Lips, ibid. 1950, **69**, 1031; see also Sixma and Wibaut, *Proc. K. Akad. Wet. Amst.* 1949, **52**, 214; Wibaut, *Experientia*, 1949, **9**, 337.

Finally, the bromination of pyridine seems to be electrophilic in character at 300°, but radical in character at 500°. The results of the many investigations with the ring system are summarized in fig. 7·3, but it can be said that electrophilic attack results in substitution mainly at the 3- or 3:5-positions. Radical attack, on the other hand, at a temperature of about 500°, results in substitution at the 2- or 2:6-positions. It will be recalled that the 3-position is the position of greatest π-electron density (p. 231), and that the 2-position is the position of greatest free valence (p. 231).

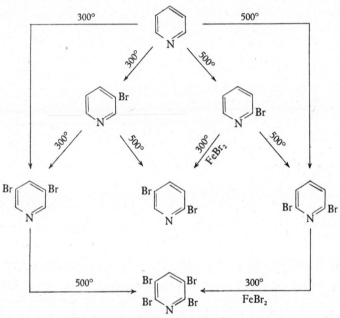

Fig. 7·3. The bromination of pyridine. (After Wibaut, *Experientia*, 1949, **9**, 337.)

THE FRIEDEL-CRAFTS AND RELATED REACTIONS

Of all the reactions involving aromatic ring systems none is more important than the Friedel-Crafts synthesis. In its original form* the reaction involved the condensation of acyl and alkyl

* Friedel and Crafts, *C.R. Acad. Sci.*, Paris, 1877, **84**, 1392, 1450; **85**, 74; *Ann. Chim.* 1884, **1**, 449: 1887, **10**, 411; **11**, 263; 1888, **14**, 433.

halides with various aromatic compounds in the presence of aluminium chloride. In this way acyl and alkyl substitution products were obtained:

$$RCl + ArH = ArR + HCl,$$
$$RCOCl + ArH = ArCOR + HCl.$$

The range of the reaction has been considerably extended during the past half-century, and it is now one of the most important synthetic methods in the whole field of organic chemistry.* Generally speaking, aluminium chloride is still the most important condensing agent, but $FeCl_3$, $SbCl_5$, BF_3, $ZnCl_2$, $TiCl_4$, HF, H_2SO_4, H_3PO_4 and P_2O_5 have also been used. In no case has the mechanism of the reaction been fully and completely established, but a variety of evidence indicates that the effective reagent in each case is probably a carbonium ion (or a polarized complex) and that the substitution is electrophilic in character.†

A few important items of evidence may be mentioned at this stage. It has been shown by Wertyporoch and his collaborators‡ that solutions of aluminium chloride, aluminium bromide and of ferric chloride in various alkyl halides are electrically conducting. This means that ionization must occur, and transference experiments have shown that the aluminium is to be found in the anions.

It has also been found that although solutions of triphenylmethyl chloride in benzene and in chloroform are colourless and non-conducting, such solutions become intensely yellow and also electrically conducting when aluminium chloride or stannic chloride is added.§ The most reasonable interpretation of these results is that ionization occurs according to equations of the type

$$R.Cl + AlCl_3 = R^+ + AlCl_4^-,$$
$$R.Cl + SnCl_4 = R^+ + SnCl_5^-.$$

---

* Calloway, *Chem. Rev.* 1935, **17**, 327–392; Thomas, *Anhydrous Aluminium Chloride in Organic Chemistry*, p. 972, Reinhold, New York, 1941.
† Cf. Pajeau and Fierens, *Bull. Soc. chim. Fr.* 1949, p. 587.
‡ Wohl and Wertyporoch, *Ber. dtsch. chem. Ges.* 1931, **64**, 1357; Wertyporoch, ibid. 1931, **64**, 1369; Wertyporoch and Firla, *Z. phys. Chem.* 1932, A, **162**, 398; Wertyporoch, Kowalski and Roeske, *Ber. dtsch. chem. Ges.* 1933, **66**, 1232; Wertyporoch and Kowalski, *Z. phys. Chem.* 1933, A, **166**, 205.
§ Norris and Sanders, *Amer. Chem. J.* 1901, **25**, 54; Gomberg, *Ber. dtsch. chem. Ges.* 1901, **34**, 2726; Kehrmann and Wentzel, ibid. 1901, **34**, 3815.

That similar ionization occurs with the other reagents is indicated by the fact that triphenylmethylcarbinol gives an intense yellow solution in concentrated sulphuric acid. The freezing-point depression indicates that four 'molecular entities' are produced, so that the reaction must be*

$$Ph_3C.OH + 2H_2SO_4 = Ph_3C^+ + OH_3^+ + 2HSO_4^-.$$

Of course, *complete* ionization need not necessarily occur in every case of Friedel-Crafts or related reactions. A polarized complex could conceivably be involved.

The experiments of Fairbrother[†] are also of significance. He found that aluminium bromide dissolves in *cyclo*hexane, and measurements of freezing-points and dielectric polarizabilities indicated that it is present almost wholly as the dimer, Al_2Br_6, and that it has a very small or zero dipole moment. The addition of ethyl bromide to such a solution, however, was found to have a profound effect. There was an increase in the apparent molar orientation polarization of aluminium bromide from about zero to more than 300 c.c., corresponding to an apparent dipole moment of about 4 D. Ethyl bromide must therefore form a highly polar complex with aluminium bromide.

Fairbrother[‡] has also obtained some significant evidence on the origin of the evolved hydrogen halide in Friedel-Crafts reactions by using aluminium chloride containing radioactive chlorine. He studied the reaction of benzene with *tert.*-butylchloride and with acetyl chloride (containing ordinary chlorine). If the chlorine in the evolved hydrogen chloride comes from the alkyl or acyl chloride, it should be non-radioactive. If it comes exclusively from the aluminium chloride, it should have the same radioactivity as the latter. However, Fairbrother found that the evolved hydrogen chloride has about 75 % of the activity expected if it comes exclusively from the aluminium chloride. The results are consistent with the view that the hydrogen chloride is formed by the interaction of a hydrogen ion with the anion $AlCl_4^-$, in which each chlorine atom has an equal chance of elimination. Fairbrother also found that an exchange of

---

* Hantzsch, *Z. phys. Chem.* 1908, **61**, 257; Hammett and Deyrup, *J. Amer. Chem. Soc.* 1933, **55**, 1900; Hantzsch, *Ber. dtsch. chem. Ges.* 1921, **54**, 2573.
† *Trans. Faraday Soc.* 1941, **37**, 763.
‡ *J. Chem. Soc.* 1937, p. 503.

chlorine atoms takes place between aluminium chloride and an organic halide even in the absence of an aromatic compound.

It seems to be fairly well established therefore that the initial Friedel-Crafts reaction can be represented:

$$RCl + AlCl_3 \rightleftharpoons R^+ + AlCl_4^-$$

or $$RCl + AlCl_3 \rightleftharpoons R^{\delta+} \ldots AlCl_4^{\delta-}.$$

By analogy with other electrophilic reactions, the attack of the reagent on an aromatic compound can be represented:

$$ArH + R^+ \rightarrow ArR + H^+$$

or $$ArH + R^{\delta+} \ldots AlCl_4^{\delta-} \rightarrow ArR + H^+ + AlCl_4^-.$$

The liberated proton evidently attacks the anion $AlCl_4^-$ as already mentioned, giving hydrogen chloride (which is evolved) and regenerating the aluminium chloride:

$$H^+ + AlCl_4^- \rightarrow HCl + AlCl_3.$$

As will be shown later, however, these reactions represent an over-simplification of the process. It is known, for example, that there are important differences between Friedel-Crafts alkylations and acylations. The alkylations require only catalytic amounts of aluminium chloride, but the acylations require slightly more than 1 mole if an acid chloride is used, and slightly more than 2 moles if an anhydride is used. Furthermore, alkylations, but not acylations, are reversible; and thirdly, the usual rules for orientation in electrophilic reactions are not always followed in alkylation experiments.* Rearrangement of carbon skeletons also takes place (in alkylations), and there are a number of related rearrangement reactions (Fries, Jacobsen, isomerization with hydrogen fluoride, etc.) which are also of interest and importance. The use of acid anhydrides for the preparation of keto-acids forms a complete subject in itself, as does the cyclization of propionic acid and butyric acid derivatives with aluminium chloride, hydrogen fluoride, sulphuric acid, or any of the other reagents.

* See also Linstead, *Ann. Rep. Chem. Soc.* 1937, **34**, 251.

ALKYLATIONS

The alkylation of aromatic compounds[*] by means of alkyl halides and aluminium chloride was first investigated by Friedel and Crafts in their classical researches. Many other alkylating agents and catalysts have since been shown to be effective. As early as 1879, Balsohn[†] found that benzene could be alkylated with olefins in the presence of aluminium chloride. Other alkylating agents which have been used include certain highly strained *cyclo*paraffins, alcohols, ethers, and even esters.[‡] The various catalysts are even more diverse in nature, ranging from aluminium chloride to stannic chloride, hydrogen fluoride and sulphuric acid. Little work has been carried out on the relative effectiveness of the various catalysts, although it is known qualitatively that some catalysts are more suitable for one type of alkylating agent than for others. It is interesting that the order of effectiveness in promoting the racemization of α-phenylethyl chloride (I) has been shown to be

$$SbCl_5 > SnCl_4 > TiCl_4 > BCl_3 > ZnCl_2 > HgCl_2.$$ [§]

The racemization probably involves a reaction of the type

$$\begin{array}{c}Ph\\ \diagdown\\ CHCl\\ \diagup\\ CH_3\\ (I)\end{array} + MCl_n \rightleftharpoons \begin{array}{c}Ph\\ \diagdown\\ CH^+\\ \diagup\\ CH_3\\ (II)\end{array} + \left[MCl_{n+1}\right]^-$$

and would therefore appear to be very similar to the Friedel-Crafts reaction. It seems likely, therefore, that the above order should also hold for the relative effectiveness of the agents as alkylation catalysts, at least for the alkyl halides.

The alkylating agents are of the type $R—X$, and, generally speaking, the ease of alkylation can be interpreted in terms of the ease of polarization of the reagent in the sense $R^{\delta+} \ldots X^{\delta-}$.

---

[*] Price, *Organic Reactions*, **3**, 1–82, John Wiley, New York, 1946.
[†] *Bull. Soc. chim. Paris*, 1879, **31**, 539.
[‡] Price, *Organic Reactions*, **3**, 1–82; Bowden, *J. Amer. Chem. Soc.* 1938, **60**, 645.
[§] Bodendorf and Böhme, *Liebigs Ann.* 1935, **516**, 1.

AROMATIC SUBSTITUTION REACTIONS

The nature of X has a profound effect, but the structure of the alkyl group is also of importance. Tertiary alkyl groups are introduced without difficulty; alkylation with secondary alkyl groups proceeds somewhat less readily, and primary alkyl groups are much less effective alkylating agents. With the powerful alkylating agents, a weak catalyst suffices; but less effective alkylating agents require the use of aluminium chloride and other strong catalysts.

One of the most interesting problems in alkylation reactions concerns the tendency for alkyl groups to undergo isomerizations. As early as 1878, Gustavson[*] found that not only did *iso*propyl bromide and benzene interact in the presence of aluminium chloride to give *iso*propylbenzene (cumene), but that the same substance is also formed if *n*-propyl bromide is used. Soon afterwards, Kekulé and Schrotter[†] found that *n*-propyl bromide is isomerized to *iso*propyl bromide in the presence of aluminium halide.

Such isomerizations of the alkylating agent do not invariably occur, but they are common in reactions of this type. Toluene and *iso*butylbromide give chiefly *meta-tert.*-butyltoluene.[‡] *iso*Amylchloride and benzene give *iso*amylbenzene, together with *tert.*-amylbenzene and 2-phenyl-3-methylbutane.[§] Moreover, Gilman and Burtner[∥] have found that the alkylation of furfural with either *n*-, *iso*- or *tert.*-butyl chlorides gives 5-*tert.*-butyl-2-furfural.

The isomerizations always result in increased branching of the alkyl chain, and this may be taken as an indication of the formation of carbonium ions in these reactions. Carbonium ions of this nature would be expected to undergo rearrangement to give the highest possible degree of branching, because the inductive effects of the methyl groups would confer the greatest stability on the most highly branched ions. All the above isomerizations, for example, can be represented as rearrangements of carbonium ions, as follows:

---

[*] *Ber. dtsch. chem. Ges.* 1878, **11**, 1251.
[†] *Bull. Soc. chim. Paris*, 1879, **34**, 485; see also Crowell and Jones, *J. Amer. Chem. Soc.* 1951, **73**, 3506.
[‡] Kelbe and Pfeiffer, *Ber. dtsch. chem. Ges.* 1886, **19**, 1723; Baur, ibid. 1891, **24**, 2832; Shoesmith and McGechen, *J. Chem. Soc.* 1930, p. 2231.
[§] Konowaloff and Egoroff, *J. Russ. Phys. Chem. Soc.* 1898, **30**, 1031.
[∥] *J. Amer. Chem. Soc.* 1935, **57**, 909.

$$CH_3.CH_2.\overset{+}{C}H_2 \longrightarrow CH_3.\overset{+}{C}H.CH_3$$

$$CH_3.CH_2.CH_2.\overset{+}{C}H_2 \longrightarrow \begin{matrix}CH_3\\CH_3\\CH_3\end{matrix}\!\!>\!\!\overset{+}{C}$$

$$\begin{matrix}CH_3\\CH_3\end{matrix}\!\!>\!\!CH.\overset{+}{C}H_2 \nearrow$$

$$\begin{matrix}CH_3\\CH_3\end{matrix}\!\!>\!\!CH.CH_2.\overset{+}{C}H_2 \longrightarrow \begin{matrix}CH_3\\CH_3\end{matrix}\!\!>\!\!CH.\overset{+}{C}H.CH_3$$

$$\xrightarrow[\text{further}]{\text{and}} \begin{matrix}CH_3\\CH_3\\C_2H_5\end{matrix}\!\!>\!\!\overset{+}{C}$$

Whether isomerization does or does not occur seems to depend on the conditions of the reaction and the type of catalyst used. As a matter of fact, n-propylbenzene can be obtained in quite good yield using n-propyl chloride and aluminium chloride provided the reaction is carried out at a low temperature. For example, Ipatieff, Pines and Schmerling* obtained a mixture consisting of 60% of n-propylbenzene and 40% of isopropylbenzene when the reaction was carried out at −6°. At 35°, however, a mixture containing 40% of n-propyl- and 60% of isopropylbenzene was obtained.

The influence of the catalyst has been illustrated by other work of Ipatieff, Pines and Schmerling.* These workers studied the alkylation of benzene in the presence of either sulphuric acid or aluminium chloride, with olefins, alcohols, with cyclopropane, and with alkyl halides. The results, which are given in table 7·2, indicate that isomerization is more likely to occur when sulphuric acid is used.

---

* *J. Org. Chem.* 1940, **5**, 253; see also Heise, *Ber. dtsch. chem. Ges.* 1891, **24**, 768; Konowaloff, *J. Russ. Phys. Chem. Soc.* 1895, **27**, 457.

This conclusion has often been confirmed. Straight-chain alcohols usually alkylate without rearrangement when aluminium chloride is used as catalyst.* The dimethylamylcarbinols give the expected octylbenzenes with aluminium chloride, although when 2:4:4-trimethylpentan-2-ol and 2:3:3-trimethylpentan-2-ol are used, some tert.-butylbenzene is also formed.†
On the other hand, rearrangement often occurs when sulphuric acid is used.‡

TABLE 7·2.§ *Alkylation products*

Alkylating agent	Sulphuric acid	Aluminium chloride
Pent-1-ene	2- and 3-Phenylpentane	—
3-Methylbut-1-ene	t.-Amylbenzene	2-Methyl-3-phenylbutane
n-Propyl alcohol	isoPropylbenzene	n-Propylbenzene
isoAmyl alcohol	t.-Amylbenzene	No alkylation
cycloPropane	n-Propylbenzene at 0°	n-Propylbenzene, regardless of temperature
	isoPropylbenzene at 65°	
n-Propylchloride	—	Mixture of n- and isopropylbenzenes

Rearrangement of olefins also occurs when sulphuric acid is used, and the product is not always that which would be obtained if the aromatic radical adds to the least hydrogenated carbon atom of the double bond. For example, the condensation of 3-methylbut-1-ene with benzene, using sulphuric acid, gave tert.-amylbenzene.‖ It has been suggested that the action of the sulphuric acid in such cases is, first, to form an addition product (the alkyl hydrogen sulphate), which subsequently undergoes rearrangement, but it seems more likely that this isomerization occurs via the carbonium ion, as shown on p. 272.¶

With pent-1-ene, a mixture of the normal product and the product of rearrangement is obtained. It is interesting that with weak acid, and at low temperatures, esters are formed, but no reaction with benzene occurs.**

---

* Bowden, *J. Amer. Chem. Soc.* 1938, **60**, 645; Ipatieff, Pines and Schmerling, *J. Org. Chem.* 1940, **5**, 253; see also Heise, *Ber. dtsch. chem. Ges.* 1891, **24**, 768; Konowaloff, *J. Russ. Phys. Chem. Soc.* 1895, **27**, 457.
† Huston, Guile, Sculati and Wasson, *J. Org. Chem.* 1941, **6**, 252.
‡ Ipatieff, Pines and Schmerling, ibid. 1940, **5**, 253; see also Heise, *Ber. dtsch. chem. Ges.* 1891, **24**, 768; Konowaloff, *J. Russ. Phys. Chem. Soc.* 1895, **27**, 457; Meyer and Bernhauer, *Mh. Chem.* 1929, **53** and **54**, 721.
§ After Ipatieff, Pines and Schmerling, *J. Org. Chem.* 1940, **5**, 254.
‖ Ipatieff, Pines and Schmerling, *J. Amer. Chem. Soc.* 1938, **60**, 353.
¶ The author is indebted to Dr R. I. Reed for this suggestion.
** Ipatieff, Corson and Pines, *J. Amer. Chem. Soc.* 1936, **58**, 919.

272 THE AROMATIC COMPOUNDS

A particularly interesting case is that of *cyclo*propane. At low temperatures this gives *n*-propylbenzene when sulphuric acid is used, but at higher temperatures *iso*propylbenzene is formed. According to Ipatieff, Pines and Schmerling* this indicates that the alkylation takes place via the ester, and that rearrangement takes place at higher but not at lower temperatures. On the other hand, when aluminium chloride is used as catalyst, *n*-propylbenzene is formed, apparently regardless of the temperature. This excludes the possibility that *n*-propyl chloride is an intermediate in the reaction, for this alkylating agent is isomerized with aluminium chloride.

$$CH_3-CH-CH=CH_2 \xrightarrow{H_2SO_4} CH_3-CH-CH_2-CH_2^+$$
$$||$$
$$CH_3CH_3$$

$$\rightleftharpoons CH_3-CH-\overset{+}{C}H-CH_3$$
$$|$$
$$CH_3$$

$$\rightleftharpoons CH_3-\overset{+}{C}-CH_2-CH_3$$
$$|$$
$$CH_3$$

Hydrogen fluoride is also interesting from this point of view. With *cyclo*propane, *n*-propyl derivatives are obtained. With *n*-propyl halides and *n*-propyl alcohol, however, the *iso*propyl derivatives are chiefly formed.†

Boron trifluoride also tends to bring about rearrangement of alkyl groups.‡ Both *n*- and *iso*propyl alcohols reacted with benzene under the influence of boron trifluoride to give identical alkylation products, viz. mono-, di- and tri-*iso*propylbenzenes. *n*-Butyl and *sec.*-butyl alcohols both gave *sec.*-butylbenzenes; and *iso*butyl and *tert.*-butyl alcohols both gave *tert.*-butylbenzenes.

* *J. Org. Chem.* 1940, **5**, 253; see also Heise, *Ber. dtsch. chem. Ges.* 1891, **24**, 768; Konowaloff, *J. Russ. Phys. Chem. Soc.* 1895, **27**, 457.
† Simons, Archer and Adams, *J. Amer. Chem. Soc.* 1938, **60**, 2955; Simons, *Industr. Engng Chem.* 1940, **32**, 178.
‡ McKenna and Sowa, *J. Amer. Chem. Soc.* 1937, **59**, 470; Toussaint and Hennion ibid. 1940, **62**, 1145.

A second interesting problem in alkylation reactions of the Friedel-Crafts type concerns the orientation of the product. There is no doubt that the reaction is electrophilic, and *ortho*- or *para*-dialkyl benzene derivatives would therefore be expected. It has long been known, however, that if aluminium chloride is used as catalyst, considerable quantities of *meta*-dialkyl derivatives are often formed. The proportion of 'abnormal' product seems to depend on the conditions of the reaction. The most vigorous conditions, using the most effective catalyst (aluminium chloride), high temperatures, and prolonged reaction periods, give rise to predominantly *meta*-disubstituted products. If much less vigorous conditions are used, however, the 'normal' products are obtained.

TABLE 7·3. *Alkylation of toluene with* MeCl

Temperature	Xylenes (%)		
	ortho	*para*	*meta*
0°	53·5	19·2	27·3
55°	12·2	0·7	87·1
106°	1·8	0·0	98·2

The effect of temperature on the orientation was shown by the experiments of Norris and Rubinstein[*] involving the alkylation of toluene with methyl chloride and aluminium chloride. At 0°, much *ortho-para* product was obtained (table 7·3). At 55° the major product was *meta*-xylene, and at 106° practically no *ortho*- or *para*-xylene was formed at all.

In the same way, further alkylation often gives significant quantities of the 1:3:5-trisubstituted product, as well as some 1:2:4- derivative.

Similar abnormal substitution has been observed in naphthalene. Using aluminium chloride as catalyst, naphthalene and *cyclo*hexanol, or *cyclo*hexene, react to give chiefly 2:6-di*cyclo*hexylnaphthalene.[†]

Aluminium chloride seems to be the most effective catalyst in promoting such abnormal substitution. Other catalysts,

[*] Ibid. 1939, **61**, 1163.
[†] Price and Tomisek, ibid. 1943, **65**, 439.

including boron trifluoride, hydrogen fluoride, ferric chloride, etc., give the normal substitution products. Thus, the methylation of toluene (III), using ferric chloride, hydrogen fluoride or boron trifluoride, gave mainly *para*-xylene (VI); but methylation using aluminium chloride as catalyst gave mainly *meta*-xylene (IV). In the same way, the methylation of *meta*-xylene using aluminium chloride gave some mesitylene (V) as well as the normal substitution product (VII).

The explanation of the abnormal orientation in the Friedel-Crafts reactions which are carried out under fairly vigorous conditions seems to lie in the fact that alkylations are reversible,* and that both intermolecular and intramolecular rearrangement occurs.† When polyalkylbenzenes are treated with aluminium chloride, the product consists of a mixture of benzene and of alkylbenzenes containing both more and fewer alkyl groups than the original. Thus pseudocumene has given a mixture of benzene, toluene, *meta*- and *para*-xylenes, durene and *iso*durene, following treatment with aluminium chloride. *meta*-Xylene has given toluene, pseudocumene, mesitylene and durene.‡ In these cir-

* Boedtker, *Bull. Soc. chim. Paris*; 1906, 35, 825; Boedtker and Halse, ibid. 1916, 19, 444; Woodward, Borcherdt and Fuson, *J. Amer. Chem. Soc.* 1934, 56, 2103.
† Baddeley, *J. Chem. Soc.* 1943, p. 527.
‡ Jacobsen, *Ber. dtsch. chem. Ges.* 1885, 18, 338.

cumstances the formation of *meta*-dialkyl derivatives may be due not so much to 'abnormal' substitution, but to the fact that isomerization of the product occurs. Thus 'normal' alkylation to give a 1:2:4-trialkyl derivative, followed by elimination of an alkyl group from the 1-position, would give a *meta*-dialkyl derivative.*

Isomerizations have very frequently been observed following treatment with aluminium chloride, and some examples may be given.

The Fries rearrangement, which involves the conversion of an O-acyl derivative into C-acyl compound, is often accompanied by isomerization. The acetate of *meta*-5-xylenol (VIII) gave 6-hydroxy-2:4-dimethylphenylalkylketone (IX) when treated with 1 mole of aluminium chloride and its 3:4-dimethyl isomeride (X) when treated with 2 moles of the catalyst. Moreover, as expected, hydroxyketones of type (IX) are isomerized to ketones of type (X) by aluminium chloride.†

(VIII)   (IX)   (X)

Similarly, *para*-cresol is isomerized to *meta*-cresol with aluminium chloride, and the xylenols all isomerize to *meta*-5-xylenol.‡

The isomerization is not confined to alkyl groups. Certain other substituents, such as the halogens, are also affected. 2-Bromo-4-methylphenol (XI), for example, has been isomerized to the 3-bromo derivative (XII); and 2:6-dibromo-4-methylphenol (XIII) has been converted into the 3:6- and 3:5-dibromo derivatives (XIV) and (XV).§

The effect of aluminium chloride on tetralin is especially noteworthy. The product is a complex mixture containing

* Ibid.
† Baddeley, *J. Chem. Soc.* 1943, p. 273.
‡ Ibid. 1943, p. 527.
§ Baddeley and Plant, ibid. 1943, p. 525.

benzene, octahydroanthracene and octahydrophenanthrene, and (probably) the cancer-producing hydrocarbon, 3:4-benzopyrene (XXIII). β-(4-Phenylbutyl)tetralin (XVI) is a probable intermediate in the formation of benzene (XVII), octahydroanthracene (XVIII) and octahydrophenanthrene (XIX), and the last two compounds are interconvertible.*

The presence of 3:4-benzopyrene has not been definitely proved, but may be inferred from the facts (i) that the reaction mixture is cancer-producing† and (ii) that the fluorescence spectrum of

* Schroeter, *Ber. dtsch. chem. Ges.* 1924, **57**, 1990.
† Kennaway, *Biochem. J.* 1930, **24**, 497.

AROMATIC SUBSTITUTION REACTIONS

the mixture shows very great similarities to that of 3:4-benzopyrene. Its formation could be explained by the following reactions:

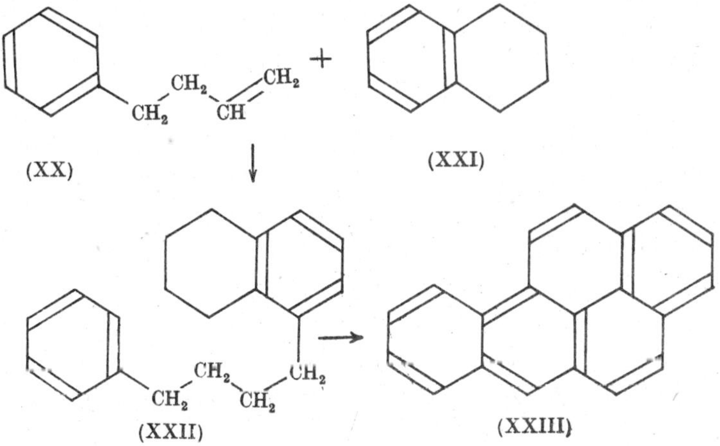

(XX) + (XXI) → (XXII) → (XXIII)

Somewhat similar rearrangements can be brought about with catalysts other than aluminium chloride. At elevated temperatures, for example, hydrogen fluoride converts xylenes into benzene, toluene and higher alkylated benzenes. Of special interest, however, are the isomerizations brought about with sulphuric acid. This type of rearrangement is known as the Jacobsen reaction.* Using this method, durene has been isomerized to a mixture of prehnitene, pseudocumene, pentamethylbenzene and hexamethylbenzene. Similarly, octahydroanthracene has been converted, in good yield, to octahydrophenanthrene.† Migration of halogen atoms takes place in the same way, for chlorodurene yields pentamethylchlorobenzene and chloropseudocumene.‡

The exact mechanisms of these reactions are not known with certainty, but some important facts have been discovered. For example, it is known that the rearrangements involve the

* Smith, *Organic Reactions*, 1, 370–84, John Wiley, New York, 1942.
† Schroeter and Götzky, *Ber. dtsch. chem. Ges.* 1937, **60**, 2035.
‡ Smith and Moyle, *J. Amer. Chem. Soc.* 1936, **58**, 1; Töhl, *Ber. dtsch. chem. Ges.* 1892, **25**, 1527.

sulphonic acid derivatives, and not the free hydrocarbons.*
Durenesulphonic acid rearranges in the presence of phosphorus
pentoxide; but this reagent has no effect on durene itself.
A second point of interest is that durenesulphonic acid (XXIV)
and *iso*durenesulphonic acid (XXVI) both isomerize to preh-
nitenesulphonic acid (XXV), but that prehnitene, although it
is sulphonated by sulphuric acid, does not rearrange.*

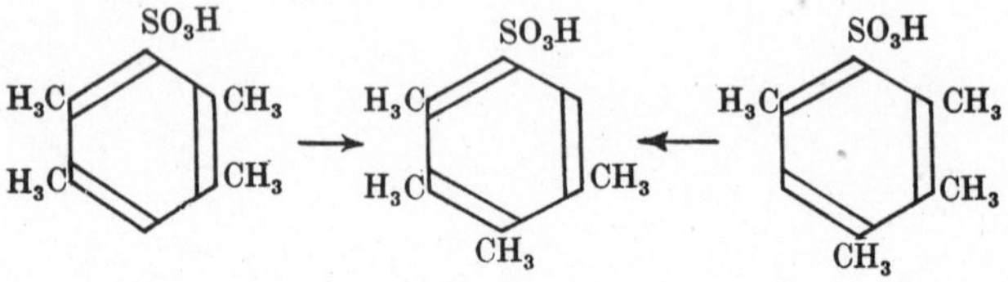

(XXIV)   (XXV)   (XXVI)

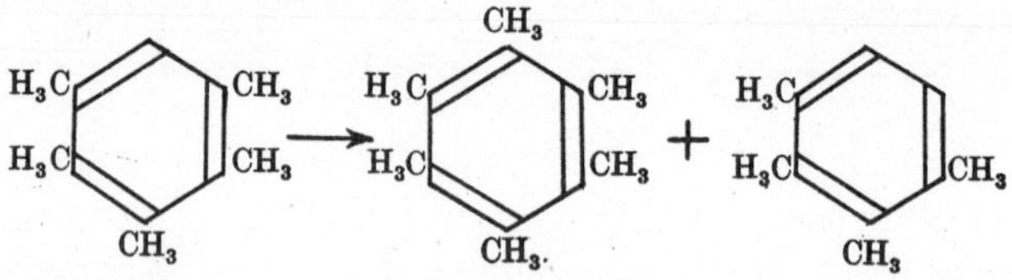

(XXVII)   (XXVIII)   (XXIX)

This and other evidence indicates that it is an alkyl group
adjacent to the sulphonic acid group which undergoes migration.
Both intramolecular and intermolecular rearrangements seem
to occur, the latter being responsible for the formation of
hexamethylbenzene (XXVIII) and prehnitene (XXIX) from
pentamethylbenzene (XXVII).†

ACYLATIONS

Acylation by the Friedel-Crafts method is perhaps of even
greater importance than alkylation. Many dyestuffs and dyestuff intermediates are produced by this reaction or one of its

* Smith and Cass, *J. Amer. Chem. Soc.* 1932, **54**, 1614.
† Jacobsen, *Ber. dtsch. chem. Ges.* 1887, **20**, 896; Smith and Lux, *J. Amer. Chem. Soc.* 1929, **51**, 2994.

variants, and it is therefore of very considerable industrial interest. In its general principles the acylation reaction closely resembles that of alkylation, and the equation may be represented as follows:

$$\text{ArH} + R\text{COX} \xrightarrow{\text{AlCl}_3} \text{ArCO}R + \text{H}X.$$

This is perhaps the most common form of acylation reaction, requiring the use of an acyl chloride, or some similar reagent, and aluminium chloride as a catalyst. A molar equivalent of aluminium chloride is needed, for the catalyst forms a complex with the product.

In spite of this superficial similarity there are many important differences. For example, both straight-chain and branched-chain acyl groups condense with aromatic compounds *without* isomerization. Indeed, this absence of isomerization of the acylating reagent renders the reaction of considerable value in synthetic chemistry. The resulting ketones have the expected structure, and the products may be reduced to give alkyl derivatives which also have the predicted structure. The method therefore serves as a useful procedure in the preparation of alkyl derivatives containing long straight chains. It is interesting that even *cyclo*propane carboxylic acid chloride reacts with benzene to give the expected phenyl*cyclo*propyl ketone in very good yield.[*]

A great variety of halides has been successfully used in this reaction. Acetyl chloride, benzoyl chloride, propionyl chloride, caproyl chloride, and many others have been used to introduce the acetyl, benzoyl, etc., radicals. Even unsaturated acyl chlorides are usually satisfactory, but acryloyl chloride ($CH_2:CH.COCl$) and benzene give indan-1-one (I), and not vinylphenyl ketone (II).[†]

Oxalyl chloride is an interesting reagent. Under various experimental conditions and with different aromatic components, the products have been ketones, diketones, or carboxylic acids. A mixture is often obtained, but under suitable conditions the reaction sometimes furnishes a satisfactory method for the

[*] Kizhner, *J. Russ. Phys. Chem. Soc.* 1911, **43**, 1163; *Chem. Abstr.* 1912, **6**, 597.
[†] Kohler, *Amer. Chem. J.* 1909, **42**, 375.

introduction of a carboxyl radical. It is significant, however, that a glyoxalic acid derivative seems never to be obtained, although such compounds can be prepared if 'chloro-oxalic ester' is used as acylating agent, followed by hydrolysis of the product.

(I)   (II)

Oxalyl chloride is not a very stable reagent. In fact, with aluminium chloride, it is known to be slowly broken down to carbon monoxide and carbonyl chloride.* Carbonyl chloride has long been known as a reagent for the preparation of mono-ketones and carboxylic acids by the Friedel-Crafts procedure:†

$$ArH + COCl_2 \rightarrow ArCOCl + HCl,$$

$$2ArH + COCl_2 \rightarrow ArCOAr + 2HCl.$$

It is reasonable to suppose, therefore, that the formation of mono-ketones and of carboxylic acids in the Friedel-Crafts reactions involving oxalyl chloride is due to the carbonyl chloride formed by degradation of the oxalyl chloride with aluminium chloride. In this connexion it is interesting that although benzene (III) reacts with oxalyl chloride to give benzoic acid (IV) and benzophenone (V), anisole (VI) reacts to give anisil (VII) in good yield.‡

Whether oxalyl chloride gives mono-ketones or diketones seems to depend on the velocity of the reaction and on the reactivity of the individual components. If the aromatic component is reactive, then oxalyl chloride reacts rapidly to give a diketone (as anisil). In other words, the oxalyl chloride reacts more rapidly with the aromatic component than it is decomposed

---

* Staudinger, *Ber. dtsch. chem. Ges.* 1908, **41**, 3558.
† Wilson and Fuller, *Industr. Engng Chem.* 1922, **14**, 406.
‡ Staudinger, *Ber. dtsch. chem. Ges.* 1908, **41**, 3558; Staudinger, Goldstein and Schlenker *Helv. chim. acta*, 1921, **4**, 342.

to carbonyl chloride by the aluminium chloride. On the other hand, if the aromatic component is unreactive, this decomposition has time to proceed to a significant extent, and monoketones and carboxylic acids are formed.

$$C_6H_6 + (COCl)_2 \longrightarrow C_6H_5.CO_2H + C_6H_5.CO.C_6H_5$$
$$\text{(III)} \qquad\qquad\qquad\qquad \text{(IV)} \qquad\qquad \text{(V)}$$

$$CH_3O\text{-}C_6H_4 + (COCl)_2 \longrightarrow CH_3O\text{-}C_6H_4\text{-}CO.CO\text{-}C_6H_4\text{-}OCH_3$$
$$\text{(VI)} \qquad\qquad\qquad\qquad\qquad\qquad \text{(VII)}$$

Carboxylic acids may also arise in another way. For example, it has been found that when glyoxylic acid chlorides are hydrolysed, they are degraded in good yield to carboxylic acids. Any glyoxylic acid chloride formed as an intermediate in these reactions would therefore yield the corresponding carboxylic acid on working up the reaction mixture, and this is evidently the reason glyoxylic acids are never observed as reaction products. Oxalyl chloride and 1:2-benzanthracene (VIII) interact under the influence of aluminium chloride to give 4:10-oxalyl-1:2-benzanthracene (X) and 1:2-benzanthracene-10-carboxylic acid (XI). The latter compound is almost certainly derived from 1:2-benzanthracene-10-glyoxylic acid chloride (IX). A large excess of aluminium chloride did not promote the formation of 1:2-benzanthracene-10-carboxylic acid as might be expected if this product arises solely from the carbonyl chloride produced as degradation product of the oxalyl chloride.*

The anhydrides also serve as very satisfactory acylating reagents. The simple aliphatic anhydrides behave as expected and give alkyl aryl ketones, two moles of aluminium chloride being required for each mole of anhydride:

$$C_6H_6 + (RCO)_2O \rightarrow C_6H_5COR.$$

Aromatic anhydrides behave similarly, and benzoic anhydride, for example, yields diaryl ketones.†

---
* Badger and Gibb, *J. Chem. Soc.* 1949, p. 799.
† Rubidge and Qua, *J. Amer. Chem. Soc.* 1914, **36**, 732.

The dibasic acid anhydrides are of special interest and importance. Phthalic anhydride reacts with benzene to give *ortho*-benzoylbenzoic acid (XII), and succinic anhydride reacts with benzene to give β-benzoylpropionic acid (XIII).

The former serves as an important method for the fusion of two additional benzene rings to an aromatic compound, as on subsequent cyclization of the benzoylbenzoic acid, and reduction of the anthraquinone formed, anthracene is obtained. The second method serves to increase the size of an aromatic compound by one benzene ring. Reduction of the aroylpropionic acid to the butyric acid derivative (XIV), followed by cyclization, and then by reduction of the ketone (XV) obtained, gives a hydroaromatic structure (XVI) which can be dehydrogenated to a fully aromatic ring system.

As two moles of aluminium chloride are required per mole of anhydride, it seems likely that the reagent-catalyst complex is of the same nature as in acylation reactions with acyl halides and that the reaction may be represented as follows:

$$(CH_3CO)_2O + 2AlCl_3 \longrightarrow \overset{\delta+}{CH_3CO} \cdots \overset{\delta-}{AlCl_4} + CH_3COOAlCl_2$$

With the dibasic acid anhydrides, it seems that the first mole of aluminium chloride must bring about the fission of the anhydride ring and the formation of the aluminium chloride salt of one carboxylic acid group and the acyl chloride of the other. The second mole of aluminium chloride evidently functions as catalyst in exactly the same manner as in normal acylations. That is:*

$$\begin{array}{c} CH_2-CO \\ | \quad\quad\quad >O \\ CH_2-CO \end{array} + 2AlCl_3 \longrightarrow \begin{array}{c} CH.COOAlCl_2 \\ | \\ CH_2-\overset{\delta+}{CO}\cdots \overset{\delta-}{AlCl_4} \end{array}$$

Of course, an unsymmetrically substituted dibasic acid anhydride can undergo fission with aluminium chloride in two possible ways, and two different keto-acids can be formed. A mono-substituted succinic anhydride, for example, can react with benzene to give either the α- acid (XVII) or the β- acid (XVIII), or both.

Ph−CO−CH₂−CH(R)−CO₂H   Ph−CO−CH(R)−CH₂−CO₂H

α-acid   β-acid
(XVII)   (XVIII)

* See Noller and Adams, *J. Amer. Chem. Soc.* 1924, **46**, 1889; Groggins and Nagel, *Industr. Engng Chem.* 1934, **26**, 1313; Groggins, *Unit Processes in Organic Synthesis*, 3rd ed. pp. 761–2, McGraw-Hill Book Co., New York, 1947; Saboor, *J. Chem. Soc.* 1945, p. 922; Berliner, *Organic Reactions*, **5**, 229–89, John Wiley, New York, 1949.

Both products are usually obtained, one isomer predominating, and it seems that the electron-attracting aluminium chloride preferentially opens the anhydride towards the carbonyl group that has the higher electron density. That is:

$$\begin{array}{c} R-CH-C=O \\ \diagdown \\ O +2AlCl_3 \\ \diagup \\ CH_2-C=O \end{array} \longrightarrow \begin{array}{c} R.CH.COOAlCl_2 \\ \delta+ \delta- \\ CH_2.CO \ldots AlCl_4 \end{array}$$

When $R = CH_3$, this effect is small, and the reaction between methylsuccinic anhydride and benzene (or toluene) gives both isomers; but with polynuclear hydrocarbons there is always a preferential formation of the α- isomer.

When $R = Ph$ or $X.C_6H_4$, however, the course of the reaction is strongly influenced both by the electronic character of the group and also by the solvent used in the reaction. For example, when toluene is used as reactant, and also as solvent, and when phenylsuccinic anhydride is used as acylating agent, the β- acid is formed predominantly. The same result is obtained when para-nitrophenylsuccinic anhydride is used. However, in the presence of an electron-donating group, the situation is reversed and there is a preferential formation of the α- isomer. The situation is again different when nitrobenzene is used as solvent. Phenylsuccinic anhydride and toluene interact to give the α- acid predominantly as also does para-methoxyphenylsuccinic anhydride. But para-nitrophenylsuccinic anhydride gives more of the β- acid. These results are summarized in table 7·4, which is due to Wali, Khalil, Bhatia and Ahmad.*

The orientation of the product in acylation reactions, either with acyl halides or with anhydrides, is of some interest, especially as anomalous orientation sometimes occurs. For example, the reaction between succinic anhydride and toluene, ethylbenzene or chlorobenzene yields only the *para* isomers, although some of the *ortho* derivatives would be expected,† and

---

* *Proc. Indian Acad. Sci.* 1941, **14A**, 139; *Chem. Abstr.* 1942, **36**, 1598; see also Berliner, *Organic Reactions*, **5**, 245.

† Krollpfeiffer and Schäfer, *Ber. dtsch. chem. Ges.* 1923, **56**, 620; Muhr, ibid. 1895, **28**, 3215; Levy, *Ann. Chim.* 1938, **9**, 5.

steric hindrance to the approach of the bulky reagent seems to be involved.

Anomalous orientation also occurs under certain conditions in the acylation of naphthalene. For example, the acetylation of naphthalene occurs predominantly in the 2-position when nitrobenzene is used as solvent, although, if carbon disulphide is used, substitution occurs in the 1-position. Similarly, succinic anhydride reacts mainly in the 2-position if nitrobenzene is used as solvent. Again, 2-methoxynaphthalene and succinic anhydride give β-(2-methoxy-1-naphthoyl)propionic acid in carbon disulphide; but in nitrobenzene the product consists of 9 parts of β-(2-methoxy-6-naphthoyl)propionic acid and 1 part β-(2-methoxy-1-naphthoyl)propionic acid.*

TABLE 7·4. *Reaction between substituted phenylsuccinic anhydrides and toluene*

Substituent in succinic anhydride	Solvent	Yield	
		α- acid	β- acid
Phenyl	Excess toluene	23	77
p-Nitrophenyl	Excess toluene	20	80
p-Methoxyphenyl	Excess toluene	82	18
Phenyl	Nitrobenzene	83	17
p-Nitrophenyl	Nitrobenzene	33	67
p-Methoxyphenyl	Nitrobenzene	Predominant	—

The problem has been carefully studied by Baddeley.[†] He found that almost pure 1-naphthyl ketones can be obtained if the reactions are carried out in solvents such as ethylene or methylene chlorides. The addition of molecular proportions of nitrobenzene, nitromesitylene, or of additional acid chloride, to the reaction mixture was found to alter the course of the reaction, and considerable quantities of the 2- isomers were produced under these conditions. These additional substances all combine with acyl chloride-aluminium chloride complexes, and the resulting double complexes must be very bulky. Baddeley therefore suggests that the *peri*-naphthalene position offers steric hindrance to the approach of such very bulky complexes to the

* Short, Stromberg and Wiles, *J. Chem. Soc.* 1936, p. 319.
† Ibid. 1949, p. S99.

1-position. No such steric hindrance affects the 2-position, with the result that under these conditions a considerable proportion of 2-substituted product is obtained. If this reasoning is valid, preferential substitution in the 1-position should occur if sufficient additional aluminium chloride is added to the reaction mixture to engage the nitrobenzene, nitromesitylene or additional acyl chloride. This has been confirmed experimentally, as table 7·5 indicates.

TABLE 7·5. *Acylation of naphthalene in methylene chloride**

Acylating agent 1 mole	Further reagent 1 mole	Composition of naphthyl ketone	
		1- isomer (%)	2- isomer (%)
AcCl—AlCl$_3$	None	98	2
AcCl—AlCl$_3$	AcCl	60	40
AcCl—AlCl$_3$	PhNO$_2$	40	60
AcCl—AlCl$_3$	2:4:6-Me$_3$C$_6$H$_2$NO$_2$	29	71
AcCl—AlCl$_3$	PhNO$_2$—AlCl$_3$	97·5	2·5

Acylation reactions, unlike alkylations, are commonly supposed to be irreversible. However, the difference seems to be one of degree rather than of kind, for intermolecular and intramolecular migrations have sometimes been observed in acylations, and such migrations may well account for some of the anomalous orientations which have been reported. In this connexion it may be mentioned that aluminium chloride converts 9-acylanthracenes into the 1- and 2- isomerides, and that the disproportionation of 1-acetylacenaphthene into acenaphthene and two isomeric diacetylacenaphthenes has been observed.[†] Furthermore, Cook and Hewett[‡] found that 1:2-benzanthracene and acetic anhydride, in nitrobenzene, gave a mixture of five ketones, including the *meso*-acetyl derivative (XIX), the 7-acetyl derivative (XX), the 6-acetyl derivative, and two ketones of undetermined orientation. The complexity of the reaction is illustrated by the fact that the *meso*-acetyl derivative (XIX) was

---

[*] After Baddeley, *J. Chem. Soc.* 1949, p. 899.
[†] See Linstead, *Ann. Rep. Chem. Soc.* 1937, **34**, 254.
[‡] *J. Chem. Soc.* 1933, p. 1408.

isomerized to 7-acetyl-1:2-benzanthracene (XX) by treatment with aluminium chloride at 40°.

Some very interesting isomerizations have been observed with alkyl aromatic ketones. For example, with at least two moles of aluminium chloride, 2-methylacetophenone (XXI) is converted, in good yield, into the 4-methyl isomeride (XXII). Such a reaction could conceivably take place either by migration of the methyl group or of the acyl group. That it takes place by the latter mechanism is indicated by the fact that if the reaction is carried out in the presence of *meta*-5-xylenol (XXIII), the yield of 4-methylacetophenone (XXII) is halved and some 2-hydroxy-4:5-dimethylacetophenone (XXIV) is also formed. It must be concluded therefore that aluminium chloride reacts with 2-methylacetophenone (XXI) to form an acylating agent.*

The effect of aluminium chloride on alkylated aromatic ketones has been extensively studied by Baddeley,* who has concluded that the isomerizations can be of two types. In type A, the mobile alkyl group moves intramolecularly into the neighbouring position. In type B, however, the acyl group migrates. With some compounds, both types of migration appear to take place simultaneously.

* Baddeley, *J. Chem. Soc.* 1944, p. 232.

In the above account, attention has been concentrated on the use of aluminium chloride as an acylation catalyst. This is reasonable, as this catalyst is certainly the one most commonly used. Nevertheless, other catalysts are also effective, as is the case in the alkylation reactions already considered.

Hydrogen fluoride is a satisfactory catalyst in some cases. For example, Fieser and Hershberg* found that acenaphthene, perinaphthene and indane are smoothly acylated by reaction with various free acids, acid anhydrides and acid chlorides in the presence of liquid anhydrous hydrogen fluoride at room temperature. Naphthalene, phenanthrene and other aromatic hydrocarbons were not acylated under these conditions, but satisfactory results were obtained at elevated temperatures. A particularly interesting acylation with hydrogen fluoride involved the action of crotonic acid ($CH_3.CH:CH.CO_2H$) on acenaphthene (XXV), the reaction product being 1'-methyl-3'-keto-2:3-*cyclo*pentenoacenaphthene (XXVI).*

(XXV)

(XXVI)

Boron trifluoride is also a suitable catalyst for acylation in some cases, but with less reactive hydrocarbons the reactions do not proceed very smoothly. The more reactive aromatic components, such as anisole and similar compounds, react readily and good yields of *para*-methoxyacetophenone are obtained using acetic anhydride as acylating agent. In general, however, there is no particular advantage in using boron tri-

---

* *J. Amer. Chem. Soc.* 1939, **61**, 1272; 1940, **62**, 49; see also Calcott, Tinker and Weinmayr, ibid. 1939, **61**, 1010.

fluoride in these condensations rather than aluminium chloride, or zinc chloride, etc.*

It has recently been found that acylations can also be brought about using perchloric acid as catalyst. It seems that mixtures of perchloric and acetic acids contain the ion $AcOH_2^+$, which is not an acylating agent; but that when acetic anhydride is added to the mixture, Ac^+ or Ac_2OH^+ ions are formed. Using this mixture, anisole has been converted into *para*-methoxyacetophenone.† The reaction is possibly

$$H^+ + Ac_2O \rightleftharpoons Ac_2OH^+ \rightleftharpoons Ac^+ + AcOH$$

or $$AcOH_2^+ + Ac_2O \rightleftharpoons Ac_2OH^+ + AcOH \rightleftharpoons Ac^+ + 2AcOH.$$

'Acetyl perchlorate', prepared *in situ* from silver perchlorate and acetyl chloride, is also an effective acetylating agent;† and acetylium ions are readily produced from acetyl chloride or acetic anhydride with zinc chloride.‡

INTRAMOLECULAR ACYLATION: CYCLIZATION

The formation of cyclic ketones by intramolecular acylation is an important synthetic process, and it can be brought about by a very wide variety of different catalysts.§ Cyclization of acids is commonly brought about with sulphuric acid, hydrogen fluoride, phosphoric acid, or with phosphorus pentoxide, etc., and cyclization of acid chlorides with aluminium chloride, stannic chloride or some similar reagent. Typical of such intramolecular acylations are the cyclization of *ortho*-benzoylbenzoic acid (I) to anthraquinone (II) with concentrated sulphuric acid, and the cyclization of γ-phenylbutyric acid chloride (III) to α-tetralone (IV) with aluminium chloride.

The mechanisms of such reactions are probably very similar to those of ordinary acylations. The cyclization of keto acids

---

* For review see Kästner in *Newer Methods of Preparative Organic Chemistry*, pp. 249–313, Interscience, New York, 1948; see also Kästner, *Angew. Chem.* 1941, 54, 273, 296.
† Burton and Praill, *J. Chem. Soc.* 1950, pp. 1203, 2034.
‡ Ibid. 1951, p. 726.
§ Johnson, *Organic Reactions*, 2, 114–77, John Wiley, New York, 1944.

with sulphuric acid almost certainly involves the intermediate formation of either a dihydroxycarbonium ion (e.g. (V)) or an oxycarbonium ion (e.g. (VI)), depending on the conditions and the acid in question. Cyclization would then occur by electrophilic substitution, with elimination of either H_3O^+ or H^+.

The cyclization of *ortho*-benzylbenzophenones (VII) to anthracene derivatives (IX), by heating with mineral acids, is evidently very similar, and proceeds via an hydroxycarbonium ion.*

Enolization does not seem to be the primary step, as has sometimes been supposed, for 2-phenylbenzophenone has been cyclized to 9-phenylfluorenone. It is also of interest that the rate of cyclization of *ortho*-benzylbenzophenones decreases with increasing length of the alkyl group, R, until *n*-butyl is reached, and remains approximately the same for *n*-pentyl and *n*-hexyl. This may be explained in terms of the increasing electron-releasing properties of the alkyl groups with increasing chain

* Berliner, *J. Amer. Chem. Soc.* 1942, **64**, 2894.

length, resulting in an increasing electron density at the positive carbon atom thereby retarding cyclization.* Steric factors are also involved in this reaction.

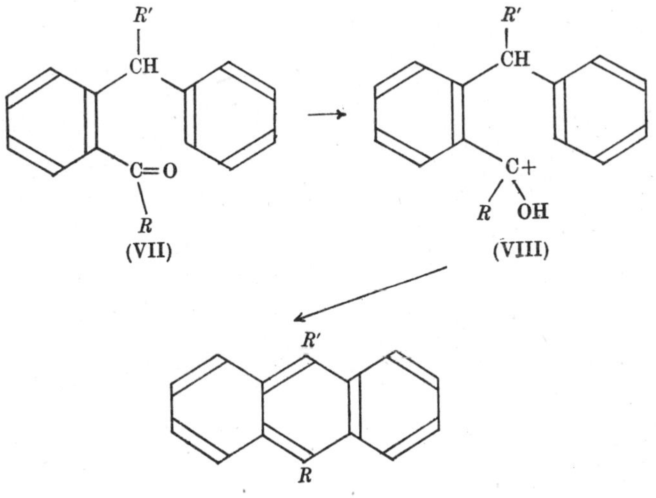

(VII) → (VIII)

(IX)

The cyclization of an acid chloride with aluminium chloride probably involves the intermediate formation of the usual complex ($RCO^{\delta+}\ldots AlCl_4^{\delta-}$), or a free oxycarbonium ion (RCO^+), ring closure being effected by electrophilic substitution with the ejection of a proton.

Another type of cyclization process is illustrated by the conversion of triarylcarbinols (X) into 9-arylfluorenes (XIII). Such cyclizations may be brought about by boiling with acetic acid containing a trace of mineral acid, and the mechanism almost certainly involves the formation of carbonium ions (XI), (XII). Cyclization would then occur by an intramolecular electrophilic substitution reaction.†

The *direction* of cyclization is often immaterial, but this is not always the case. With unsymmetrical compounds cyclization

* Ibid. 1944, **66**, 533; Bradsher and Vingiello, ibid. 1949, **71**, 1434.
† Ibid. 1942, **64**, 2894.

can occur in two, or sometimes more, possible ways. In practice, however, one direction is nearly always favoured, and a number of factors may contribute to this. For example, *meta*-methoxy-hydrocinnamic acid (XIV) can conceivably cyclize to yield either

(X)   (XI)

(XII)   (XIII)

(XIV)   (XV)   (XVI)

5-methoxyindan-1-one (XV) or 7-methoxyindan-1-one (XVI), depending on whether the reaction occurs either *para* or *ortho* to the substituent. In fact, both products are obtained in this reaction, the major one being 5-methoxyindan-1-one.*

* Ingold and Piggott, *J. Chem. Soc.* 1923, **123**, 1469.

As such intramolecular acylations are electrophilic reactions, it is reasonable to suppose that cyclization will take place most readily at the position of greater, or greatest, π-electron density, or at the position at which two π-electrons can be provided most readily, and this has been generally confirmed. The cyclization of β-tolyl-β-phenylpropionic acid (XVII), for example, can conceivably take place to either of the aromatic rings, but a methyl group increases the π-electron densities of the annular carbon atoms, and the cyclization proceeds to give a phenylmethylindanone (XVIII) rather than the alternative tolylindanone.* The formation of (XVIII) involves cyclization *para* to the methyl group.

Substitution at an α-position in naphthalene proceeds more readily than at a benzene position, for two π-electrons can be provided more easily at an α-position. Naphthylphenylpropionic acid (XIX) cyclizes to form a phenylbenzindan-1-one (XX).*

Again, cyclization takes place more readily into an α-position in naphthalene than into a β-position, for γ-2-naphthylbutyric acid (XXI) cyclizes to give 4-keto-1:2:3:4-tetrahydrophenanthrene (XXII). The alternative cyclization into the β-position to give the linear compound does not occur.†

* von Braun, Manz and Reinsch, *Liebigs Ann.* 1929, **468**, 277.
† Haworth, *J. Chem. Soc.* 1932, p. 1125; Bachmann and Edgerton, *J. Amer. Chem. Soc.* 1940, **62**, 2219.

Other factors may also affect the direction of cyclization. Other things being equal, it is found that six-membered rings are formed in preference to either five-membered or seven-membered rings. This is shown by the fact that α-benzyl-γ-phenylbutyric acid (XXIII) gives 2-benzyltetral-1-one (XXIV), and none of the isomeric indanone seems to be formed.*

(XXI)   (XXII)

(XXIII) → (XXIV)

Steric factors also control the direction of cyclization in some cases, and examples are also known in which the nature of the product is governed by the catalyst and/or solvent used in the reaction (see pp. 121-3).

DIAZO COUPLING

Phenols, amines, and certain other classes of aromatic compounds rapidly react with various diazo compounds, in aqueous alkaline solution, to give highly coloured azo compounds:

$$ArH + Ar'.N_2X = Ar.N:N.Ar' + HX.$$

This reaction is known as 'diazo coupling'.†

* von Braun, *Ber. dtsch. chem. Ges.* 1928, **61**, 441; Leuchs, ibid. 1928, **61**, 144.
† For a review see Saunders, *The Aromatic Diazo-Compounds and their Technical Applications*, 2nd ed., Edward Arnold, London, 1949.

The mechanism of this important reaction has not yet been established with complete certainty. For many years a mechanism involving prior reaction of the reagent with the functional group of the aromatic compound was favoured.* The coupling of phenols, for example, was supposed to occur by attachment of the diazo group to the phenolic oxygen, and subsequent rearrangement of the diazo-ether to the diazo-phenol. Intermediates of the nature postulated have sometimes been isolated. For example, Dimroth and Hartmann[†] obtained a labile diazo-ether (II) from *para*-nitrophenol (I) and *para*-bromobenzenediazonium hydroxide, and showed that it is converted into an azo compound (III) on gentle warming.

In the same way, the coupling of amines was supposed to occur by rearrangement of the diazoamino compounds first formed. It is, of course, well known that aryldiazonium chlorides and amines (IV) interact under certain conditions to form relatively stable diazoamino compounds (V) which are readily converted into aminoazo compounds (VI).[‡]

However, although azo compounds *can* be formed by a rearrangement of this nature, it does not seem that this is the

* See Fieser in Gilman's *Organic Chemistry*, 2nd ed. 1, 117–213, John Wiley, New York, 1943.
† *Ber. dtsch. chem. Ges.* 1908, **41**, 4012.
‡ See Goldschmidt and Reinders, *Ber. dtsch. chem. Ges.* 1896, **29**, 1369, 1899; Goldschmidt and Salcher, *Z. phys. Chem.* 1899, **29**, 89.

normal mechanism of diazo coupling, and Meyer* has shown that the coupling of amines can occur without the preliminary formation of a diazoamino compound. Moreover, phenolic ethers and tertiary amines can couple with diazo compounds, although intermediates of the type postulated are clearly impossible in such cases. Furthermore, certain very reactive hydrocarbons (such as acenaphthene and benzpyrene) also undergo the coupling reaction. It seems that the diazo-coupling reaction must be considered in the same light as any other aromatic substitution reaction, and that amino and hydroxy groups facilitate the reaction.

In preparative work it is well known that the high acidity necessary for the preparation of a diazonium compound must be decreased before coupling. In general, reactions are carried out in alkaline or at least in buffered solution. The substituents —O⁻ and —NH_2 are strongly electron-donating, the substituent —OH is less strongly electron-donating, and the —NH_3^+ group is electron-attracting, so that electrophilic substitution of phenols and amines would be expected to proceed most rapidly in alkaline conditions such that the concentration of phenate ions or of free base is a maximum. Kinetic studies have in fact confirmed that phenols couple as phenate ions, and amines as free bases.†

(VII)   (VIII)

The effects of other substituents also indicate that the reaction is electrophilic. Benzene does not couple, but mesitylene (VII), in which the position attacked is activated by two *ortho-* and one *para*-standing methyl groups, does react. An azo compound (VIII) is obtained in good yield from mesitylene (VII) and diazotized picramide.‡

---

* *Ber. dtsch. chem. Ges.* 1921, **54**, 2265.
† Wistar and Bartlett, *J. Amer. Chem. Soc.* 1941, **63**, 413; Hauser and Breslow, ibid. p. 418.
‡ Meyer and Tochtermann, *Ber. dtsch. chem. Ges.* 1921, **54**, 2283.

Pentamethylbenzene and *iso*durene also couple, but durene does not.* Only in the latter compound is there no *para*-methyl group for activation. In the same way, *meta*-xylenol ether (IX) couples more readily than *meta*-cresol ether (X), because all the groups activate the position of substitution. Under the same conditions *ortho*-cresol ether (XI) does not couple, for the methyl group is now *meta* to the probable position of attack and therefore exerts practically no activating action.† Moreover, steric hindrance probably prevents the methoxy group assuming a coplanar configuration with the ring, thereby reducing its $+T$ effect.

For the same reasons, dimethyl-*meta*-toluidine (XII) is very much more reactive in diazo coupling and other electrophilic reactions than dimethyl-*ortho*-toluidine (XIII). In the former compound, the $+T$ effects of both substituents reinforce one another. In the latter, the two $+T$ effects act independently and, furthermore, steric hindrance must inhibit the electron-donating effect of the dimethylamino group, as explained elsewhere (pp. 214–23).

Both phenols and amines are predominantly substituted in the *para* positions in this reaction. With phenol itself the major product is the *para*-hydroxyazo compound, but a little of the

---

* Smith and Paden, *J. Amer. Chem. Soc.* 1934, **56**, 2169.
† von Auwers and Michaelis, *Ber. dtsch. chem. Ges.* 1914, **47**, 1275; von Auwers and Borsche, ibid. 1915, **48**, 1716.

ortho-hydroxyazo compound is also formed. If the *para* position is already occupied by a stable substituent, such as a methyl group, the attack is transferred to the *ortho* position. In other words, the *ortho*:*para* ratio is exceptionally low for the diazo-coupling reaction. As a matter of fact, the tendency for reaction to occur *para* to a hydroxy group is so great that some substituents in this position may be removed. *para*-Hydroxy-benzoic acid (XIV) and phenol-*para*-sulphonic acid (XVI) behave in this way, and couple with benzene diazonium salts to give *para*-hydroxyazobenzene (XV).

That there is no inherent resistance to *ortho* coupling is shown by the fact that bisazo compounds can often be obtained by increasing the concentration of reactants, and especially if the aromatic component is activated, e.g. by an alkyl group *meta* to the hydroxy group. Under the right conditions phenol itself gives a bisazo compound (XVII), and even a trisazo compound (XVIII); and resorcinol also gives a trisazo compound (XIX).

In the naphthalene series both *ortho* and *para* positions are often substituted. 1-Naphthol couples at both the 2- and 4-positions, the proportion at which the two positions are attacked depending both on the pH and on the diazo compound. 2-Naphthol couples in the 1-position, and not in the 3-position. In this and in similar compounds it seems that the activating

AROMATIC SUBSTITUTION REACTIONS

influence of the hydroxy group is transferred more strongly along a bond of high 'order' than along a bond of low 'order'.

The inherent 'reactivity' of a position in a polycyclic compound is also of importance. Thus 4-hydroxypyrene (XX) couples in the 3-position, but 3-hydroxypyrene (XXI) does not couple. The 3-position in pyrene is inherently much more 'reactive' *vis-à-vis* electrophilic reagents than is the 4-position.

(XX) (XXI)

(XXII)

Para substitution is likewise favoured in amines of the benzene series. Aniline itself does not couple with weak diazo compounds, for it is less strongly activated than the phenate ion; but it does couple in the 4-position with powerful coupling reagents.* Similarly, predominant *para* substitution is observed with the more strongly activated alkylarylamines and arylalkylamines, etc. As with the hydroxy compounds, 1-naphthylamine couples in both the 2- and 4-positions; and 2-naphthylamine is attacked at the 1-position.

The results of coupling experiments with various aminoquinolines (see (XXII)) are also of interest in this connexion. Except in the case of 2-aminoquinoline, which yields a triazen,

* Gattermann and Rolfes, *Liebigs Ann.* 1921, **425**, 135.

the products are benzeneazo- compounds, and, as can be seen from table 7·6, substitution always occurs at the positions which are activated to the greatest extent by the amino group. Such activating influences are exerted more strongly through the αβ- than through the ββ-bonds.

TABLE 7·6. *Coupling positions of aminoquinolines**

NH_2 at	2	3	4	5	6	7	8
Coupling at	Triazen	4	Fails	8, 6	5	8	5

The different diazonium compounds are not all equally effective. Qualitatively, it is well known that *para*-nitrobenzenediazonium chloride is a much stronger coupler than is the unsubstituted benzenediazonium chloride, and the diazo compound formed from picramide is exceptionally powerful, reacting with compounds (such as hydrocarbons) which are unaffected by less powerful diazo compounds. Similarly, the introduction of electron-releasing substituents into the diazo compound reduces its coupling effectiveness. Quantitatively, Conant and Peterson[†] have investigated a number of different diazo compounds by measuring the rate of coupling with certain phenols. As the reaction is electrophilic, it seems likely that the effective reagent is the weakly cationoid diazonium ion:

$$\text{C}_6\text{H}_5\text{—N}\overset{+}{\equiv}\text{N}$$

The relative rates[‡] for the various substituted ions are:

O_2N—C$_6$H$_4$—N_2^+ 1300 Br—C$_6$H$_4$—N_2^+ 13 H_3C—C$_6$H$_4$—N_2^+ 0·4

^-O_3S—C$_6$H$_4$—N_2^+ 13   C$_6$H$_5$—N_2^+ 1 CH_3O—C$_6$H$_4$—N_2^+ 0·01

* Renshaw, Friedman and Gajewski, *J. Amer. Chem. Soc.* 1939, **61**, 3322.
† Ibid. 1930, **52**, 1220.
‡ See Dewar, *The Electronic Theory of Organic Chemistry*, p. 183, Oxford University Press. 1949.

AROMATIC SUBSTITUTION REACTIONS

These figures support the suggestion that the diazonium ion is the effective reagent, for electron-attracting substituents would decrease the electron density on the coupling nitrogen, thereby increasing its electron affinity and electrophilic activity. Similarly, electron-donating substituents, such as CH_3- and CH_3O-, would reduce the positive charge on the coupling nitrogen, making it a less effective coupler. On the other hand, these figures would also support the contention that the actual reagent is a polarized complex $Ar.N_2^{\delta+}-X^{\delta-}$. It seems not unlikely that both mechanisms may operate under certain conditions.

SULPHONATION

The sulphonation of aromatic compounds is usually brought about by heating with sulphuric acid, but other reagents have also been used. The most common variant is probably a solution of sulphur trioxide in an inert solvent or in sulphuric acid (oleum):

$$ArH + H_2SO_4 = ArSO_3H + H_2O,$$
$$ArH + SO_3 = ArSO_3H.$$

Sulphur trioxide complexes with various organic substances have also been used; so have chlorosulphonic acid and the alkaline bisulphates.[*]

The reagents are not all equally effective, for the reaction of aromatic hydrocarbons with sulphur trioxide occurs very much more rapidly and under milder conditions than with other sulphonating agents. As a matter of fact, the rate of sulphonation is intimately connected with the concentration of sulphur trioxide; it increases with increasing SO_3 content, both in sulphuric acid of less than 100% concentration, also in oleum.[†] Certain catalysts, such as boron trifluoride, increase the rate of sulphonation, and the sulphates of sodium, mercury, cadmium, aluminium, lead, arsenic, bismuth and iron are also effective in this respect. On the other hand, certain substances

---

[*] Suter and Weston, *Organic Reactions*, 3, 141, John Wiley, New York, 1946.
[†] Martinsen, *Z. phys. Chem.* 1908, 62, 713; Pinnow, *Z. Elektrochem.* 1915, 21, 380; 1917, 23, 243.

inhibit sulphonation. For example, in cyclizations using sulphuric acid, boric acid is often added to decrease the amount of sulphonation which occurs.

In spite of the importance of the reaction remarkably little is known of the mechanism. It is known, however, that the reaction is electrophilic (cationoid) in character. As indicated by table 7·7, electron-releasing substituents in benzene facilitate the reaction, and electron-attracting substituents retard it. Furthermore, there is a linear relationship between the dipole moments of benzene derivatives and the energies of activation for this reaction, so that reactivity must be directly related to the electron density at the point of attack.

TABLE 7·7. *Sulphonation in nitrobenzene at* $40°$*

Compound	SO_3		H_2SO_4	
	k	E, kcal./mole	$10^6 k$	E, kcal./mole
Benzene	48·8	4·8	15·5	7·5
Toluene	—	—	78·7	6·8
Chlorobenzene	2·4	7·7	10·6	8·9
Bromobenzene	2·1	7·8	9·5	8·9
Nitrobenzene	$7·8 \times 10^{-6}$	11·4	0·24	11·0
p-Nitrotoluene	$9·5 \times 10^{-4}$	11·0	3·3	9·8

By analogy with other electrophilic reactions it is reasonable to suppose that the actual sulphonating agent is the cation, —SO_3H^+, and under certain experimental conditions this does seem to be the case. The kinetics of sulphonation by sulphuric acid in nitrobenzene as solvent, and the kinetics of sulphonation in fuming sulphuric acid, were both found to be consistent with this assumption.[†]

On the other hand, sulphur trioxide also seems to be a sulphonating agent. This is indicated by the fact that sulphur trioxide complexes with dioxan, and with pyridine, can sulphonate benzene and other aromatic compounds.[‡] Hinshelwood

---

* Vicary and Hinshelwood, *J. Chem. Soc.* 1939, p. 1372; Wadsworth and Hinshelwood, ibid. 1944, p. 469; Dresel and Hinshelwood, ibid. 1944, p. 649; Stubbs, Williams and Hinshelwood, ibid. 1948, p. 1065.

† Brand, ibid. 1950, pp. 997, 1004.

‡ Suter, Evans and Kiefer, *J. Amer. Chem. Soc.* 1938, **60**, 538; Wagner, *Ber. dtsch. chem. Ges.* 1886, **19**, 1158; Baumgarten, ibid. 1926, **59**, 1976; Burkhardt and Lapworth, *J. Chem. Soc.* 1926, p. 684.

and his collaborators found that sulphonation with sulphur trioxide in nitrobenzene is initially of the first order with respect to the aromatic compound, and of the second order with respect to sulphur trioxide, and that the reaction is retarded by the sulphonic acid formed. The results were consistent with the view that the sulphur trioxide dimer, S_2O_6, is the effective reagent:

$$2SO_3 \rightleftharpoons S_2O_6,$$

$$ArH + S_2O_6 = ArS_2O_6H,$$

$$ArS_2O_6H \rightleftharpoons ArSO_3H + SO_3.$$

Alternatively, sulphur trioxide may be the effective reagent, a second sulphur trioxide molecule being required as a proton-acceptor:[*]

$$ArH + SO_3 \rightleftharpoons ArH.SO_3,$$

$$ArH.SO_3 + SO_3 \rightarrow ArSO_3^- + SO_3H^+,$$

$$ArSO_3^- + SO_3H^+ \rightleftharpoons ArSO_3H + SO_3.$$

It has long been known that the sulphonation reaction is reversible. When heated with water or with dilute acids, aromatic sulphonic acids are hydrolysed to the parent aromatic components, so that both sulphonation and desulphonation reactions occur at measurable rate in dilute sulphuric acid. It is to be expected that an equilibrium mixture of the aromatic component and of its sulphonic acid will be formed in certain circumstances, and this is found to be the case. For example, the sulphonation of benzene in the temperature range 100–200° attains equilibrium with 73–78 % sulphuric acid.[†] In general, the sulphonic acids which are most readily formed are those which are most readily hydrolysed, and this fact has sometimes been used to separate two aromatic compounds. The mixture of sulphonic acids obtained on sulphonation is heated with water or with dilute acid. The sulphonic acid of the more reactive substance is hydrolysed, and the more stable sulphonic acid of the less reactive substance remains relatively unaffected.

---

[*] See also Braude, *Ann. Rep. Chem. Soc.* 1949, **46**, 135.
[†] Guyot, *Chim. et ind.* 1919, **2**, 879; Zakharov, *J. Chem. Ind. U.S.S.R.* 1929, **6**, 1648; *Chem. Abstr.* 1931, **25**, 5154.

The reversible character of the reaction has some important consequences with regard to the orientation of the sulphonic acids in sulphonation reactions. For example, in the sulphonation of toluene (I) at 0°, the three isomers are formed in the following proportions: *ortho*, 43%; *meta*, 4%; *para*, 53%. That is, the quantity of *para* isomer formed is only slightly greater than that of *ortho*. At 100°, however, (see (II)) the three isomers are formed in the proportions: *ortho*, 13%; *meta*, 8%; *para*, 79%. At this temperature the *para* isomer is predominantly formed.* That this effect is due to desulphonation-sulphonation is indicated by the fact that both *ortho* and *para* isomers can be partially transformed into one another by heating with sulphuric acid and a little water. The *meta* isomer, however, is stable under the same conditions.

$$\begin{array}{cc} \text{CH}_3 & \text{CH}_3 \\ \text{43\%} & \text{13\%} \\ \text{4\%} & \text{8\%} \\ \text{53\%} & \text{79\%} \\ 0°\text{ C} & 100°\text{ C} \\ (\text{I}) & (\text{II}) \end{array}$$

The situation with naphthalene is similar. Naphthalene is more readily sulphonated than benzene, and at temperatures below 40° (III) the product consists of 96% naphthalene-1-sulphonic acid and 4% of the 2-sulphonic acid.† At 165° (IV), however, the product consists of only 15% naphthalene-1-sulphonic acid, about 85% being the 2- acid, and traces of other products are also formed.‡ Naphthalene-1-sulphonic acid is readily converted into an equilibrium mixture containing mainly the 2- isomer by heating with sulphuric acid. There is little doubt that the conversion involves a desulphonation-sulphonation process, and that it is not a direct rearrangement. In this connexion it is significant that if a mixture of the two acids is heated

---

* Holleman and Caland, *Ber. dtsch. chem. Ges.* 1911, **44**, 2504; Bradfield and Jones, *Trans. Faraday Soc.* 1941, **37**, 731.
† Fierz-David, *J. Soc. Chem. Ind., Lond.*, 1923, **42**, 421 T.
‡ Witt, *Ber. dtsch. chem. Ges.* 1915, **48**, 743.

with water to 150°, the 1- isomer is completely hydrolysed; but there is relatively little loss of the 2- isomer.*

Other things being equal, it seems that in this and in other cases the more stable isomer is the one in which the sulphonic acid group occupies a less hindered position. There is therefore a close relationship between apparent rearrangements of this type and rearrangements of the Friedel-Crafts and Jacobsen type (pp. 275–8).

96% 4%   15% 85%

below 40°   165°
(III)   (IV)

Steric hindrance seems to play a very important part in this reaction, and this is not surprising in view of the relatively large dimensions of the —SO_3H group. *tert.*-Butylbenzene is sulphonated exclusively in the *para* position under conditions in which toluene gives a fair proportion of the *ortho* isomer.† Similarly, *para*-cymene is sulphonated predominantly *ortho* to the less bulky methyl group.‡ Perhaps of even greater interest is the fact that, even at low temperatures, anthracene-9-sulphonic acid seems never to be formed. In this case the *peri* positions offer considerable steric hindrance, and this seems to be the explanation for the fact that reaction occurs exclusively at the 1- and 2-positions. It is possible, however, that preliminary reaction occurs at the 9-position, but that desulphonation occurs very readily, subsequent sulphonation giving the more stable 1- or 2- acids. It is also of interest that sulphonation at low temperatures favours the 1- rather than the 2-position, but that substitution in the 2-position is favoured by high temperatures. This is *not* due to desulphonation-sulphonation,

* Masters, U.S. Patent, 1,922,813; *Chem. Abstr.* 1933, **27**, 5085; Vorozhtzov and Krasova, *Anilin. Prom.* 1932, **2**, 15; *Chem. Abstr.* 1933, **27**, 5321.
† Senkowski, *Ber. dtsch. chem. Ges.* 1890, **23**, 2412.
‡ Phillips, *J. Amer. Chem. Soc.* 1924, **46**, 686; Le Fèvre, *J. Chem. Soc.* 1934, p. 1501; it must be admitted, however, that this result can also be explained by the superior hyperconjugation effect of the methyl group.

for anthracene-1-sulphonic acid (V) is not converted into the 2- acid on heating with sulphuric acid, but into the 1:5-disulphonic acid (VI) and 1:8-disulphonic acid (VII).* It must therefore be assumed that primary β- substitution of anthracene is favoured at high temperatures.

The orientation of the phenanthrene-sulphonic acids is also affected by the temperature of the reaction. At low temperatures (VIII) the 9- acid is formed; at 60° (IX) the product was found to be a mixture of the 1-, 2-, 3- and 9- acids, the 2- and 3- acids being isolated in the greater quantity. At 120–125° (X) much of the phenanthrene was found to be converted into disulphonic acids, but the 2- and 3-monosulphonic acids were also isolated in good yield.† Steric hindrance may also be a factor here, favouring β- substitution at high temperatures.

Apparently anomalous orientation also occurs in some sulphonation reactions which are carried out in the presence of mercury compounds. For example, the sulphonation of anthraquinone with oleum gives almost entirely β- substitution; but it was discovered independently by Iljinsky and by Schmidt, in 1903, that if a trace of mercury is added, predominant α- substitution occurs. Fierz-David‡ found that the pure β- acid is formed with 30% oleum at 145°, but in the presence of mercury

---

* Battegay and Brandt, *Bull. Soc. chim. Fr.* 1923, **33**, 1667.
† Fieser, *J. Amer. Chem. Soc.* 1929, **51**, 2460; *Organic Syntheses*, Coll. vol. 2, 482 (1943).
‡ *Helv. chim. acta*, 1927, **10**, 197; 1928; **11**, 197.

the product consisted of 97 % of the α- isomer and 3 % of the β- acid.

The sulphonation of naphthalene and of anthracene is unaffected by the addition of mercury,* but the sulphonation of *ortho*-xylene, *ortho*-dichlorobenzene and of *ortho*-dibromobenzene is affected to some extent. In the absence of mercury, the

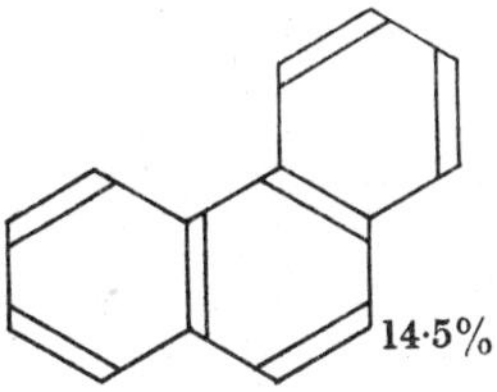

14·5%   19% 18%   13%
         4%
20°   60°
(VIII)   (IX)

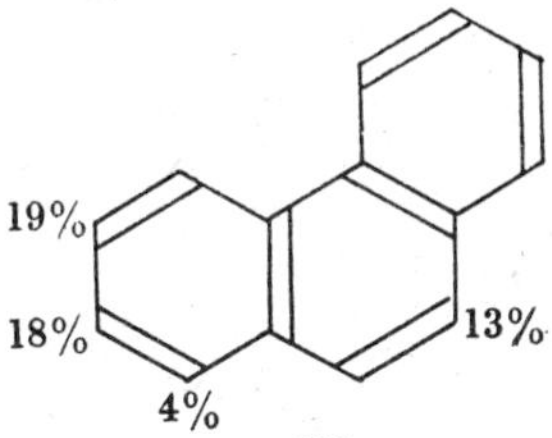

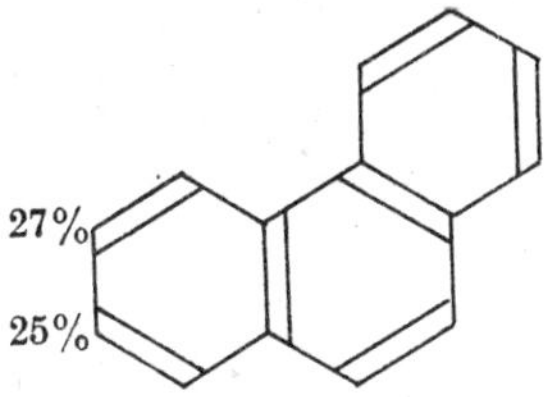

27%
25%
120–125°
(X)

4-sulphonic acid is the exclusive product in each case, but the addition of mercury to the reaction mixture gives a product containing some 20–25 % of the 3-sulphonic acid.† It seems likely that in all these cases the anomalous orientation is due to mercuration, followed by replacement with a sulphonic acid group, and the problem is therefore one of the orientation in mercuration reactions.

HYDROXYLATION

Following the discovery that hypobromous acid becomes a powerful brominating agent in the presence of mineral acids, Derbyshire and Waters‡ suggested that hydrogen peroxide may

* Euwes, *Rec. trav. chim. Pays-Bas*, 1909, **28**, 298.
† Lauer, *J. prakt. Chem.* 1933, **138**, 81.   ‡ *Nature, Lond.*, 1950, **165**, 401.

react similarly with mineral acids, to give hydroxyl cations, OH^+ or $H_3O_2^+$:

$$HO\text{---}OH + H^+ \rightleftharpoons H_2O + OH^+.$$

That this reaction does occur was indicated by the observation that hydrogen peroxide becomes a very effective hydroxylating agent in the presence of mineral acid. In this way mesitylene (I) was oxidized to mesitol (II).

(I)   + $\overset{+}{OH}$ ⟶   (II)

Hydrogen peroxide does not always undergo fission to give an electrophilic reagent. In the presence of ferrous salts it apparently decomposes to give free hydroxyl radicals:*

$$HO\text{---}OH + Fe^{++} \rightarrow HO^- + \cdot OH + Fe^{+++}.$$

Such a reaction mixture also hydroxylates aromatic compounds, the proportions of the various isomers formed being in good agreement with its characterization as a radical reaction, and not as an electrophilic reaction.

The oxidation of nitrobenzene under these conditions was found to give all three possible isomers in the approximate proportions: *ortho*, 0·56; *meta*, 0·44; *para*, 1·0. It is noteworthy that, in spite of the fact that of the five available positions of substitution only one is a *para* position, the amount of *para* isomer formed is considerably in excess of either the *ortho* or *meta*.†

Free hydroxyl radicals also seem to be formed by the action of various radiations (X-rays, neutrons, α-rays) on aqueous solutions. Thus the irradiation of dilute aqueous solutions of benzene in the absence of oxygen has been shown to give phenol and diphenyl. Similarly, the irradiation of dilute solutions of

* Fenton, *J. Chem. Soc.* 1894, **65**, 899; Haber and Weiss, *Proc. Roy. Soc.* A, 1934, **147**, 332.
† Loebl, Stein and Weiss, *J. Chem. Soc.* 1949, p. 2074.

benzoic acid has been shown to give *ortho*-, *meta*- and *para*-hydroxybenzoic acids.* Nitrobenzene has also been hydroxylated with free hydroxyl radicals produced by X-ray irradiation of an aqueous solution. It is interesting that the ratio of the *ortho*, *meta* and *para* isomers formed in this way is essentially the same as by hydroxylation with hydrogen peroxide and ferrous salts.†
Stein and Weiss‡ have also investigated the proportions of isomers formed in the hydroxylation of benzoic acid, and phenol, by means of X-rays. In the case of benzoic acid the three isomers were all formed, in the ratio: *para*, 10; *ortho*, 5; *meta*, 2. In the case of phenol no *meta* isomer was formed, the *para*:*ortho* ratio being approximately 2:1 in neutral solution, and rising to approximately 8:1 in acid or alkaline solutions. In all cases *para* substitution seems to be favoured in radical substitution.

A third type of direct hydroxylation is also known, and occurs with deactivated aromatic compounds, such as nitrobenzene. The reaction is effected by heating the substance with hot potash in the presence of air. Nitrobenzene has been converted into *ortho*- and a trace of *para*-nitrophenol in this way;§ pyridine has given α-pyridone;‖ 2:6-dimethylpyridine has given 2:6-dimethyl-γ-pyridone. In the same way, alizarin has been prepared from 1- or 2-hydroxyanthraquinone.¶ These reactions are probably nucleophilic, and the reagent is doubtless the hydroxyl anion.

AMINATION: THE TSCHITSCHIBABIN REACTION

The direct introduction of an amino substituent can sometimes be achieved by treatment of an aromatic compound with sodium amide, a reaction discovered by Tschitschibabin and Seide:**

$$ArH + NaNH_2 \rightarrow ArNH_2 + NaH,$$
$$ArNH_2 + NaH \rightarrow ArNHNa + H_2.$$

---

* Loebl, Stein and Weiss, *J. Chem. Soc.* 1951, p. 405; Stein and Weiss, ibid. 1949, pp. 3245, 3254; Weiss, *Nature, Lond.*, 1944, **153**, 748; *Trans. Faraday Soc.* 1947, **43**, 314.
† Loebl, Stein and Weiss, *J. Chem. Soc.* 1950, p. 2704.
‡ *Nature, Lond.*, 1950, **166**, 1104.
§ Wohl, *Ber. dtsch. chem. Ges.* 1899, **32**, 3486.
‖ Tschitschibabin, ibid. 1923, **56**, 1879.
¶ Baeyer and Caro, ibid. 1874, **7**, 968.
** *J. Russ. Phys. Chem. Soc.* 1914, **46**, 1216; *Chem. Abstr.* 1915, **9**, 1901.

Hydrogen is evolved and the intermediate metal derivative formed must be hydrolysed to give the free amine.*

Benzene and alkylated benzenes do not undergo this reaction; indeed, these compounds are often used as solvents. The reaction proceeds most readily with heterocyclic bases such as pyridine and quinoline. Pyridine (I) itself reacts to give mainly 2-aminopyridine (II) and a little 4-aminopyridine (III); but under more vigorous conditions, 2:6-diaminopyridine and 2:4:6-triaminopyridine (IV) are formed.† 2:6-Disubstituted pyridines give the corresponding 4-amino compounds.

Similarly, quinoline gives a mixture of 2-aminoquinoline (V) and 4-aminoquinoline (VI).‡

*iso*Quinoline gives 1-amino*iso*quinoline (VII),§ and phenanthridine very readily gives 6-aminophenanthridine (VIII).∥ The available evidence indicates that the amino group enters the

---

* For a review see Leffler, *Organic Reactions*, 1, 91–104, John Wiley, New York, 1942.
† Ger. Patent, 663,891; U.S. Patent, 1,789,022.
‡ Bergstrom, *J. Org. Chem.* 1937, 2, 411; Shreve, Riechers, Rubenkoenig and Goodman, *Industr. Engng Chem.* 1940, 32, 173; Tschitschibabin and Zataepina, *J. Russ. Phys. Chem. Soc.* 1920, 50, 553.
§ Bergstrom, *Liebigs Ann.* 1934, 515, 34; Tschitschibabin and Oparina, *J. Russ. Phys. Chem. Soc.* 1920, 50, 543.
∥ Morgan and Walls, *J. Chem. Soc.* 1932, p. 2225.

AROMATIC SUBSTITUTION REACTIONS 311

position having the least negative (most positive) charge in heterocyclic compounds.

The reaction is not confined to heterocyclic bases, although it has so far received most attention in this field. Naphthalene reacts with sodium amide to give 1-naphthylamine and 1:5-diaminonaphthalene.*

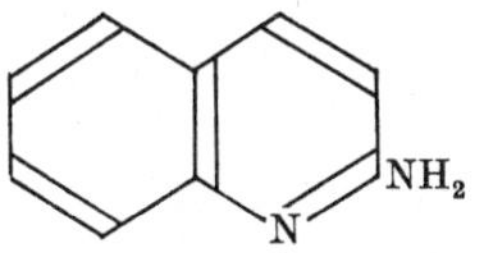

(V) 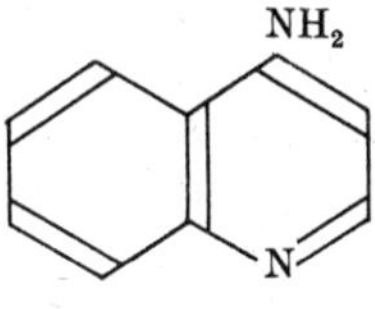 (VI)

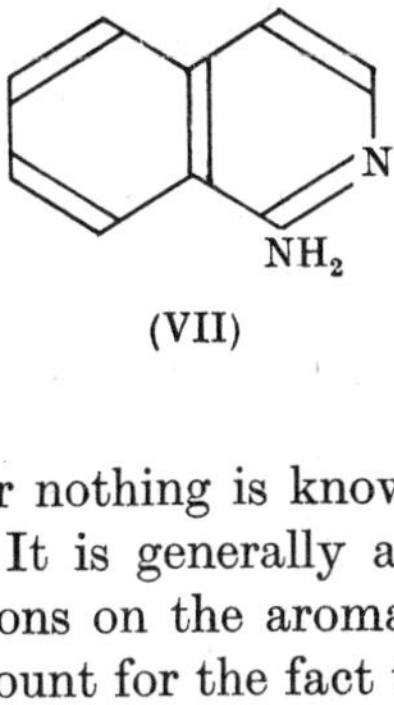

(VII) 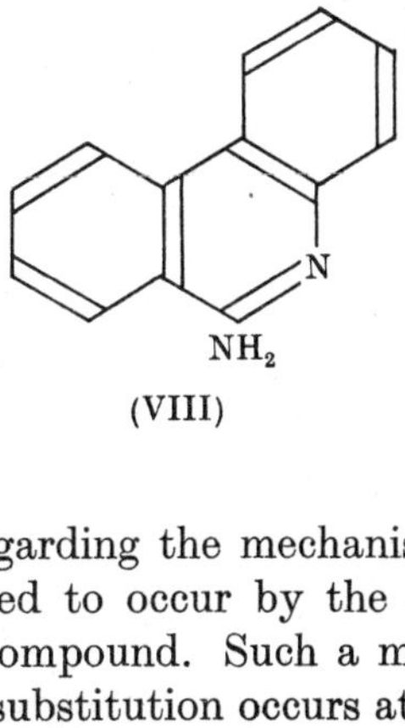 (VIII)

Little or nothing is known regarding the mechanism of the reaction. It is generally assumed to occur by the attack of amine anions on the aromatic compound. Such a mechanism would account for the fact that substitution occurs at the least negative carbon atom in heterocycles, and at the α-position in naphthalene (see pp. 244, 248).

INTRODUCTION OF THE CYANO GROUP

There are several examples in the literature of the direct introduction of a cyano group into an aromatic ring system. Possibly the best known of these reactions is the conversion of *meta*-dinitrobenzene (I) into 2-nitro-6-methoxybenzonitrile (III) by

* Sachs, *Ber. dtsch. chem. Ges.* 1906, **39**, 3006.

heating with potassium cyanide in methanol. It seems likely that this is a nucleophilic reaction brought about by cyano anions, followed by a nucleophilic replacement of one of the nitro groups by methoxyl anions.* In the same way, 5-cyano-6-methoxyquinoline is produced from 6-nitroquinoline, the point of attack by the cyano anion again being doubly 'deactivated'.

(I) → $\overset{\text{CN}^-}{\longrightarrow}$ (II) → $\overset{\text{OCH}_3^-}{\longrightarrow}$ (III)

Another reaction which seems to involve the participation of cyano anions is due to von Richter† and has recently been more thoroughly investigated by Bunnett, Cormack and McKay.‡ With alcoholic potassium cyanide a wide variety of aromatic nitro compounds has been converted into carboxylic acids, the carboxylic acid groups being *ortho* to the position previously occupied by the nitro groups. Thus *para*-nitrobromobenzene (IV) gave *meta*-bromobenzoic acid (V), *meta*-nitrobromobenzene gave a mixture of *ortho*- and *para*-bromobenzoic acids, and 2:5-dibromonitrobenzene (VI) gave 2:5-dibromobenzoic acid (VII).

(IV) (V) (VI) (VII)

These reactions probably proceed in three steps:‡ (i) attack by a cyano anion at an unoccupied position *ortho* to the nitro group; (ii) loss of the nitro group; and (iii) hydrolysis of the cyano group to the carboxylic acid.

* de Bruyn, *Rec. trav. chim. Pays-Bas*, 1883, **2**, 205, 235; de Bruyn and van Geuns, ibid. 1904, **23**, 26.
† *Ber. dtsch. chem. Ges.* 1874, **7**, 1145; 1875, **8**, 1418.
‡ *J. Org. Chem.* 1950, **15**, 481.

THE GOMBERG REACTION

The Gomberg reaction provides a very useful method for the introduction of aryl groups into aromatic compounds, and therefore for the synthesis of unsymmetrical diaryls.*

The reaction can be carried out in several ways. According to the method of Gomberg and Bachmann[†] the substituting reagent is prepared from an aromatic amine, which is diazotized in the usual way and then treated with alkali in the presence of the aromatic compound to be attacked. The substitution of benzene with a *para*-bromophenyl group, for example, can be represented by the following equations:

$$Br.C_6H_4.NH_2 \rightarrow Br.C_6H_4.N_2Cl,$$

$$Br.C_6H_4.N_2Cl + C_6H_6 + NaOH$$
$$\rightarrow Br.C_6H_4.C_6H_5 + N_2 + NaCl + H_2O.$$

There do not seem to be any indifferent solvents for this reaction, which is therefore limited to aromatic compounds which are liquid at the desired reaction temperature. Nevertheless, very many aromatic substances, including benzene, pyridine, nitrobenzene, toluene and so on, have been successfully substituted with a great variety of aryl groups.

Alternatively, the substituting reagent can be produced from an aromatic amine by acetylation, followed by nitrosation.[‡] The resulting nitrosoacetamido compound is then decomposed in the presence of the liquid aromatic compound to be attacked. For example, the substitution of benzene with a *meta*-nitrophenyl group can be represented by the following equations:

$$NO_2.C_6H_4.NH_2 \rightarrow NO_2.C_6H_4.NH.COCH_3,$$

$$NO_2.C_6H_4.NH.COCH_3 \rightarrow NO_2.C_6H_4.N(NO).COCH_3,$$

$$NO_2.C_6H_4.N(NO).COCH_3 + C_6H_6$$
$$\rightarrow NO_2.C_6H_4.C_6H_5 + N_2 + CH_3COOH.$$

---

* See Bachmann and Hoffman, *Organic Reactions*, **2**, 224–61, John Wiley, New York, 1944.
† *J. Amer. Chem. Soc.* 1924, **46**, 2339.
‡ Grieve and Hey, *J. Chem. Soc.* 1934, p. 1797; France, Heilbron and Hey, ibid. 1940, p. 369.

This method is really very similar to that of Gomberg and Bachmann, because nitrosoacetanilides are known to be tautomers of phenyldiazoacetates:*

$$\phi.N(NO).COCH_3 \rightleftharpoons \phi.N_2.O.COCH_3.$$

A third method for the introduction of aryl groups involves the use of aroyl peroxides. The decomposition of benzoyl peroxide in benzene, for example, gives diphenyl, together with some benzoic acid, phenyl benzoate, terphenyl and quaterphenyl.†

All these methods have been used extensively in synthetic work. Even *meta*-substituted diphenyls are readily available by the Gomberg reaction, and complex diaryls of large molecular weight can also be obtained without difficulty. For example, 1-phenylnaphthalene can be obtained from 1-naphthylamine and benzene, and *para*-terphenyl can be obtained from 4-aminodiphenyl and benzene.

With the limitation already mentioned, all classes of simple aromatic substance can be substituted with an aryl group in this way.

The orientation of the product is of considerable importance when considering the mechanism of the reaction, and it is noteworthy that benzene derivatives substituted with either electron-attracting or electron-releasing groups are attacked. Moreover, nitrobenzene is attacked more rapidly than toluene.

In some cases all the possible isomers are obtained, but, generally speaking, *ortho-para* derivatives are formed predominantly, whatever the nature of the substituent group. The phenylation of pyridine, for example, has been shown to give a mixture of 2-, 3- and 4-phenylpyridines, the former predominating.‡ Similarly, the phenylation of ethyl benzoate gave a mixture of 2-, 3- and 4-carbethoxydiphenyl, indicating substitution *ortho*, *meta* and *para* to the carbethoxy group.§

DeTar and Scheifele[||] have studied the phenylation of nitro-

* Bamberger, *Ber. dtsch. chem. Ges.* 1894, **27**, 914; von Pechmann and Frobenius, ibid. 1894, **27**, 651; Hantzsch and Wechsler, *Liebigs Ann.* 1902, **325**, 226.
† Gelissen and Hermans, *Ber. dtsch. chem. Ges.* 1925, **58**, 285, 476, 984.
‡ Haworth, Heilbron and Hey, *J. Chem. Soc.* 1940, p. 349; Hey and Walker, ibid. 1948, p. 2213; Adams, Hey, Mamalis and Parker, ibid. 1949, p. 3181.
§ Grieve and Hey, ibid. 1934, p. 1797; France, Heilbron and Hey, ibid. 1940, p. 369.
|| *J. Amer. Chem. Soc.* 1951, **73**, 1442.

AROMATIC SUBSTITUTION REACTIONS

benzene using all three types of reagent, namely, benzenediazonium chloride and alkali, N-nitrosoacetanilide, and benzoyl peroxide. In each case the product was found to be a mixture of *ortho*-nitrodiphenyl (60–70 %) and *para*-nitrodiphenyl (30–40 %) with only a little *meta*-nitrodiphenyl (0–10 %). In other words, although all three isomers are formed, the nitro group is predominantly *ortho-para*-directing in this reaction.

These facts clearly indicate that the reaction cannot be electrophilic, but that it is probably of a free radical character. Much evidence which supports this conclusion has accumulated.*

The decomposition of diazohydroxides and similar compounds may be represented

$$Ph.N_2.X \rightarrow Ph\cdot + X\cdot + N_2.$$

This is supported by the fact that benzenediazonium chloride decomposes in acetone, in the presence of excess calcium carbonate, to give mainly benzene and chloroacetone:

$$Ph\cdot + CH_3COCH_3 \rightarrow Ph.H + CH_3COCH_2\cdot,$$
$$Cl\cdot + CH_3COCH_3 \rightarrow CH_3COCH_2Cl + H\cdot,$$
$$Cl\cdot + CH_3COCH_2\cdot \rightarrow CH_3COCH_2Cl.$$

Furthermore, metals such as antimony, bismuth and lead are attacked under these conditions, giving metallic chloride. It is also of interest that the decomposition of benzenediazonium hydroxide in carbon disulphide gives diphenyldisulphide, and that decomposition in *cyclo*hexane gives benzene.†

Waters† has also shown that the metals lead, tin, copper and mercury are converted into the metal acetates when nitrosoacetanilide is decomposed in their presence. This proves the intermediate formation of acetate radicals, for these metals are not attacked by acetate ions, or by acetic acid, under the same conditions. Moreover, carbon dioxide is sometimes evolved during reactions of this nature, and this evidently arises by decomposition of the acetate radicals:

$$CH_3COO\cdot \rightarrow CO_2 + CH_3\cdot.$$

---

* Hey and Waters, *Chem. Rev.* 1937, **21**, 169; Hey, *Ann. Rep. Chem. Soc.* 1940, **37**, 278; Waters, *The Chemistry of Free Radicals*, Oxford University Press, 1948.
† Waters, *J. Chem. Soc.* 1937, pp. 2007, 2014; Makin and Waters, ibid. 1938, p. 843.

Additional evidence supporting the radical nature of the processes arises from the fact that when nitrosoacetanilide is decomposed in the presence of hexane, benzene is obtained, and that when it is decomposed in the presence of carbon tetrachloride, chlorobenzene is formed. These observations indicate that nitrosoacetanilide must decompose to give phenyl radicals:

$$Ph.N(NO).COCH_3 \rightarrow Ph\cdot + CH_3COO\cdot + N_2,$$

and that these react with the hexane or carbon tetrachloride as follows:

$$Ph\cdot + RH \rightarrow PhH + R\cdot,$$
$$Ph\cdot + CCl_4 \rightarrow PhCl + CCl_3\cdot.$$

Finally, it is reasonable to suppose that benzoyl peroxide also decomposes to give phenyl radicals:

$$PhCO.O.O.COPh \rightarrow Ph\cdot + CO_2 + PhCO.O\cdot,$$
$$PhCO.O\cdot \rightarrow Ph\cdot + CO_2.$$

The available evidence therefore indicates that the Gomberg reaction is almost certainly a substitution reaction involving an attack by free aryl radicals.

THE PSCHORR SYNTHESIS

Closely allied to the Gomberg reaction is the Pschorr synthesis of phenanthrene and other polycyclic and heterocyclic compounds. As originally developed by Pschorr* the reaction involved the diazotization of α-phenyl-*ortho*-aminocinnamic acid (I), and subsequent treatment of the diazo solution with copper powder. In this way, phenanthrene-9-carboxylic acid (III) was obtained in 93% yield. In some cases the addition of copper powder has been found to be unnecessary, the decomposition being effected simply by warming the aqueous solution on the water-bath.†

---

* *Ber. dtsch. chem. Ges.* 1896, **29**, 496.
† Pschorr and Tappen, ibid. 1906, **39**, 3115; Pschorr *et al. Liebigs Ann.* 1912. **391**, 23; Mayer and Balle, ibid. 1914, **403**, 167.

As indicated by structures (I)–(III), the synthesis clearly requires a *cis* configuration of the aryl nuclei, and it is noteworthy that *cis-ortho*-aminostilbene gives phenanthrene directly, but that *trans-ortho*-aminostilbene does not do so. As a matter of fact, the success of the method introduced by Pschorr depends on the circumstance that the carboxyl group induces a *cis* configuration of the aryl nuclei in the Perkin reaction used to prepare the nitro acids.*

The Pschorr synthesis has received very wide application in the preparation of substituted phenanthrene derivatives, especially those of interest in the field of natural products. Pschorr and Sumuleanu[†] used the method to confirm the structure of dimethylmorphol, which was obtained as a degradation product from morphine.

Other methoxyphenanthrenes have been prepared by Pschorr[‡] and, more recently, by Buchanan, Cook and Loudon.[§]

The method has also been applied to the synthesis of fluorenes and fluorenones. Thus *ortho*-aminodiphenylmethane (IV) gives fluorene (V),[ǁ] and *ortho*-aminobenzophenone (VI) gives fluorenone (VII).[¶] Similarly, 3-nitro-6-aminobenzophenone has been

---

* See Ruggli and Staub, *Helv. chim. acta*, 1936, **19**, 1288; 1937, **20**, 37; Ruggli and Dinger, ibid. 1941, **24**, 173.
† *Ber. dtsch. chem. Ges.* 1900, **33**, 1810.
‡ *Liebigs Ann.* 1912, **391**, 40.
§ *J. Chem. Soc.* 1944, p. 325.
ǁ Fischer and Schmidt, *Ber. dtsch. chem. Ges.* 1894, **27**, 2786.
¶ Graebe and Ullmann, ibid. 1894, **27**, 3483.

cyclized to 2-nitrofluorenone,* and many other similar reactions have been successfully carried out.†

In the preparation of substituted phenanthrenes and fluorenes, and also in the synthesis of certain polycyclic compounds, it is important to recognize that cyclization can sometimes give two isomerides. Indeed, two isomerides are almost invariably formed

unless substituents are present only in the position *para* to the amino group and *para* to the ethylenic link. For example, in the preparation of substituted phenanthrenes, two isomerides are formed if there is a substituent *meta* to the ethylenic link in the ring which does not carry the amino group. Thus *meta*-methyl-α-*ortho'*-aminophenylcinnamic acid (VIII) gives a mixture of 2- and 4-methylphenanthrene-9-carboxylic acids ((IX) and (X)).‡

* Ullmann and Mallet, *Ber. dtsch. chem. Ges.* 1898, **31**, 1694.
† Ullmann and Bleier, ibid. 1902, **35**, 4273; Miller and Bachman, *J. Amer. Chem. Soc.* 1935, **57**, 2443.
‡ Mayer and Balle, *Liebigs Ann.* 1914, **403**, 167.

Similarly, application of the Pschorr synthesis to the diamino compound (XI) resulted in cyclization in both possible directions, giving a mixture of 1.2.5:6 dibenzanthracene-4:8-dicarboxylic acid (XII) and of 3:4:5:6-dibenzophenanthrene-1:8-dicarboxylic acid (XIII).*

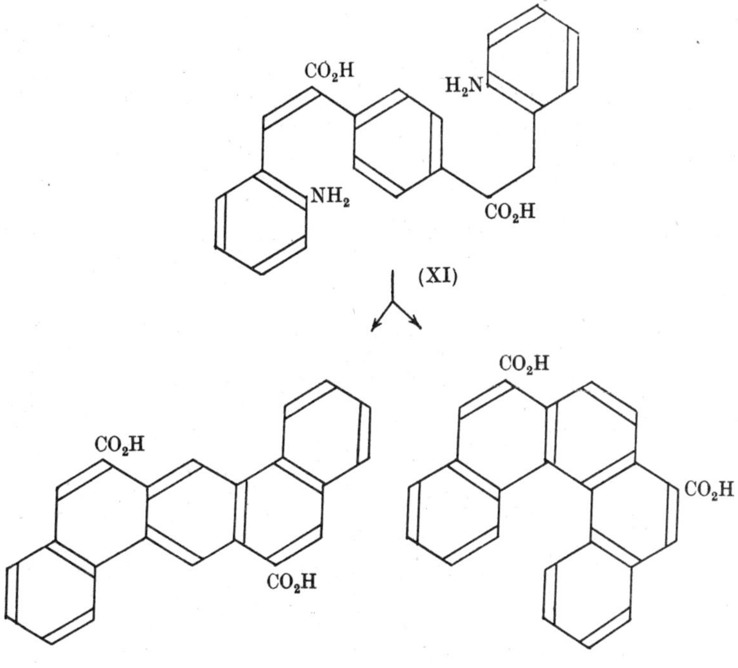

(XI)

(XII)   (XIII)

Again, cyclization of the acid (XIV) was found† to take place mainly at the β-position of the naphthalene ring to give the 1:2-benzanthracene derivative (XV), and only to a lesser extent at the more 'reactive' α-position to give the 3:4-benzophenanthrene derivative (XVI). This is an important result, for preferential β- substitution in naphthalene is very rare, and suggests that the process is of a free-radical character.

* Weitzenböck and Klingler, *Mh. Chem.* 1918, **39**, 315.
† Cook, *J. Chem. Soc.* 1931, p. 2524.

Recent investigations, particularly by Hey and his co-workers, have provided further evidence for the free-radical character of the cyclization and, moreover, have introduced certain modifications and improvements which were suggested by analogy with the Gomberg reaction. Hey and Osbond* converted *ortho*-amino-α-phenylcinnamic acid into phenanthrene-9-carboxylic

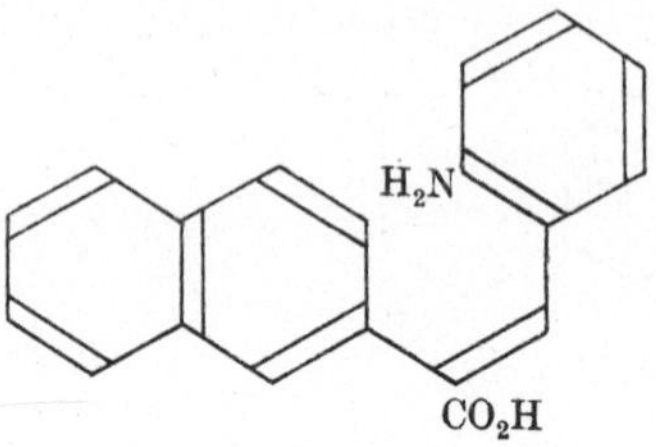

(XIV)

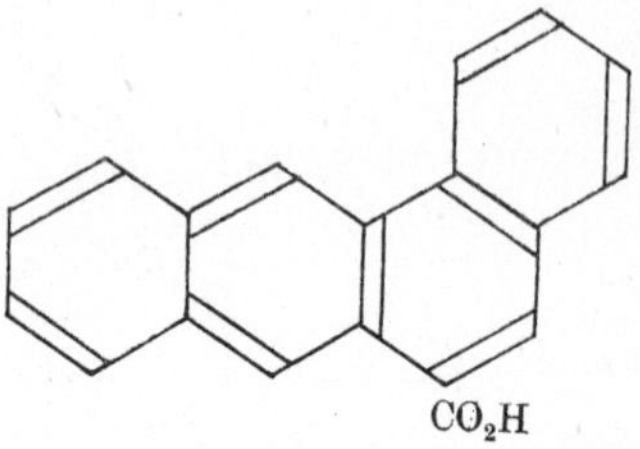

(XV)

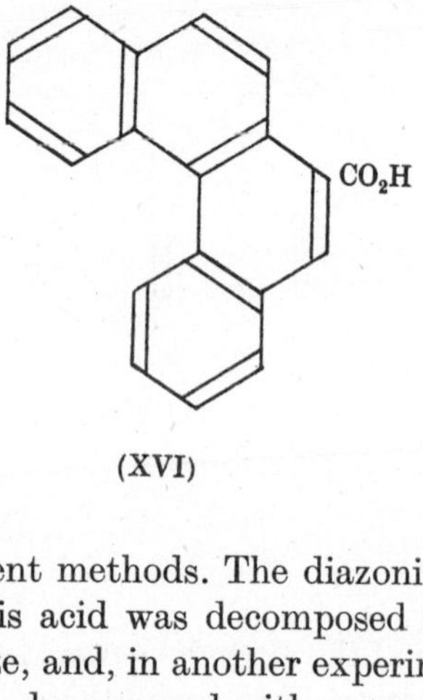

(XVI)

acid by six different methods. The diazonium chloride (XVII) prepared from this acid was decomposed in aqueous solution with copper bronze, and, in another experiment, the dry diazonium chloride was decomposed with copper under acetone. In the latter case the diazonium chloride appeared to be stable until a trace of copper was added. Both methods gave the required product in satisfactory yield. In a third modification the diazo solution was decomposed by the gradual addition of aqueous sodium hydroxide; and the reaction was also effected by the addition of aqueous sodium acetate in place of the sodium

* *J. Chem. Soc.* 1949, p. 3164.

AROMATIC SUBSTITUTION REACTIONS

hydroxide, at room temperature. The fifth method was analogous to the nitrosoacetanilide method in the Gomberg reaction. The *ortho*-amino-α-phenylcinnamic acid was acetylated and then nitrosated, to give *ortho*-nitrosoacetamido-α-phenylcinnamic acid (XVIII). On warming in benzene this gave phenanthrene-carboxylic acid (XX). The sixth method involved the conversion

of the amino acid into the triazen (XIX), and subsequent treatment with hydrogen chloride. The success of all these methods provides strong circumstantial evidence for the participation of free radicals.

In this connexion it is also significant that Hey and Osbond have successfully carried out a Pschorr type of synthesis involving cyclization into a pyridine ring, as in (XXI)–(XXII).

Finally, Hey and Osbond* have demonstrated that Pschorr-type reactions can be carried out on a number of *ortho*-amino-α-phenylcinnamic acids which are substituted with deactivating groups in the α-phenyl ring. It seems that the facility of inter-nuclear cyclization is not dependent to any appreciable extent on either the nature or positions of substituent groups in the α-phenyl ring; and this behaviour is also reminiscent of radical substitution. As examples of this cyclization into a deactivated ring, it may be mentioned that *ortho*-amino-α-(*ortho*-nitro-

phenyl)cinnamic acid (XXIII) was successfully cyclized to 1-nitrophenanthrene-10-carboxylic acid (XXIV), and that *ortho*-amino-α-(*para*-cyanophenyl)cinnamic acid (XXV) was converted into 3-cyanophenanthrene-10-carboxylic acid (XXVI).

In spite of these successes it is noteworthy that DeTar and Sagmanli† found that the decomposition of the diazonium salt from 2-aminobenzophenone in *acidic* media gave fluorenone in 80% yield, but that the decomposition in alkaline media (which would be expected to favour the free-radical mechanism) gave only traces of fluorenone. Similar results were obtained in other

* J. Chem. Soc. 1949, p. 3172.
† J. Amer. Chem. Soc. 1950, 72, 965.

cases, and DeTar and Sagmanli interpreted the results as indicating that the reaction proceeds by an ionic mechanism *in acidic media*. This observation is of some significance and further work is clearly required.

THE ELBS REACTION

The Elbs reaction* involves the pyrolysis of *ortho*-methyl diarylketones (as (I)), whereby water is split out and cyclization occurs to give aromatic hydrocarbons (as (II)). It may therefore be regarded as a special type of substitution reaction.

(I) (II)

(III) (IV)

The reaction has proved of very great value in the synthesis of polycyclic aromatic hydrocarbons, especially in the hands of Clar, Fieser and Cook.† As a general rule the yields are not good, averaging about 10–50 %, but this is usually offset by the ready availability of the ketones. The pyrolysis temperature of 400–450° often results in extensive charring, and for this reason zinc dust is sometimes added to the melt; but it is doubtful whether this improves the yield. It is interesting that the ketones (III) and (IV) both give 1:2-benzanthracene, but the cyclization of

* Elbs *et al. Ber. dtsch. chem. Ges.* 1884, **17**, 2847; 1885, **18**, 1797; 1886, **19**, 408; *J. prakt. Chem.* 1886, **33**, 180; 1887, **35**, 465; 1890, **41**, 1; 1890, **41**, 121.
† For a review see Fieser, *Organic Reactions*, **1**, 129–54, John Wiley, New York, 1942.

the former, involving reaction at a β-naphthalene position, takes place much more readily, and in better yield, than the latter, which involves cyclization into the phenyl ring.

A free methyl group is not essential to the reaction, for many aroylindanes have been cyclized without difficulty. For example, 7-methyl-4-(α-naphthoyl)-indane (V) gave the highly carcinogenic hydrocarbon, methylcholanthrene (VI), in about 50% yield.*

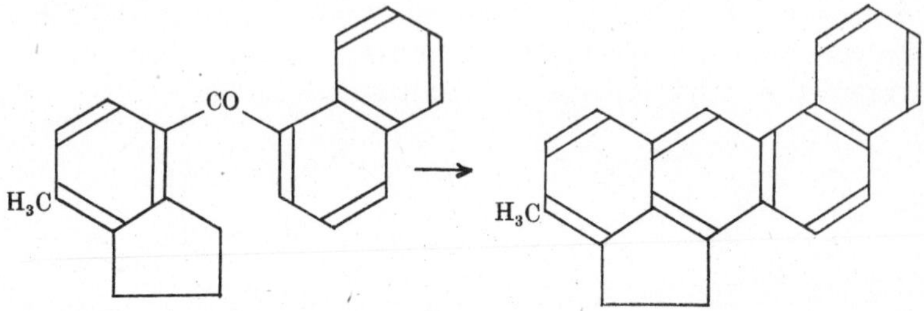

(V)   (VI)

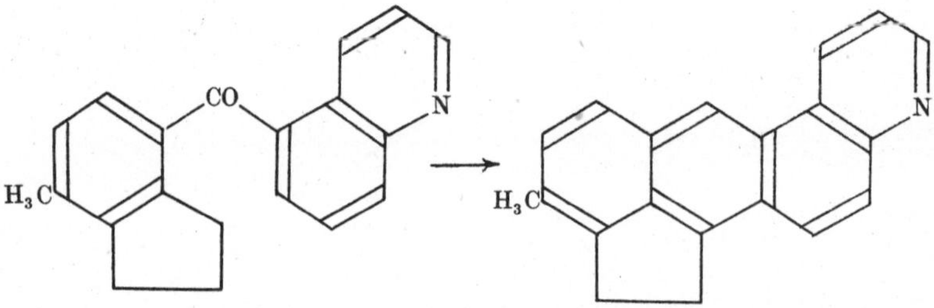

(VII)   (VIII)

Many other similar reactions have also been carried out, an interesting one being described by Fieser and Hershberg,† and involving cyclization of (VII), to give 20-methyl-4-azacholanthrene (VIII).

The utility of the reaction is not confined to mono-ketones, but double cyclizations can also be brought about. For example,

* Fieser and Seligman, *J. Amer. Chem. Soc.* 1935, **57**, 942; 1936, **58**, 2482; Bachmann, *J. Org. Chem.* 1937, **1**, 347.
† *J. Amer. Chem. Soc.* 1940, **62**, 1640.

2:6-dimethyl-1:5-di(2'-naphthoyl)naphthalene (IX) has been converted into the complex high-melting hydrocarbon 2:3:8:9-di-(naphtho-1':2')chrysene (X), in good yield;* and 4:6-dibenzoyl-1:3-xylene (XI) also undergoes double cyclization with the formation of dihydropentacene (XII).† The formation of the dihydride, rather than the expected fully aromatic structure,

is interesting. The additional hydrogen is no doubt furnished by other molecules which undergo decomposition at the elevated temperature of the reaction, the formation of the dihydride being facilitated by the enhanced reactivity of the aromatic compound, pentacene.

One of the most interesting applications of the Elbs reaction is to the preparation of the carcinogen 1:2:5:6-dibenzanthracene

---

* Fieser and Dietz, *Ber. dtsch. chem. Ges.* 1929, **62**, 1827.
† Clar and John, ibid. 1929, **62**, 3021; 1931, **64**, 981.

(XV). Cook* showed that 2-methyl-1:1′-dinaphthyl ketone (XIII) and 2-methyl-1:2′-dinaphthyl ketone (XIV) do not yield isomeric hydrocarbons on pyrolysis, as expected, but that 1:2:5:6-dibenzanthracene (XV) is the major product in each case. The first ketone (XIII) must suffer rearrangement before cyclization, and this is no doubt associated with the fact that, in this case, the methylnaphthoyl group is attached to an α-naphthalene

(XI)

(XII)

position. 1-Acyl- and 1-aroyl-naphthalenes are known to rearrange to the β- isomers under the influence of catalysts, and it is not unlikely that similar rearrangements can occur at especially elevated temperatures. It has also been found that the pyrolysis of both ketones gives some of the yellow-orange hydrocarbon, 1:2-benzonaphthacene (XVI), and this can only arise *directly* from the ketone (XIV), by cyclization into the free β- rather than into the free α-position.

The major product can be obtained in a pure colourless condition, freed from the yellow benzonaphthacene, by taking advantage of the enhanced reactivity of the latter. It is sulphonated, or oxidized, more rapidly than 1:2:5:6-dibenzanthracene; it reacts more rapidly with maleic anhydride; and it is more strongly adsorbed on alumina, thereby facilitating chromatographic separation.

* *J. Chem. Soc.* 1931, p. 489; see also Clar, *Ber. dtsch. chem. Ges.* 1929, **62**, 350; Fieser and Dietz, ibid. 1929, **62**, 1827.

AROMATIC SUBSTITUTION REACTIONS

Other rearrangements of a similar type are also known,* and the usefulness of the reaction is also limited by the fact that alkyl groups are sometimes eliminated, either partially or completely, during pyrolysis. Such methyl groups are lost most readily from *meso* and α-positions of the formed hydrocarbons.

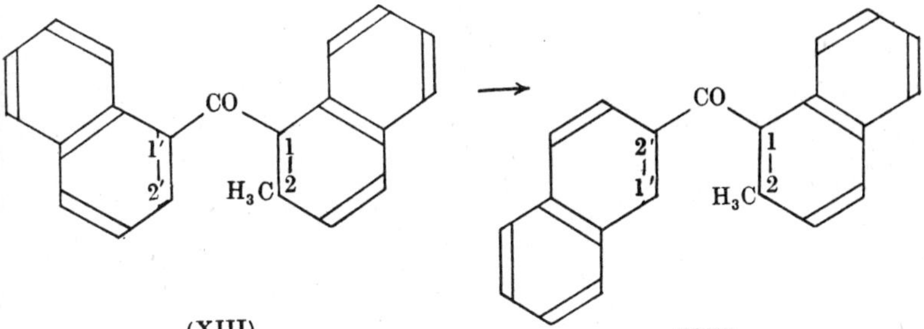

(XIII) → (XIV)

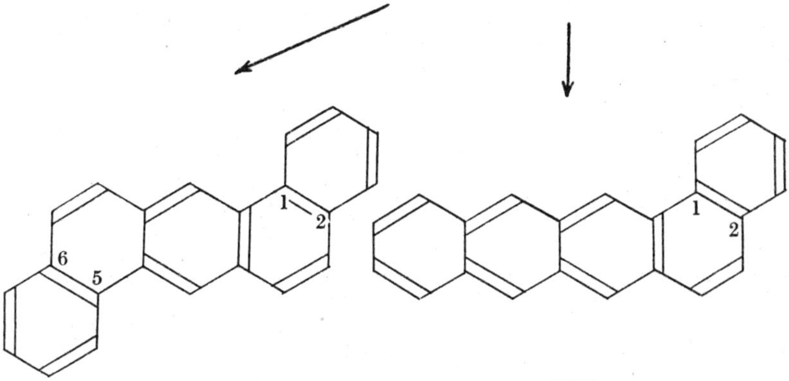

(XV) major product

(XVI) minor product

Pyrolysis of the ethyl ketone (XVII), for example, gave 1:2:5:6-dibenzanthracene by loss of a methyl group,† and the methyl groups indicated by asterisks in the ketones (XVIII) and (XIX) were also eliminated during pyrolysis.‡ Moreover, *iso*propyl groups are sometimes degraded to methyl.

---

* See Clar, *Aromatische Kohlenwasserstoffe*, 2nd ed., Springer-Verlag, Berlin, 1952.
† Fieser and Newman, *J. Amer. Chem. Soc.* 1936, **58**, 2376.
‡ Cook, *J. Chem. Soc.* 1932, p. 456; 1931, p. 489.

The mechanism of the Elbs reaction is entirely unknown. Cook* postulated primary enolization of the ketone (XX), and intramolecular cyclization of the enol (XXI) to give a dihydroanthranol (XXII), which would then suffer dehydration at the elevated temperature.

(XVII)

(XVIII)

(XIX)

(XX)

(XXI) → (XXII)

* J. Chem. Soc. 1931, p. 487.

AROMATIC SUBSTITUTION REACTIONS

Alternatively, Fieser and Dietz* suggested that the reaction proceeds by forced 1:4 addition of the methyl group to the conjugated system of the ketone, again with the formation of a dihydroanthranol, as in (XXIII)–(XXV).

(XXIII)   →   (XXIV)   →

(XXV)

One aspect of the reaction mechanism has been investigated in some detail by Hurd and Azorlosa.† Using deuterium-substituted *ortho*-methylbenzophenones, it has been shown that the hydrogen atom appearing at the 9-position in the anthracene formed comes from the *ortho* nuclear position, and not from the methyl group. Assuming the mechanism of Cook, therefore, the reaction can be illustrated by the scheme (XXVI)–(XXIX).

It is interesting to record the various by-products which have been encountered. Anthrones are sometimes found in the reaction mixtures.‡ Phenanthrene has been obtained from a phenanthrylaryl ketone§ and anthracene from an anthrylaryl ketone.∥ Benzoic acid and benzaldehyde have been detected after pyrolysis of 1:5-dibenzoyl-2:6-dimethylnaphthalene.¶ Heidelberger, Brewer and Dauben** found that 2-methyl-1:2′-di-

---
* Fieser and Dietz, *Ber. dtsch. chem. Ges.* 1929, **62**, 1827.
† *J. Amer. Chem. Soc.* 1951, **73**, 37.
‡ For example, Morgan and Coulson, *J. Chem. Soc.* 1929, p. 2551; Fieser and Peters, *J. Amer. Chem. Soc.* 1932, **54**, 3742.
§ Fieser and Dietz, *Ber. dtsch. chem. Ges.* 1929, **62**, 1827.
∥ Clar, John and Hawran, *Ber. dtsch. chem. Ges.* 1929, **62**, 940.
¶ Clar, Wallenstein and Avenarius, ibid. 1929, **62**, 950.
** *J. Amer. Chem. Soc.* 1947, **69**, 1389.

330 THE AROMATIC COMPOUNDS

naphthyl ketone labelled at the carbonyl carbon atom with radioactive carbon pyrolysed to give radioactive 1:2:5:6-dibenzanthracene, but they found that the effluent gas from the pyrolysis was also highly radioactive, indicating the volatilization of some of the carbonyl carbon.

All these facts indicate that considerable and extensive decomposition occurs during the pyrolysis, and it seems likely that a radical mechanism is involved. The fact that all attempts to find a catalyst for the reaction have failed also supports this viewpoint; but further work is necessary before any conclusion can be reached.

THE *ORTHO*:*PARA* RATIO

The basic principles of aromatic substitution seem to be well established, but the observed variations in the *ortho*:*para* ratio are still somewhat obscure.

If a given substituent in benzene has the same activating (or deactivating) influence on the *ortho* and *para* positions, then the amount of the *ortho*-disubstituted product formed should be twice that of the *para* product, as there are two available *ortho* positions and only one *para* position. The *ortho*:*para* ratio should be 2, or, alternatively, the ½ *ortho*:*para* ratio should be 1. This, however, is hardly ever observed; for some substituents the ratio is less than unity (*para* position favoured), and for others it is greater than unity (*ortho* position favoured).

It is true that the inductive and inductomeric effects are short-range effects, and may be expected to influence the *ortho* positions to a greater extent than the more distant *para* positions. Early attempts to explain the variations in the *ortho*:*para* ratio were largely based on this fact.[*] On the other hand, all the observed variations (table 7·8) cannot be explained in this way.

TABLE 7·8. *Ratio of* ortho *to* para *substitution accompanying nitration*[†]

Compound	% ortho	% meta	% para	½ ortho:para ratio
Ph.CH$_3$	58·8	4·4	36·8	0·80
Ph.CH$_2$Cl	40·9	4·2	54·9	0·37
Ph.CHCl$_2$	23·3	33·8	42·9	0·27
Ph.CCl$_3$	6·8	64·5	28·7	0·12
Ph.I	41·1	—	58·7	0·35
Ph.Br	37·6	—	62·4	0·30
Ph.Cl	30·1	—	69·9	0·22
Ph.F	12·4	—	87·6	0·07
Ph.Ph	53	—	47	0·56
Ph.NHAc	40·7	—	59·3	0·34
Ph.OMe	20	—	80	0·12
Ph.CO$_2$Et	28·3	68·4	3·3	4·3
Ph.CO$_2$H	18·5	80·2	1·3	7·1
Ph.NO$_2$	6·4	93·2	0·25	12·8

The interpretation of the problem is to some extent complicated by the fact that the ratio may depend on the nature of the attacking reagent. For example, de la Mare[‡] has pointed out that although bromination of acetanilide gives *para*-bromoacetanilide almost exclusively, chlorination also gives much of the *ortho* derivative. Again, halogenation of toluene proceeds about 400 times faster than that of benzene, whereas the nitration of toluene is only about 25 times more rapid than that of benzene.[§] Similarly, *para*-halogenation of fluorobenzene

[*] Lapworth and Robinson, *Mem. Manch. Lit. Phil. Soc.* 1927, **72**, 43.
[†] Data collected by Dewar, *J. Chem. Soc.* 1949, p. 463.
[‡] de la Mare, ibid. 1949, p. 2871.
[§] de la Mare and Robertson, ibid. 1943, p. 279.

332   THE AROMATIC COMPOUNDS

is more rapid than that of benzene, but *para*-nitration of fluorobenzene is less rapid.

The variations in the *ortho*:*para* ratio are almost certainly dependent on several factors, and the following may be mentioned:

(i) The π-electron densities at the *ortho* and *para* positions may not be the same. This may arise from the fact that $+I$ substituents may be expected to activate the *ortho* positions to a greater extent than the *para* positions, and that $-I$ substituents will preferentially deactivate the *ortho* positions. In the absence of other factors then, $+I$ substituents should lead to preferential *ortho* substitution, and $-I$ substituents to *para* substitution. This is evidently the explanation for the fact that in nitration of halogenated benzenes the $\frac{1}{2}$ *ortho*:*para* ratio decreases from iodobenzene to fluorobenzene.

(ii) In the case of *ortho-para*-directing substituents, the quasiquinonoid transition state for *para* substitution (II) is more stable than that for *ortho* substitution (I), so that preferential *para* substitution is to be expected.*

(iii) On the other hand, in the case of the *meta*-directing substituents, the preferential stability of the *para*-quinonoid resonance structures, such as (III), may lead to greater de-

* Waters, *J. Chem. Soc.* 1948, p. 727.

activation of the *para* position, and hence preferential substitution in the *ortho* position.*

(iv) Steric effects may be expected to favour *para* substitution. Even small groups, such as methyl groups, appear to exert a slight effect on the *ortho* positions, and with larger substituents the effect probably becomes of paramount importance. The size of the attacking reagent may also be an important factor; *ortho* substitution should be favoured by a small rather than by large reagents.

(v) Chelation may also favour *ortho* substitution in the transition state between the reagent and the substituent. This would seem to be most likely in the case of sulphonation or nitration of amines, phenols, etc.†

* de la Mare, ibid. 1949, p. 2871.   † Waters, ibid. 1948, p. 727.

CHAPTER 8

THE DIELS-ALDER REACTION

SURVEY OF THE REACTION WITH AROMATIC DIENES

The addition of ethylenic or acetylenic compounds to the 1:4-positions of a conjugated diene according to the equation (I)–(III) is known as the 'Diels-Alder reaction' or the 'diene synthesis'. A great variety of dienes and dienophiles undergo the reaction, and the method is therefore a very useful synthetic tool. Several excellent reviews of the reaction have been published.*

Of the ethylenic components, maleic anhydride is possibly the most important; but many other dienophilic reagents have been used, including dibromomaleic anhydride, citraconic anhydride, acrylic acid, acrolein, cinnamic acid, *cyclo*hexene, indene and *para*-benzoquinone. Similarly, many dienes have been used. With acyclic dienes, the reaction usually proceeds with great facility, and excellent yields of the adducts are obtained. Even with less reactive dienes, however, the reaction proceeds at a useful rate. With a given dienophile, the ease of reaction appears to be governed by the 'character' of the double bonds forming the diene. This is illustrated by the fact that although 1:2-dihydrobenzene functions as a diene in the Diels-Alder synthesis, benzene does not. In the former case the bonds are ethylenic in character and constitute a reactive diene; but in benzene the bonds are all aromatic and relatively unreactive.

* Alder in *Newer Methods of Preparative Organic Chemistry*, pp. 381–511, Interscience, 1948; Kloetzel, *Organic Reactions*, **4**, 1–59, John Wiley, New York; Holmes, ibid. pp. 60–173; Butz and Rytina, *Organic Reactions*, **5**, 136–92, John Wiley, New York; Norton, *Chem. Rev.* 1942, **31**, 319–523.

Compounds having one ethylenic double bond conjugated with an aromatic double bond can often function as dienes in the Diels-Alder reaction, particularly if the aromatic double bond has pronounced double-bond character. With styrene and its simple derivatives the reaction is usually complicated by the tendency to polymerize. Moreover, styrene (V) also functions as a dienophile (rather than as a diene), for it condenses with 2:3-dimethylbutadiene (IV) to give 3:4-dimethyl-1:2:5:6-tetrahydrodiphenyl (VI). Certain derivatives of styrene do, however, function satisfactorily as dienes. This is particularly the case when there is a *para*-alkoxy group, and when there is a β-alkyl group on the double bond. These conditions are fulfilled in *iso*eugenol methyl ether (VII) and also in *iso*safrole (VIII),* both of which function satisfactorily as dienes. On the other hand, adducts are not formed with *meta*-methoxystyrene, *para*-methoxystyrene or with 3:4-methylenedioxystyrene.

It is of some interest that 1:1-diphenylethylene (IX) adds *two* molecules of maleic anhydride, to give (XI) probably via the intermediate (X).†

The phenyl ring in these aromatic-aliphatic dienes may be replaced by other aromatic ring systems, often with a pronounced increase in the 'reactivity' of the resulting diene. Fairly reactive dienes are obtained, for example, by utilization of the 1:2 bond in naphthalene, or the 9:10 bond in phenanthrene, as the aromatic component of the diene. 1-Vinylnaphthalene (XII), for example,

* Hudson and Robinson, *J. Chem. Soc.* 1941, p. 715.
† Wagner-Jauregg, *Liebigs Ann.* 1931, **491**, 1.

has been used extensively as a diene. In boiling toluene or xylene, it adds maleic anhydride to give a 1:2:3:11-tetrahydrophenanthrene derivative (XIII), which is isomerized to a 1:2:3:4-tetrahydrophenanthrene derivative (XIV) by boiling with acetic acid containing hydrogen chloride.* If the reaction is carried out in acetic acid instead of in a hydrocarbon solvent, the aromatic structure is obtained directly.

Bachmann and Deno† have studied the addition of a number of $\alpha\beta$- and $\alpha\beta\gamma\delta$- unsaturated acids to 1-vinylnaphthalene. In most of the reactions two structural isomers were possible, and the isomer actually formed was found to be that expected by

* Cohen, *Nature, Lond.*, 1935, **136**, 869; Cohen and Warren, *J. Chem. Soc.* 1937, p. 1315.
† *J. Amer. Chem. Soc.* 1949, **71**, 3062.

application of the general principle* that the most nucleophilic carbon atom of the diene becomes linked to the most electrophilic carbon atom of the dienophile. That is, 1-vinylnaphthalene (XV) reacted with unsaturated acids (such as (XVI)) in acetic acid or in propionic acid solution, to give 1:2:3:4-tetrahydrophenan-

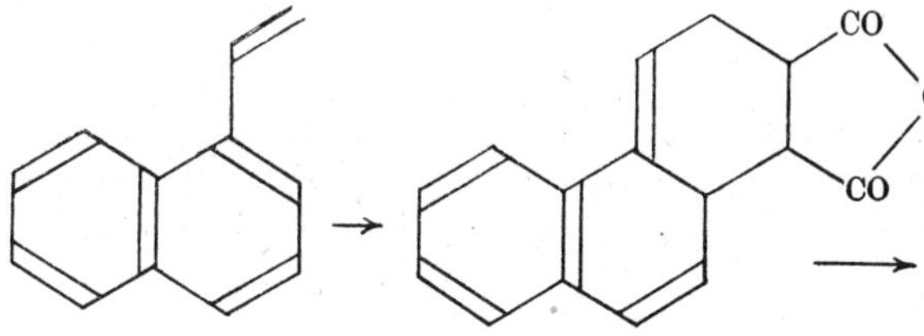

(XII) (XIII)

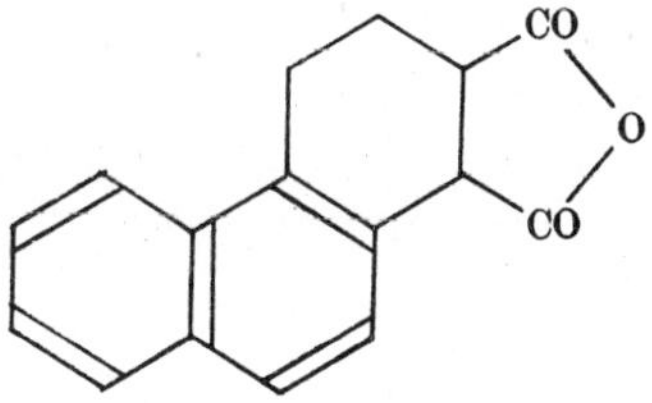

(XIV)

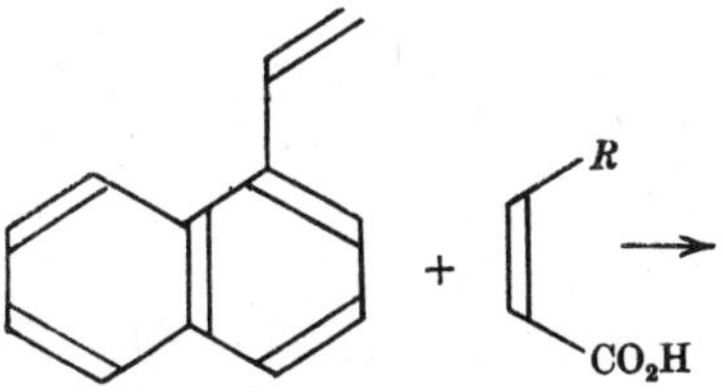

(XV) (XVI)

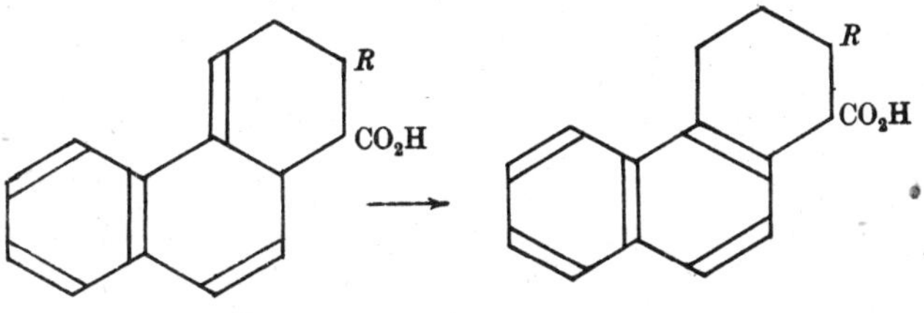

(XVII) (XVIII)

* Hudson and Robinson, *J. Chem. Soc.* 1941, p. 715.

338   THE AROMATIC COMPOUNDS

threne-1-carboxylic acids (XVIII), and *not* the alternative corresponding 2-carboxylic acids.

Bachmann and Deno also found that 1-vinylnaphthalene reacts both as a diene (XIX) and as a dienophile (XX), for when

refluxed alone in propionic acid, dimerization occurred, and 1-(1'-naphthyl)-1:2:3:4-tetrahydrophenanthrene (XXI) was isolated.

2-Vinylnaphthalene also functions as a diene.* In this case either the 1:2 or 2:3 bond of naphthalene could function as the

* Cohen, *Nature, Lond.*, 1935, **136**, 869; Cohen and Warren, *J. Chem. Soc.* 1937, p. 1315.

THE DIELS-ALDER REACTION

'double bond', but only the former does so, for the product was found to be a tetrahydrophenanthrene derivative. This result may be cited as evidence that the 1:2 bond of naphthalene has considerably more double-bond character than the 2:3 bond.

Other vinyl- derivatives of aromatic compounds also behave as dienes. These include 9-vinylphenanthrene (XXII), 9-*iso*propenylphenanthrene (XXIII), *cyclo*pentenylphenanthrene and 1-α-naphthyl-1-*cyclo*pentene (XXIV). The latter compound reacts nearly quantitatively with maleic anhydride.

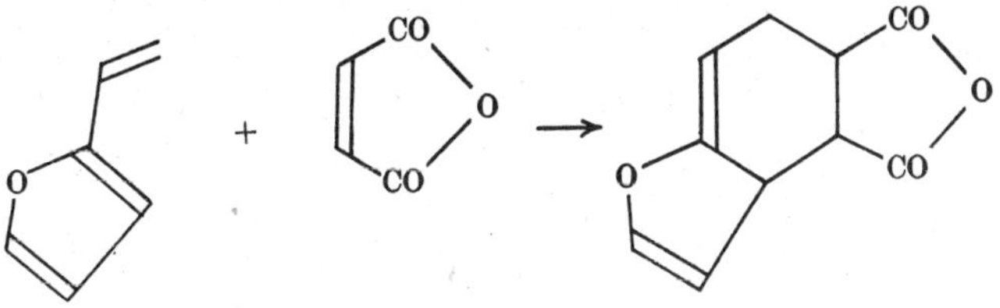

(XXV)   (XXVI)   (XXVII)

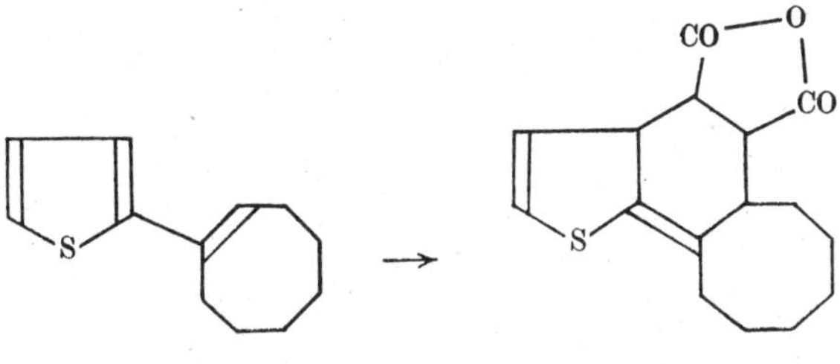

(XXVIII)   (XXIX)

Vinyl aromatic compounds in which the aromatic double bond is provided by a heterocyclic compound are of particular interest. α-Vinylfuran (XXV), for example, reacts with maleic anhydride (XXVI) without difficulty,[*] to give the expected adduct (XXVII).

Dienes composed of an alicyclic double bond and one double bond of the thiophen nucleus are also effective.[†] Maleic anhydride adds normally to 1-(2'-thienyl)-*cyclo*oct-1-ene (XXVIII) for example, to give the adduct (XXIX).

[*] Paul, *C.R. Acad. Sci., Paris*, 1939, **208**, 1028.
[†] Szmuszkovicz and Modest, *J. Amer. Chem. Soc.* 1950, **72**, 571.

Heterocyclic compounds such as furan, thiophen and pyrrole may be considered as dienes, and furan does react as such in this synthesis. It is the simplest aromatic substance to function as a diene, and this indicates that the two 'double bonds' in this compound have pronounced double-bond character. The interaction of furan (XXX) and maleic anhydride under mild conditions gives 3:6-*endo-oxo*-1:2:3:6-tetrahydrophthalic anhydride (XXXI), which on hydrogenation yields 3:6-*endo-oxo*-hexahydrophthalic anhydride (XXXII).*

(XXX)   (XXXI)   (XXXII)

(XXXIII)   (XXXIV)

Many furan derivatives, including alkyl- and halogeno-substituted compounds, yield similar adducts, but the presence of carbethoxy, cyano and nitro substituents appears to prevent the reaction.† It is interesting that 2-ethylfuran (XXXIII) undergoes normal addition with maleic anhydride to give 3-ethyl-3:6-*endo-oxo*-1:2:3:6-tetrahydrophthalic anhydride (XXXIV),‡ for, as already mentioned, 2-vinylfuran (XXV) undergoes the alternative reaction to give an adduct which does not contain an oxygen bridge.

The adducts formed from furan derivatives are often very labile, and readily dissociate into the components when warmed in solvents, or when heated alone. This is not surprising, because

---

* von Bruchhausen and Bersch, *Arch. Pharm.* 1928, **266**, 697; *Chem. Abstr.* 1929, **23**, 1647; Diels, Alder and Naujoks, *Ber. dtsch. chem. Ges.* 1929, **62**, 554.
† van Campen and Johnson, *J. Amer. Chem. Soc.* 1933, **55**, 430.
‡ Paul, *Bull. Soc. chim. Fr.* 1943, **10**, 163.

the Diels-Alder synthesis has been shown to be reversible, and the instability of certain adducts merely means that the equilibrium

$$\text{diene} + \text{dienophile} \rightleftharpoons \text{adduct}$$

lies on the side of the cleavage products. Stable adducts are formed when the equilibrium lies far over on the right-hand side. With furan (XXXV) and pyrocinchonic anhydride (XXXVI), the equilibrium evidently lies far over on the left-hand side. This reaction has been attempted under a wide variety of conditions (for the expected adduct would give cantharidin (XXXVIII) on hydrogenation); but no adduct has been obtained. That the equilibrium lies too far on the side of the components rather than the adduct is supported by the observation that dehydrogenation of cantharidin yields furan and pyrocinchonic anhydride.

(XXXV)   (XXXVI)   (XXXVII)   (XXXVIII)

The adducts from furans and dienophiles such as maleic anhydride are readily converted into benzene derivatives by treatment with hydrogen bromide in acetic acid, and this process is valuable both to prove the structure of the adduct and as a synthetic procedure.* For example, 3-bromofuran and maleic anhydride give an adduct which is easily converted into 4-bromophthalic acid.

With maleic anhydride and other dienophiles, the *iso*benzofurans yield adducts which may be converted into derivatives of naphthalene. Dufraisse and Priou,† for example, prepared 1:4-diphenylnaphthalene (XLII) by addition of maleic anhydride to 1:3-diphenyl*iso*benzofuran (XXXIX) to give the adduct (XL), elimination of H_2O to give (XLI), and subsequent decarboxylation.

---

* van Campen and Johnson, *J. Amer. Chem. Soc.* 1933, **55**, 430.
† *Bull. Soc. chim. Fr.* 1938, **5**, 502; Barnett, *J. Chem. Soc.* 1935, p. 1326.

1:3-Diphenyl*iso*benzofuran (XXXIX) also forms adducts with *para*-benzoquinone and with 1:4-naphthaquinone. The adduct (XLIII) with the latter quinone was found to be unstable in the presence of hydrogen ions, dehydration taking place giving diphenylnaphthacenequinone (XLIV).* This synthesis is of interest, for treatment of the quinone with phenylmagnesium bromide, followed by reduction of the resulting diol (XLV), gave the important hydrocarbon, rubrene (XLVI).

Thiophen is intermediate between benzene and furan in reactivity, and does not undergo the Diels-Alder reaction. However, certain thiophen derivatives are reactive. Clapp[†] found that maleic anhydride reacts with 2:3:4:5-bis(1:8-naphthylene)thiophen (XLVII). The addition product (XLVIII)

---

* Bergmann, *J. Chem. Soc.* 1938, p. 1147; Dufraisse and Compagnon, *C.R. Acad. Sci., Paris*, 1938, **207**, 585.
† *J. Amer. Chem. Soc.* 1939, **61**, 2733.

THE DIELS-ALDER REACTION

evidently lost hydrogen sulphide for the expected substituted phthalic anhydride (XLIX) was obtained. Similarly, Allen and Gates* found that 1:3:5:6-tetraphenylisobenzothiophen (L) adds maleic anhydride to give an adduct (LI) having a sulphur bridge.

Pyrroles might be expected to undergo the Diels-Alder addition in the same way as furans, especially as the former are so very reactive, but this is not the case. Reaction between pyrroles and dienophiles does occur in many cases, but the normal adducts are not formed.† Pyrrole (LII) itself reacts with maleic anhydride (LIII), in aqueous medium, to give ammonia, carbon dioxide and a diketo acid (LIV). In addition, a small amount of 2:5-pyrrolylenedisuccinic acid is formed.

Similarly, N-methylpyrrole (LV) reacts with maleic anhydride to form N-methyl-α-pyrrylsuccinic anhydride (LVI).

In addition to these heterocyclic compounds which have been used as dienes, many aromatic *hydrocarbons* are sufficiently

---

* Ibid. 1943, **65**, 1283.
† Diels and Alder, *Liebigs Ann.* 1929, **470**, 62; 1931, **486**, 211; 1932, **498**, 1.

344 THE AROMATIC COMPOUNDS

reactive to react with maleic anhydride and other dienophiles. For example, benzanthrene (LVII) reacts readily with maleic anhydride to give an adduct (LVIII) which suffers rearrangement of hydrogen to give a hydrogenated *peri*-benzopyrene (LIX).*

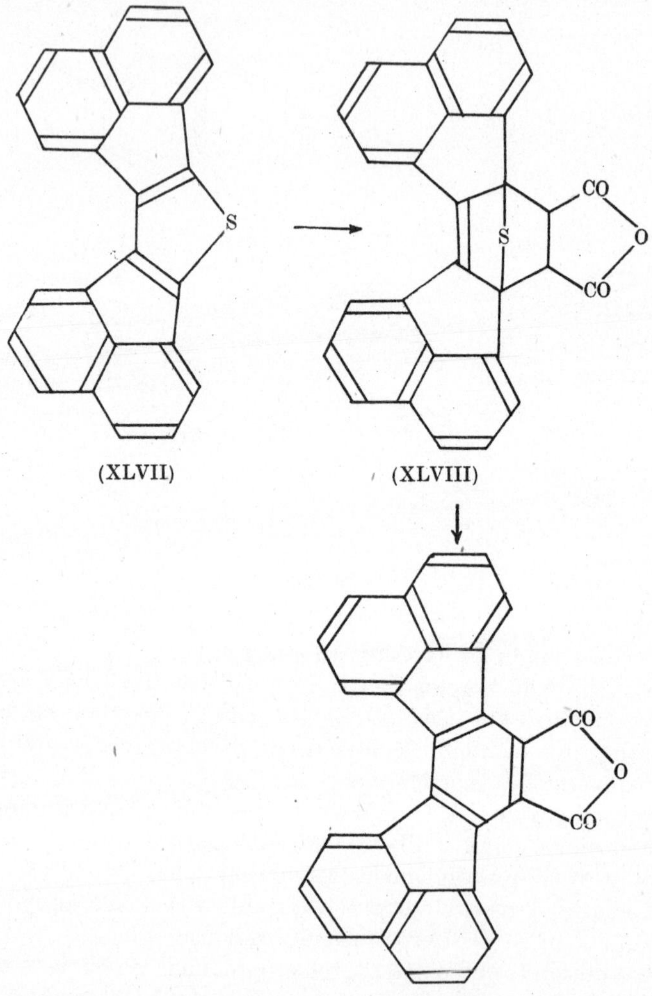

(XLVII)   (XLVIII)

(XLIX)

* Clar, *Ber. dtsch. chem. Ges.* 1932, **65**, 1425.

The addition of maleic anhydride to perylene (LX) takes place in weakly oxidizing solution (e.g. in boiling nitrobenzene), the adduct (LXI) evidently formed being converted into 1:12-benzoperylenedicarboxylic anhydride (LXII) by loss of hydrogen. 1:12-Benzoperylene was obtained by decarboxylation.*

It is of some interest that perylene does not add *two* molecules of maleic anhydride, and that 1:12-benzoperylene does not react with dienophiles at all. Evidently the attachment of an additional benzene ring in this way reduces the reactivity of the carbon atoms involved.

On the other hand, *meso*-anthradianthrene (LXIII) reacts with maleic anhydride in boiling nitrobenzene to give the

* German Patent, 651,677; Clar, *Ber. dtsch. chem. Ges.* 1932, **65**, 846; 1940, **73**, 351.

anhydride (LXIV); and *meso*-naphtha-dianthrene (LXV) reacts with two molecules of maleic anhydride to give the dianhydride (LXVI). Both anhydrides have been decarboxylated to the same hydrocarbon, ovalene (LXVII), which may be considered as a naphthalene ring system surrounded by eight benzene rings.*

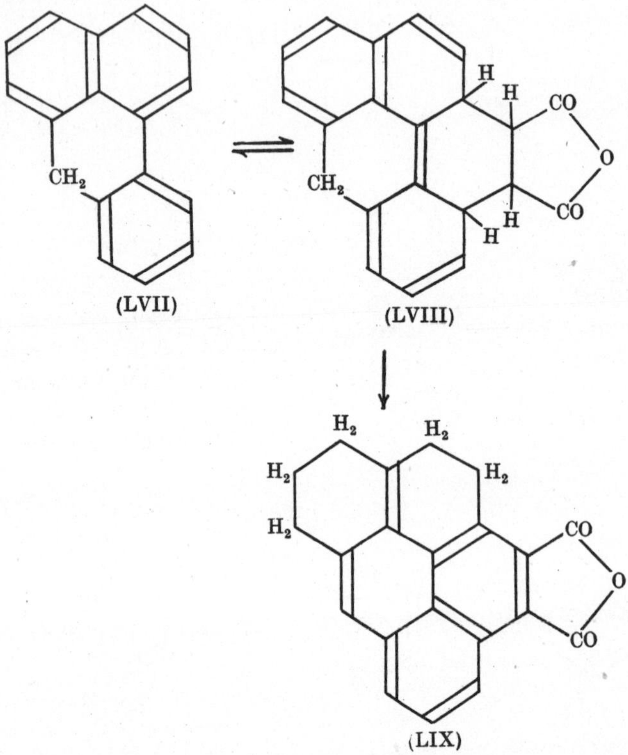

(LVII)   (LVIII)

(LIX)

Simple aromatic dienes such as diphenyl and phenanthrene are not sufficiently reactive to form adducts with maleic anhydride. The terminal positions of the diene in diphenyl (LXVIII), for example, have a free-valence number of 0·118.†

* Scholl and Meyer, *Ber. dtsch. chem. Ges.* 1934, **67**, 1236; Clar, ibid. 1932, **65**, 846; *Nature, Lond.*, 1948, **161**, 238.
† Baldock, Berthier and Pullman, *C.R. Acad. Sci., Paris*, 1949, **228**, 931; Buu-Hoï, Coulson, Daudel, Daudel, Martin, Pullman and Pullman, *Rev. Sci.* 1947, **85**, 1041; Coulson and Longuet-Higgins, ibid. 1947, **85**, 929.

In phenanthrene (LXIX), the terminal positions have a free-valence number of 0·122. The terminal positions in perylene (LXX) have a free-valence number of 0·135, and even here reaction can only be brought about under conditions such that the adduct is continuously removed (by oxidation) as a stable aromatic complex. Aliphatic dienes such as butadiene (LXXI)

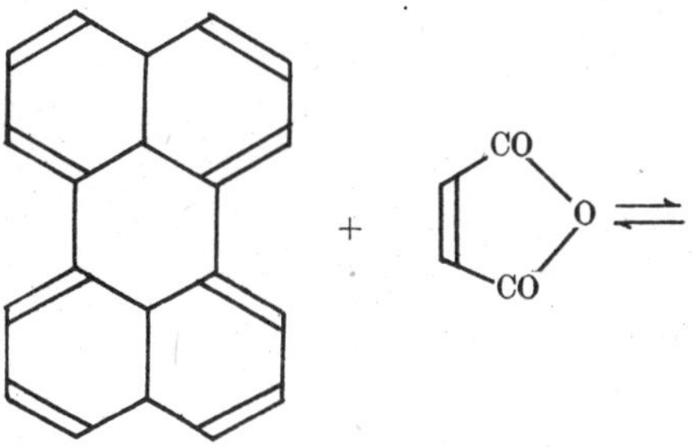

(LX)

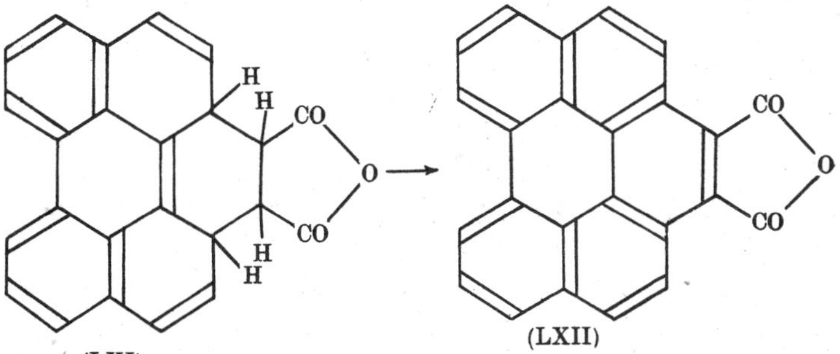

(LXI)   (LXII)

have smaller free-valence numbers, but they also have smaller polarization energies and these compounds are therefore very reactive.

As already mentioned, benzene does not undergo the diene synthesis. *para*-Xylene likewise is ineffective in this reaction, although maleic anhydride reacts with one of the methyl groups of this compound to give a substituted succinic anhydride.*
When naphthalene is fused with 30 molecular proportions of maleic anhydride at 100° for 24 hours, reaction occurs to the

* Bickford, Fisher, Dollear and Swift, *J. Amer. Oil Chem. Soc.* 1948, **25**, 251.

extent of less than 1%. In the presence of suitably disposed alkyl groups, however, good yields of adducts can be obtained. Kloetzel, Dayton and Herzog* found that 1:2:3:4-tetramethyl-

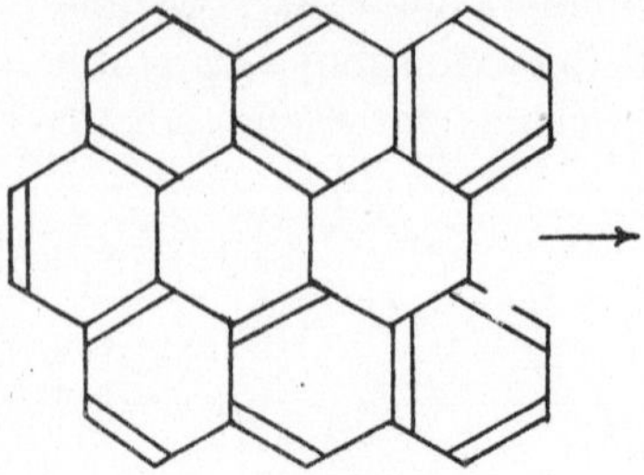

(LXIII)

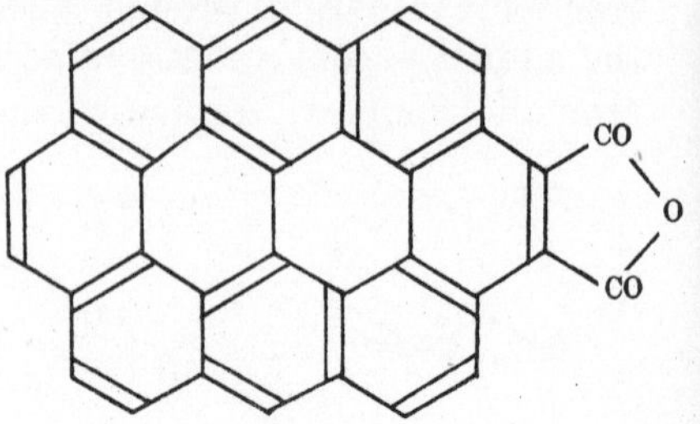

(LXIV)

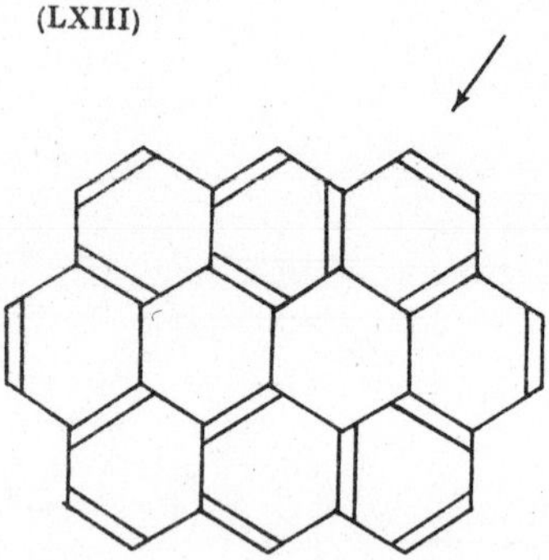

(LXVII)

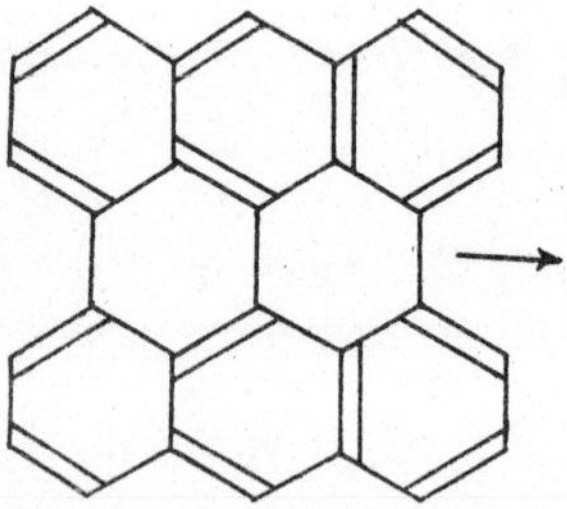

(LXV)

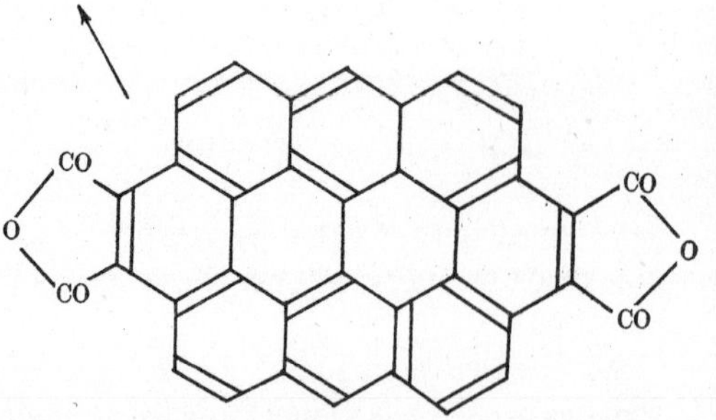

(LXVI)

naphthalene (LXXII) reacts with maleic anhydride to give 1:2:3:4-tetramethyl-1:4-dihydronaphthalene-1:4-*endo*-αβ-succinic anhydride (LXXIII). With equimolar quantities, the

* *J. Amer. Chem. Soc.* 1950, **72**, 273.

equilibrium is established (in boiling xylene) within 46 hours, the mixture containing 4·6–6·4% adduct. With a large excess of maleic anhydride, yields of up to 90% were obtained. In the same way, Kloetzel and Herzog* have shown that 1:2:4-triethylnaphthalene also gives an adduct with maleic anhydride.

(LXVIII)   (LXIX)   (LXX)   (LXXI)

(LXXII) ⇌ (LXXIII)

No positive results have been obtained with phenanthrene, chrysene or with pyrene, but anthracene (LXXIV) and its derivatives are reactive and undergo reaction with a variety of dienophiles. With maleic anhydride, 9:10-dihydroanthracene-9:10-endo-αβ-succinic anhydride was obtained,† and the pronounced reactivity of the meso-anthracene positions is indicated by the fact that 2-isopropenylanthracene undergoes addition exclusively at the 9:10 positions.‡ With anthracene itself, reaction also occurs with a variety of maleic anhydride deriva-

* J. Org. Chem. 1950, **15**, 370.
† Clar, Ber. dtsch. chem. Ges. 1931, **64**, 2194; Diels and Alder, Liebigs Ann. 1931, **486**, 191.
‡ Bergmann and Bergmann, J. Amer. Chem. Soc. 1940, **62**, 1699.

tives, with *para*-benzoquinone, with acetylenedicarboxylic ester, with acrolein, with allyl chloride and with vinyl acetate.

Many anthracene derivatives and anthracene benzologues undergo the reaction, although the position of the additional benzene rings has a profound effect on the reactivity, and on the velocity of the addition. 1:2-Benzanthracene (LXXV), for example, reacts with maleic anhydride at the positions indicated, but less readily than does anthracene (LXXIV) itself. 1:2:5:6-Dibenzanthracene (LXXVI) reacts even less readily. On the other hand, naphthacene (LXXVII) reacts very readily indeed with maleic anhydride, and pentacene (LXXVIII) and hexacene react almost instantaneously.*

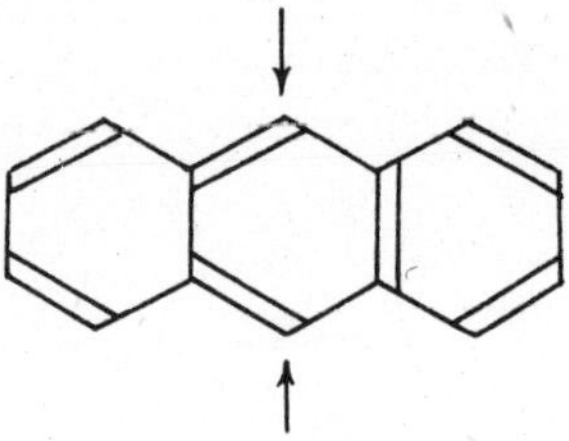

(LXXIV) 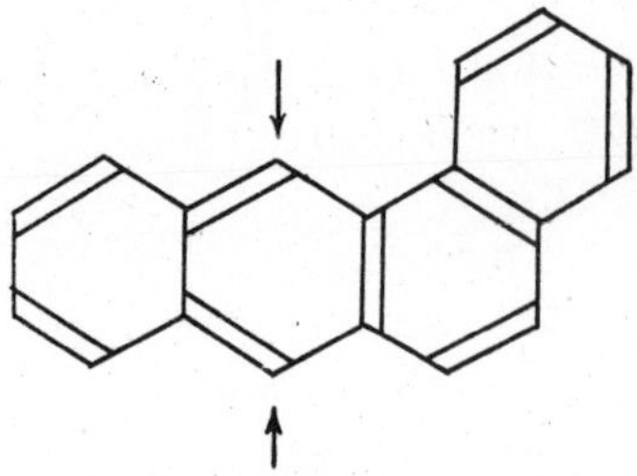 (LXXV)

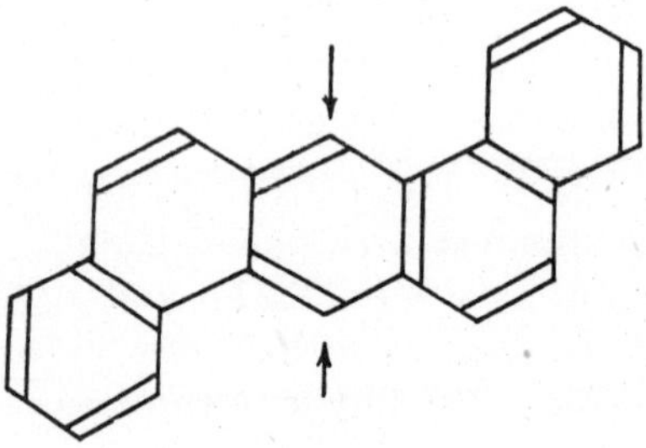

(LXXVI) 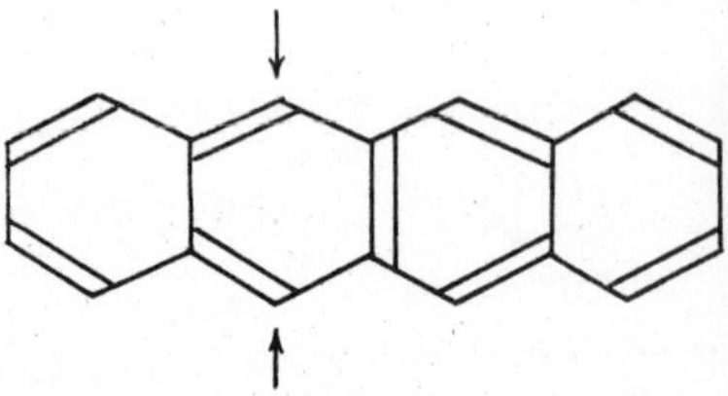 (LXXVII)

In the pentacyclic compounds (LXXIX)–(LXXXI) the reactivity increases so greatly in the sequence given that the reaction with maleic anhydride can be used for their separation.†
Similarly, the reaction can be used with advantage to remove

* Clar, *Aromatische Kohlenwasserstoffe*, 2nd ed., Springer-Verlag, Berlin, 1952.
† Clar and Lombardi, *Ber. dtsch. chem. Ges.* 1932, **65**, 1411.

THE DIELS-ALDER REACTION

the very reactive 'chrysogen' (naphthacene) from commercial anthracene.

The presence of substituents, especially in the *meso* positions of the anthracene ring system, has a marked effect both on the position of the equilibrium and on the 'reactivity' of the diene.

(LXXVIII)   (LXXIX)

(LXXX)   (LXXXI)

In the case of 9:10-dimethylanthracene the equilibrium is in favour of the adduct, and almost quantitative yields may be obtained when the reactants are present in equimolar proportions. With 9:10-diphenylanthracene, however, the equilibrium is in favour of the constituent molecules, and the equilibrium mixture contains only 16 % adduct.* These and other results are summarized in table 8·1.

The presence of methyl substituents increases the velocity of the addition. 9-Methylanthracene reacts much faster than anthracene, and the reaction between 9:10-dimethylanthracene and maleic anhydride proceeds rapidly even at room temperature. Again, 9:10-dimethyl-1:2-benzanthracene and 5:9:10-trimethyl-1:2-benzanthracene react more rapidly with maleic

* Bachmann and Kloetzel, *J. Amer. Chem. Soc.* 1938, **60**, 481.

TABLE 8·1. *Equilibrium mixtures from anthracene derivatives and maleic anhydride in boiling xylene**

Hydrocarbon	Percentage adduct in equilibrium mixture
Anthracene	99
9-Methylanthracene	99
9:10-Dimethylanthracene	98
1:2-Benzanthracene	84
9-Phenylanthracene	75
1:2:5:6-Dibenzanthracene	30
20-Methylcholanthrene	22
9:10-Diphenylanthracene	16

anhydride than does 1:2-benzanthracene. On the other hand, 9:10-diethyl-1:2-benzanthracene is less reactive than the 9:10-dimethyl- derivative.[†] *Meso*-phenyl groups retard the reaction enormously. The reaction between equimolar proportions of 9:10-diphenylanthracene and maleic anhydride is incomplete (does not reach equilibrium) even after boiling, in benzene, for several days. That *meso*-phenyl groups should retard the reaction is of some interest, for in other reactions (e.g. photo-oxidation) these groups have an activating influence. The reaction between diphenylpentacene and maleic anhydride provides an interesting example of the retarding influence of the phenyl group. Allen and Bell[‡] isolated both a mono- and a di- addition product from 6:13-diphenylpentacene (LXXXII) and maleic anhydride. The mono- addition product (LXXXIII) was found to be unchanged after further treatment with maleic anhydride, so it cannot be an intermediate in the formation of the di- addition product, and it seems probable that the *meso*-phenyl groups so hinder the addition at the 6:13 positions of (LXXXII) that addition also takes place to some extent at the alternative 5:14 and 7:12 positions, to give the di- adduct (LXXXIV).

Not only may substituents markedly affect the ease of formation of the adduct with maleic anhydride or other dienophile, but the formation of the adduct may affect the 'reactivity' of the substituent itself. In an adduct derived from an anthracene

---

[*] After Bachmann and Kloetzel, *J. Amer. Chem. Soc.* 1938, **60**, 481.
[†] Bachmann and Chemerda, ibid. 1938, **60**, 1023.
[‡] Ibid. 1942, **64**, 1253.

THE DIELS-ALDER REACTION 353

derivative, for example, the *meso* carbon atoms become aliphatic in character. This is illustrated by the fact that the chlorine atoms in the adduct (LXXXV) are 'reactive', and are capable of displacement by aryl groups in the Friedel-Crafts reaction, giving (LXXXVI) and (LXXXVII).*

* Clar, *Ber. dtsch. chem. Ges.* 1931, **64**, 2194.

On the other hand, in other reactions, *meso* halogen atoms have been found to be extremely unreactive. The adduct from 9-bromoanthracene and maleic anhydride, for example, was found to give no ionized halogen after prolonged boiling with alcoholic potash, and the adduct from 9:10-dichloroanthracene was similarly found to be unreactive.*

QUINONES AS DIENOPHILES; TRIPTYCENE

The use of *para*-benzoquinone as a dienophile has already been mentioned, but this reaction is worthy of special treatment because it gives rise to some extremely interesting compounds.

It has been shown that equimolecular quantities of anthracene (I) and benzoquinone (II) give the adduct (III), in 93 % yield;† and on cautious oxidation this adduct loses two atoms of hydrogen to yield 9:10-dihydro-9:10-*endo-ortho*-phenylene-1:4-anthraquinone (IV). Using two molecules of anthracene per molecule of benzoquinone, however, a different adduct (V) is obtained.

* Barnett, Goodway, Higgins and Lawrence, *J. Chem. Soc.* 1934, p. 1224; Bachmann and Kloetzel, *J. Org. Chem.* 1938, 3, 55; Bartlett and Cohen, *J. Amer. Chem. Soc.* 1940, **62**, 1183.
† Clar, *Ber. dtsch. chem. Ges.* 1931, **64**, 1676.

THE DIELS-ALDER REACTION

Tetrachloro-*para*-benzoquinone also forms an adduct with anthracene. In this case two chlorine atoms are easily lost, and a dichloro derivative (VI) is obtained.*

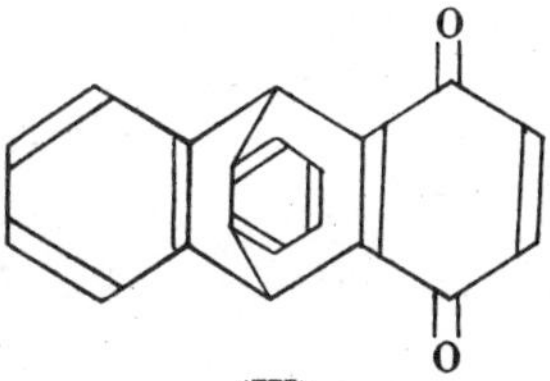

(IV) 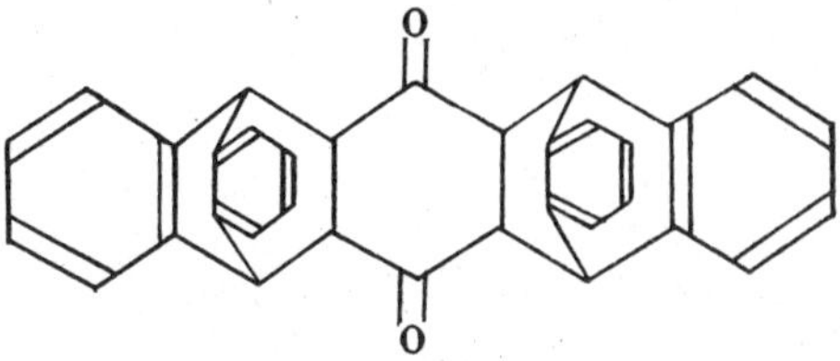   (V)

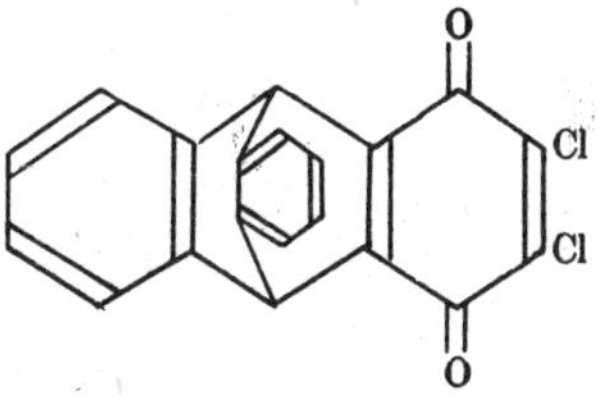

(VI)

In the same way, pentacene forms adducts with benzoquinone, and also with tetrachlorobenzoquinone. The latter adduct (VII) loses two chlorine atoms very easily to give the compound (VIII).†

The adduct from anthracene and *para*-benzoquinone has also been converted, by the series of transformations indicated in the formulae (IX)–(XIV), into the symmetrical hydrocarbon, 9:10-*ortho*-benzenoanthracene (XIV).‡ It has been called 'triptycene' from the triptych of antiquity, a book having three leaves hinged together. It is an analogue of the triphenylmethyl radical, but unlike the latter it is a very stable substance. Oxidation of triphenylmethane with chromic anhydride readily gives triphenylcarbinol, but triptycene is oxidized under similar conditions to anthraquinone and carbon dioxide.

---

* Ibid.
† Ibid.; Clar and John, ibid. 1930, **63**, 2967.
‡ Bartlett, Ryan and Cohen, *J. Amer. Chem. Soc.* 1942, **64**, 2649.

1-Bromotriptycene has also been synthesized by a similar series of reactions starting from 9-bromoanthracene.*

Triptycene may be called tribenzobi*cyclo*(2, 2, 2)octatriene, and Bartlett, Ryan and Cohen† suggest that the internal bond angles of the bi*cyclo*(2, 2, 2)octatriene system must always approach the value 109° 28′. There is no evidence to support this conclusion,

(VII)

(VIII)

but the molecule must certainly be strained. The two saturated carbon atoms (see (XV)) are each linked to three benzene rings, and it is impossible for *both* the significant angles (α and β) to have the 'normal' values of 109° 28′ and 120° respectively. It is more probable that α is increased and that β is reduced in magnitude.

It is noteworthy that Bartlett and Lewis‡ have shown that the redox potential of triptycenequinone ((XI), 0·666 V.) is very close to that of indanequinone (0·641 V.; cf. benzoquinone = 0·711 V.), suggesting a similar strain in both compounds.

* Bartlett *et al. J. Amer. Chem. Soc.* 1950, **72**, 1003.
† Ibid. 1942, **64**, 2649.
‡ Ibid. 1950, **72**, 1005.

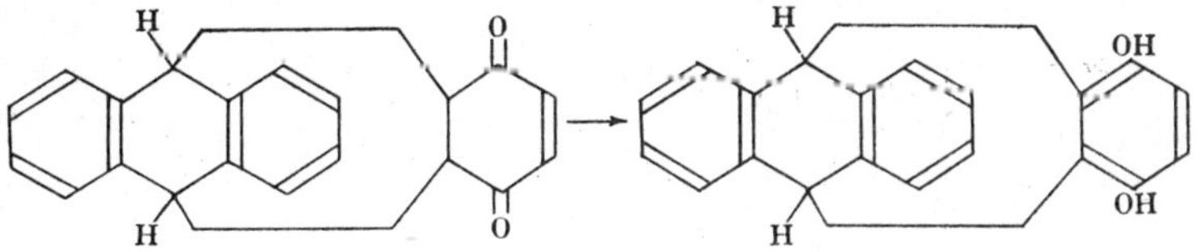

(IX)　(X)

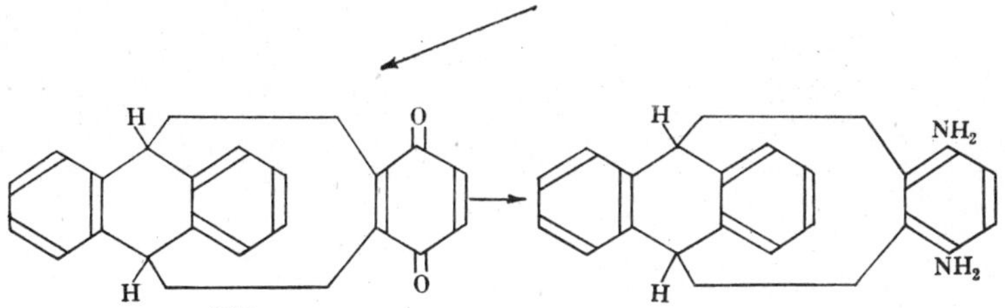

(XI)　(XII)

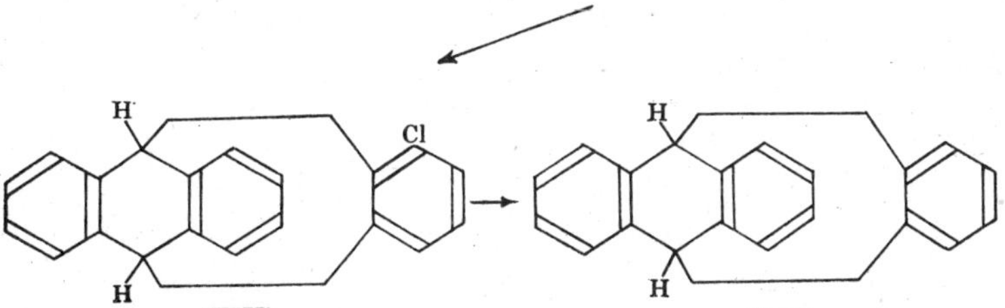

(XIII)　(XIV)

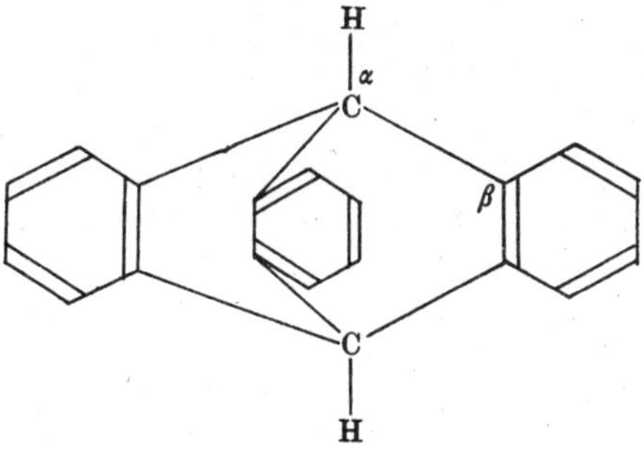

(XV)

SOME STEREOCHEMICAL CONSIDERATIONS

The diene synthesis is remarkably stereo-specific. When several stereoisomers of an adduct appear to be possible, only one is actually formed, for the reaction appears always to involve a *cis* addition. This is indicated by the fact that butadiene (I) and maleic acid (II) give *cis*-1:2:3:6-tetrahydrophthalic acid (III), while butadiene and fumaric acid (IV) give *trans*-1:2:3:6-tetrahydrophthalic acid (V).

Similar results have been obtained using aromatic dienes. Anthracene and maleic anhydride have been shown to give *cis*-9:10-dihydroanthracene-9:10-*endo*-αβ-succinic anhydride, which on hydrolysis gives the *cis*-DL- acid (VI). This *cis*-DL- acid loses water and forms the anhydride even on attempted recrystallization, or on standing in a vacuum desiccator. In the same way, the *cis*-dimethyl ester has been obtained by addition of dimethylmaleate to anthracene, and the product is identical with that obtained by methylation of the *cis*-DL- acid. On the other hand, fumaric acid reacts with anthracene to give the *trans* adduct (VII), and dimethylfumarate gives the *trans*-dimethyl ester. Unlike the *cis* compound, however, the *trans*-DL- acid was found to be remarkably stable. It showed no tendency to lose water to form an anhydride, and could even be recrystallized unchanged from a mixture of acetic acid and acetic anhydride.

THE DIELS-ALDER REACTION

The methyl derivatives of the *cis* and *trans* adducts have likewise been obtained from citraconic anhydride and mesaconic acid respectively,* and it is significant that in all cases studied the addition of the *trans* acids was found to be slower than that of the corresponding *cis* acids or anhydrides.

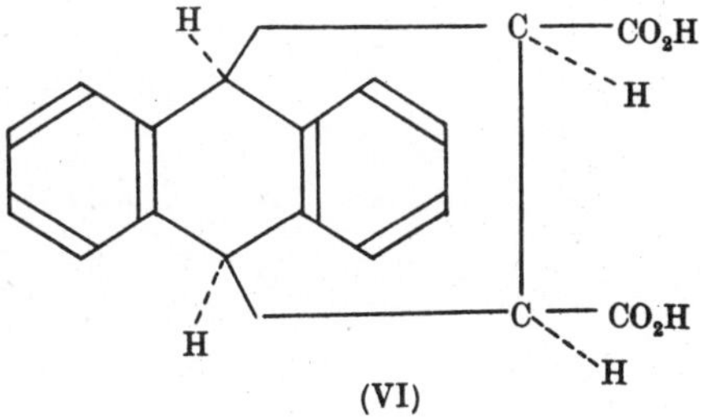

(VI)

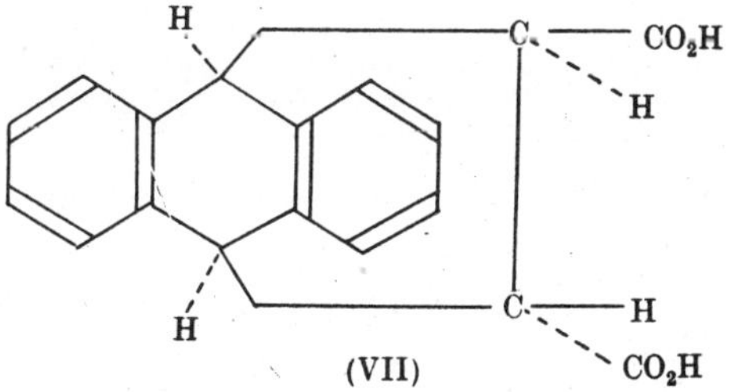

(VII)

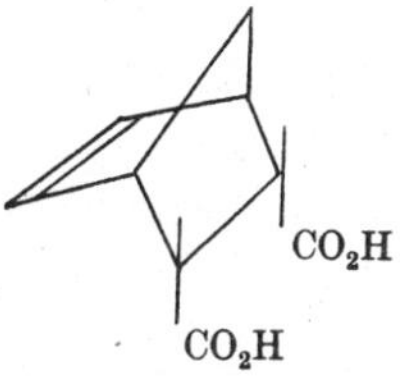

(VIII)
endo-cis

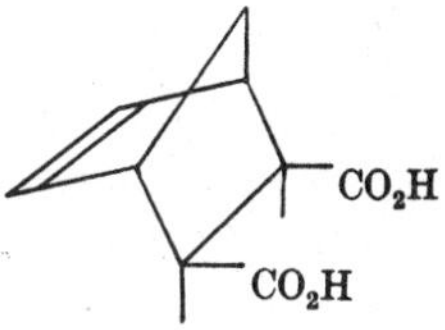

(IX)
exo-cis

The rule of *cis* addition has been well established. With cyclic dienes, such as *cyclo*pentadiene or furan, however, two *cis* configurations are theoretically possible, namely, *endo-cis* (VIII) and *exo-cis* (IX).

* Bachmann and Scott, *J. Amer. Chem. Soc.* 1948, **70**, 1458.

It was formerly believed that the *endo-cis* configuration is produced exclusively in reactions of this type,* but some exceptions have been found. It has been shown that although the reaction between furan and maleic acid in water gives an adduct of the *endo-cis* type, that formed from furan and maleic anhydride in ether has the *exo-cis* configuration.† Furthermore, Alder and Stein‡ have found that at low temperatures *cyclo*pentadiene gives mainly the *endo* form, but that at high temperatures more of the *exo* dimer is formed.§

THE MECHANISM OF THE DIELS-ALDER REACTION

The mechanism of the diene synthesis is still largely unknown, but the effect of substituents on the course (and velocity) of the addition gives important information. The work of Hudson and Robinson‖ indicated that the tendency of a styrene derivative to react with maleic anhydride is enhanced by the presence of an electron-donating substituent *para* (but not *meta*) to the unsubstituted side chain. More recently, Bergmann and Szmuszkowicz¶ have examined the addition of maleic anhydride to unsymmetrically substituted 1:1-diarylethylenes (I). With such compounds the dienophile was found to give only one of the two possible isomeric addition products, which (after dehydrogenation) have either of the two structures (II) or (III).

The work showed that aromatic substituents could be divided into two classes, first, those which promote participation of the substituted ring in the addition, and secondly, those which prevent participation of the substituted ring. Their results appear to be explicable only if it is assumed that the reaction takes place by electrophilic attack at the β-carbon atom of the side chain, and nucleophilic attack at the 2- or 2'-position, as in (IV) and (V). In this way the course of the reaction must be dependent not only on the electronic influence of a substituent on the β-position,

* Alder and Stein, *Angew. Chem.* 1937, **50**, 510; *Liebigs Ann.* 1934, **514**, 1.
† Woodward and Baer, *J. Amer. Chem. Soc.* 1948, **70**, 1161.
‡ *Liebigs Ann.* 1933, **504**, 219.
§ See also Wassermann, *J. Chem. Soc.* 1935, pp. 828, 1511; *Trans. Faraday Soc.* 1939, **35**, 841.
‖ *J. Chem. Soc.* 1941, p. 715.
¶ *J. Amer. Chem. Soc.* 1948, **70**, 2748.

THE DIELS-ALDER REACTION

but also on the 2-position. Using Hammett's σ constants, Bergmann and Szmuszkowicz showed that the course of the reaction is dependent on the increased positive charge in the 2-position as well as the increased electron density at the β-carbon atom.

(I) [structure with labels R, 2, α, β]

Dehydrogenation of adduct →

(II) [structure]

or

(III) [structure]

This mechanism is also supported by the orientation of the adducts formed from 1-vinylnaphthalene and unsymmetrical dienophiles already mentioned.* This may be summarized in the form: the most electrophilic carbon atom of the dienophile reacts with the most nucleophilic carbon atom of the diene.

It also seems to be established by stereochemical and kinetic evidence† that the two new C—C bonds are formed simul-

* Ibid. 1949, **71**, 3062.
† Bergmann and Eschinazi, ibid. 1943, **65**, 1405; Wassermann, *J. Chem. Soc.* 1942, p. 612.

taneously. This is supported by the principle of *cis* addition and by the fact that *non-reactive* dienes sometimes cause isomerization of maleic into fumaric acid.*

Although it is likely that the great majority of Diels-Alder reactions occur by a polar mechanism, it seems that a non-polar addition can also occur in certain circumstances, such as in the gas phase.

(IV)   (V)

In either case, as Brown[†] has pointed out, the formation of an adduct must involve the eventual 'localization' of two of the π-electrons of the diene. According to the polar mechanism the two electrons would have to be localized on one of the carbon atoms; according to the non-polar mechanism one electron would be localized on each of the terminal atoms of the diene. The energy required to bring about this localization, called the *para*-localization energy, can be obtained from the equation

$$P = E_r - E + 2\alpha,$$

where E is the π-electron energy of the original conjugated system, E_r is the total π-electron energy of the one or more separate conjugated systems when the two π-electrons are localized, and 2α is the energy of two isolated π-electrons. Alternatively, P may be expressed in terms of the resonance energies of the two systems.

* Bergmann and Bergmann, *J. Amer. Chem. Soc.* 1937, **59**, 1443.
† *J. Chem. Soc.* 1950, p. 691.

THE DIELS-ALDER REACTION

The ease of reaction with a dienophile may be assumed to depend mainly on the magnitude of the energy required to bring about the required localization (table 8·2). For benzene the energy requirement is calculated to be 91 kcal./mole, which is evidently too large for reaction to occur. For the *meso* positions of anthracene, however, the energy is 70 kcal./mole, and reaction occurs readily with a variety of dienophiles. As the series of linear benzologues is ascended, the magnitude of the localization energy becomes smaller, and the reaction proceeds with increasing facility.

TABLE 8·2. para-*Localization energies and free-valence numbers for the* meso *positions of the polyacenes*

Acene	para-Localization energy* (kcal./mole)	Free-valence number†
Benzene	91	0·081
Naphthalene	82	0·134
Anthracene	70	0·202
Naphthacene	68	0·212
Pentacene	66	0·222
Hexacene	65	—
Heptacene	64	—

An approximate indication of the 'reactivity' of an acene is also given by the free-valence numbers of the positions involved in any Diels-Alder addition. These are also reported in table 8·2.

It will be noted that the free-valence number increases with the number of linear benzene rings.

* Brown, ibid. 1950, p. 691.
† Molecular orbital method. See also Badger, *J. Aust. Chem. Inst.* 1950, **17**, 14; and Burkitt, Coulson and Longuet-Higgins, *Trans. Faraday Soc.* 1951, **47**, 553.

CHAPTER 9

PHOTO-OXIDATION AND PHOTO-DIMERIZATION

THE STRUCTURE OF THE PHOTO-OXIDES

The addition of a molecule of oxygen to an aromatic compound, in the presence of light, with the formation of a 'photo-peroxide', was first demonstrated by Dufraisse and his collaborators with the brilliant red hydrocarbon rubrene. It was found that solutions of this hydrocarbon (R) soon became decolorized and lost their characteristic fluorescence when exposed to sunlight, or to artificial light, in the presence of air or oxygen. The resulting colourless peroxide (RO_2) decomposed at about 180° to give the original hydrocarbon and oxygen.*

$$R + O_2 \underset{\text{heat}}{\overset{\text{light}}{\rightleftarrows}} RO_2$$

Rubrene was first prepared by elimination of hydrogen chloride from phenylethinyldiphenylchloromethane (I), and its structure was not known for several years after the discovery of the photo-addition reaction. It was, however, eventually proved to be 5:6:11:12-tetraphenylnaphthacene (II).†

$$2\text{Ph}.\text{C} \equiv \text{C}.\text{C}.\text{Ph}_2 \quad \xrightarrow{-2\text{HCl}}$$
$$|$$
$$\text{Cl}$$

(I)　　　　　　　　(II)

The recognition of rubrene as a polynuclear aromatic hydrocarbon was of great importance, and the photo-oxidation reaction was soon extended to include a wide variety of compounds of this type. Naphthalene and phenanthrene do not add

---

* Moureu, Dufraisse and Dean, *C.R. Acad. Sci., Paris*, 1926, **182**, 1440, 1584; Moureu, Dufraisse and Girard, ibid. 1928, **186**, 1027.

† Dufraisse, *Bull. Soc. chim. Fr.* 1936, **3**, 1857; Dufraisse and Velluz, *C.R. Acad. Sci., Paris*, 1935, **201**, 1394.

oxygen in this way,* but the reaction is now known to be a fairly general attribute of anthracene derivatives. Anthracene itself gives a colourless photo-oxide under suitable conditions,† and many substituted compounds react in the same way.

It was early recognized that there seems to be an intimate relationship between the ability of a polynuclear aromatic hydrocarbon to add oxygen in the presence of light, and its ability to add dienophilic reagents such as maleic anhydride.‡ However, the analogy is by no means complete. The addition of oxygen can *only* be induced in the presence of light, and certain substances which are photo-oxidized with great facility are somewhat resistant to the addition of maleic anhydride. Nevertheless, the *structural* analogy seems to be close. For example, the absorption spectra of the photo-oxide and of the maleic anhydride adduct of anthracene are both similar to that of 9:10-dihydroanthracene, but very different from that of anthracene itself.§

The most important evidence bearing on the structure of the photo-oxides is provided by the hydrogenation experiments. It has been shown for several photo-oxides that catalytic hydrogenation yields the corresponding *meso*-dihydroxy derivatives.‖ This evidence indicates that the photo-oxides cannot be molecular compounds, and that the oxygen cannot be held by 'polarization forces', but must be held by normal bonds. Moreover, although the heat of formation of rubrene photo-oxide is small, it is too great for that of a molecular compound.¶ There seems little doubt, therefore, that the oxygen is bound to the two *meso* positions by ordinary chemical bonds. The photo-oxides of an anthracene derivative and of rubrene may therefore be formulated as (III) and (IV). On catalytic hydrogenation, these photo-oxides pass to the diols (V) and (VI).

The representation of the photo-oxides in this way is, however,

---

* Dufraisse and Priou, *Bull. Soc. chim. Fr.* 1938, **5**, 611.
† Dufraisse and Gérard, *C.R. Acad. Sci., Paris*, 1935, **201**, 428; 1936, **202**, 1859; *Bull. Soc. chim. Fr.* 1937, **4**, 2052.
‡ Dufraisse, ibid. 1939, **6**, 422; Bergmann and McLean, *Chem. Rev.* 1941, **28**, 367.
§ Dufraisse and Gillet, *C.R. Acad. Sci., Paris*, 1947, **225**, 191.
‖ Dufraisse and Houpillart, ibid. 1937, **205**, 740; Cook and Martin, *J. Chem. Soc.* 1940, p. 1125; see also Badger and Cook, *Chem. & Ind.* 1937, 1949, p. 353.
¶ Dufraisse and Enderlin, *C.R. Acad. Sci., Paris*, 1930, **191**, 1321.

somewhat misleading, for the structures represent derivatives of dihydroanthracene and dihydronaphthacene, respectively. The structures are therefore non-coplanar: the planes of the terminal rings in anthracene photo-oxide are inclined to one another, and the two oxygen atoms are disposed away from this angle, as in (VII).

Assuming the O—O distance to be 1·47 Å.,* the distance between the two *meso*-carbon atoms to be 2·80 Å., and the distance between each *meso*-carbon atom and the nearer oxygen atom to be 1·43 Å.,† the angle between the valency bonds of the two oxygen atoms may be calculated to be approximately 118°. This is only slightly greater than the figures (*c.* 105–115°) which have been determined experimentally for various oxygen

* Cf. Giguère and Schomaker, *J. Amer. Chem. Soc.* 1943, **65**, 2025.
† Cf. Pauling, *The Nature of the Chemical Bond*, 2nd ed., Cornell University Press, 1940.

compounds of the type C—O—C and Cl—O—Cl.* It would seem, therefore, that no unusual strain or distortion is involved in the formation of transannular peroxides of the type assumed.

PHOTO-OXIDATION OF POLYACENES

As already indicated, the photo-oxidation of anthracene was not discovered until 1935, and the successful isolation of the product was very largely due to the use of carbon disulphide as solvent, and to the use of dilute, rather than concentrated, solutions. In other solvents, and in concentrated solution, anthracene more readily forms the photo-dimer, dianthracene, which will be discussed later.

As with the Diels-Alder reaction, the higher benzologues of anthracene are photo-oxidized with increasing facility as the number of linear benzene rings is increased. Naphthacene, the parent hydrocarbon of rubrene, forms a photo-oxide quite readily.† Pentacene is very rapidly photo-oxidized,‡ and it is inadvisable to work with solutions of this hydrocarbon in broad daylight. Hexacene is even more readily oxidized.§ Solutions of this hydrocarbon are almost instantly photo-oxidized on exposure to light and air. These facts are in conformity with the hypothesis formulated by Dufraisse that the ability to undergo photo-oxidation is associated with the reactivity of the *meso* positions.‖ It will be shown in a subsequent section that substituents also have a very marked effect on the ease of formation, and stability, of photo-oxides.

PHOTO-OXIDATION OF SIMPLE HETEROCYCLIC COMPOUNDS

Benzene does not form a photo-oxide, but some of its heterocyclic analogues are so much more reactive that the formation of photo-oxides might be anticipated. No photo-oxide of a simple

---

* Ibid.; Sutton and Brockway, *J. Amer. Chem. Soc.* 1935, **57**, 473.
† Dufraisse and Horclois, *Bull. Soc. chim. Fr.* 1936, **3**, 1880.
‡ Clar and John, *Ber. dtsch. chem. Ges.* 1930, **63**, 2967.
§ Clar, ibid. 1939, **72**, 1817.
‖ Dufraisse, *Bull. Soc. chim. Fr.* 1939, **6**, 422.

heterocyclic compound, such as furan, has, however, been isolated, although there is some evidence for their existence in solution. For example, 2-methylfuran (I), when photo-sensitized with eosin, evidently gives a peroxide (II), which on hydrogenation affords the aldehyde (III), and the acid (IV), and which, on disproportionation in ethanol, gives 2-keto-5-ethoxy-5-methyl-2:5-dihydrofuran (V).*

In the same way, 1:3-diphenyl*iso*benzofuran (VI) is readily photo-oxidized to *ortho*-dibenzoylbenzene (VIII), and the intermediate formation of a photo-oxide (VII) is usually postulated. Such a peroxide has actually been isolated by Dufraisse and Ecary,† but it proved to be very unstable, even at low temperatures.

* Schenck, *Angew. Chem.* 1948, **60**A, 244.
† *C.R. Acad. Sci., Paris*, 1946, **223**, 735.

THE INFLUENCE OF SUBSTITUENTS ON THE EASE OF FORMATION OF PHOTO-OXIDES

meso-Alkyl groups very greatly facilitate the photochemical addition of oxygen. 9:10-Dimethylanthracene readily forms a photo-oxide, even in the ordinary light of the laboratory. Similarly, Cook and Martin[*] found that *meso*-alkyl-substituted 1:2-benzanthracenes are very readily photo-oxidized even with the aid of an ordinary gas-filled lamp. Methoxy groups facilitate the addition to an even greater degree, for 9:10-dimethoxyanthracene is photo-oxidized in a few seconds in sunlight.[†] Phenyl groups, both in anthracene and naphthacene, also promote the addition. In a manometric study, Dufraisse and Le Bras[‡] found that rubrene is more readily photo-oxidized than naphthacene, and that 9:10-diphenylanthracene is more rapidly attacked than anthracene.

The effect of some other *meso* substituents has also been studied. 9:10-Diacetoxyanthracene is oxidized only very slowly, and, in general, the presence of *meso* substituents of the *meta*-directing type either inhibits or entirely prevents the reaction.[§] Substituents in the 'outer' rings also affect the rate of photo-oxidation, but the effect is less marked than when the substituents are present in the *meso* positions.

As might be expected, certain aza- analogues of anthracene derivatives also photo-oxidize. In some cases the photo-oxide cannot be isolated, but the presence of phenyl substituents in the *meso* positions facilitates the addition and seems to stabilize the product. Thus 9:10-diphenyl-1-azanthracene, 9:10-diphenyl-2-azanthracene and even 2-phenyl-1-azanthracene form photo-oxides (I)–(III).[‖]

In an attempt to prepare *bis*-photo-oxides, Dufraisse, Velluz and Velluz[¶] found, contrary to expectation, that 9:9'-dianthryl

---

[*] Dufraisse and Houpillart, ibid. 1937, **205**, 740; Cook and Martin, *J. Chem. Soc.* 1940, p. 1125; Badger and Cook, *Chem. & Ind.* 1949, p. 353; see also Cook, Martin and Roe, *Nature, Lond.*, 1939, **143**, 1020.
[†] Dufraisse and Priou, *Bull. Soc. chim. Fr.* 1939, **6**, 1649.
[‡] *C.R. Acad. Sci., Paris*, 1943, **216**, 60.
[§] Dufraisse and Mathieu, *Bull. Soc. chim. Fr.* 1947, p. 307.
[‖] Étienne, *C.R. Acad. Sci., Paris*, 1943, **217**, 694; Étienne and Robert, ibid. 1946, **223**, 331; Étienne, ibid. 1944, **219**, 622.
[¶] *Bull. Soc. chim. Fr.* 1938, **5**, 600.

and 10:10′-diphenyl-9:9′-dianthryl (IV) do not photo-oxidize. These compounds do not react with maleic anhydride either, and steric hindrance may be a factor here. On the other hand, 9:9′:10:10′-tetraphenyl-1:1′-dianthryl (V) forms a *bis*-photo-oxide, each anthracene residue taking up a molecule of oxygen.*

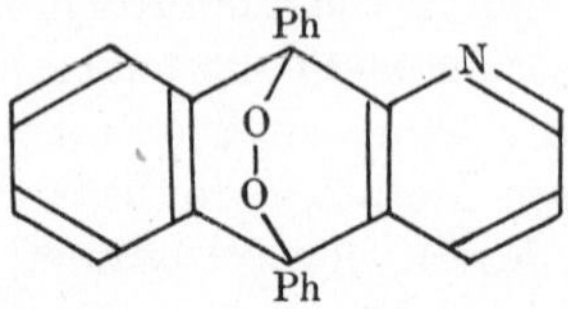

(I) 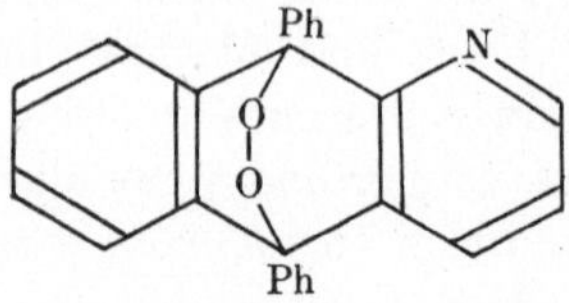   (II)

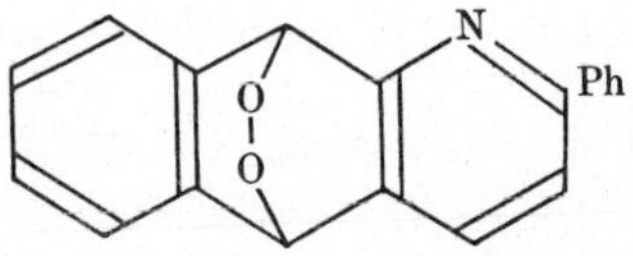

(III)

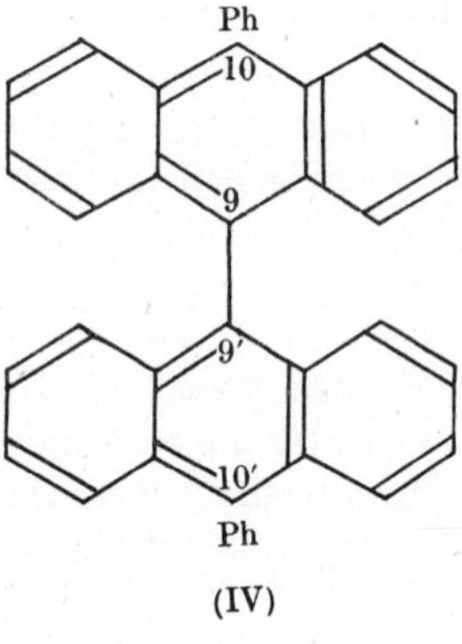

(IV)

These results are of interest, for *meso*-diphenylhelianthrene (VI), which differs from diphenyldianthryl (IV) by only two hydrogen atoms, has been found to form a *mono*-photo-oxide.†
This photo-oxide is yellow, and is the only coloured compound

* Sauvage, *Ann. Chim.* 1947, **2**, 844; Dufraisse and Sauvage, *C.R. Acad. Sci., Paris*, 1945, **221**, 665.
† Sauvage, *Ann. Chim.* 1947, **2**, 844; Dufraisse and Sauvage, *C.R. Acad. Sci., Paris*, 1947, **225**, 126.

of this nature so far discovered. The reaction proceeds very readily, and the intense violet colour of the hydrocarbon in carbon disulphide solution is completely discharged in less than a second by exposure to direct sunlight. The structure of the product has not been determined with certainty, although certain speculations have been made. Clearly it cannot have a structure directly comparable with that of the anthracene photo-oxides.

(V) (VI)

Dufraisse and Sauvage* postulate the structure (VII) and disregard a structure of type (VIII) for the reason that phenanthrene does not form a similar photo-oxide. This reasoning is not, however, convincing. For example, phenanthrene does not add maleic anhydride, although the closely related perylene molecule does do so under certain conditions, and the difference is probably due to the fact that the comparable positions in perylene are more 'reactive' than in phenanthrene.

An important feature of the helianthrene molecule, however, is that it must be strained, for the hydrogen atoms at positions 8 and 8' must interfere with one another. It seems likely, therefore, that the molecule would tend to revert to a 10:10'-dihydride with great ease. The structure (IX) for the photo-oxide of *meso*-diphenylhelianthrene would therefore seem to be probable. The analogous hydrocarbon, *meso*-naphthodianthrene, formed by

* Ibid.; see also Brockmann and Mühlmann, *Chem. Ber.* 1948, **81**, 467.

cyclization across the 8:8' positions, is not strained, and has no similar tendency to revert to a dihydride. It is therefore significant that diphenyl-*meso*-naphthodianthrene does not form a photo-oxide.*

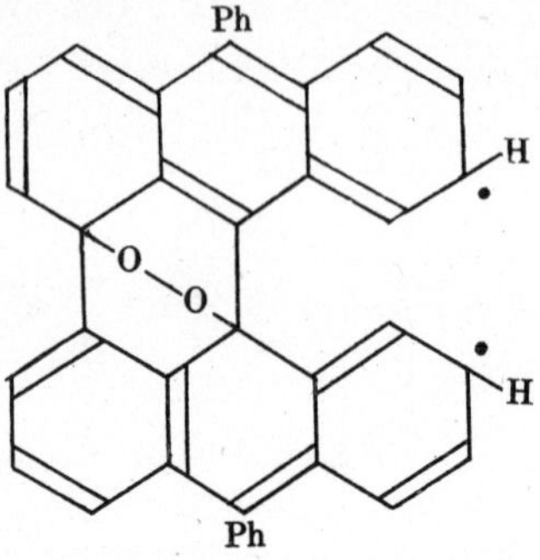

(VII) 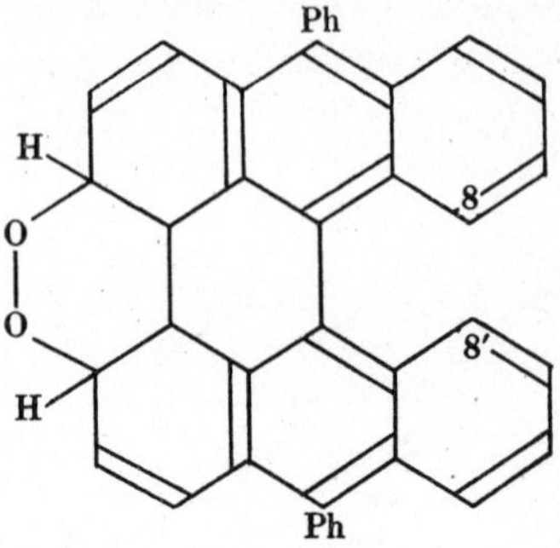 (VIII)

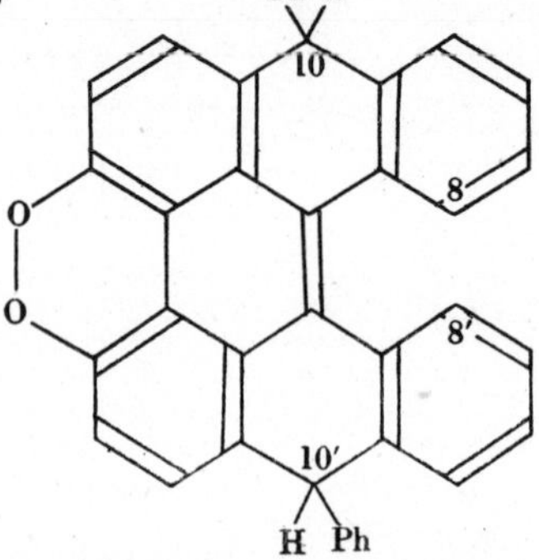

(IX)

THE INFLUENCE OF SUBSTITUENTS ON THE THERMAL STABILITY OF PHOTO-OXIDES

When rubrene photo-oxide is heated it dissociates into rubrene and oxygen, and in quantitative experiments it is possible to recover over 80% of the oxygen.† All the photo-oxides of aromatic compounds decompose on heating, but the yield of

* Sauvage, *Ann. Chim.* 1947, **2**, 844.
† Moureu, Dufraisse and Dean, *C.R. Acad. Sci.*, Paris, 1926, **182**, 1584; Moureu, Dufraisse and Girard, ibid. 1928, **186**, 1027.

oxygen which can be recovered varies considerably with the structure. On the one hand, the photo-oxides of anthracene and naphthacene decompose explosively at about 120°, and no free oxygen can be detected;* but the photo-oxides of 9:10-diarylanthracenes dissociate with the liberation of up to 98 % of their combined oxygen.†

Substituents have a profound effect on the yield of oxygen which can be recovered. Thus, although the 9:10-diarylanthracene photo-oxides give almost quantitative yields of oxygen on heating, the 9-alkyl-10-arylanthracene photo-oxides normally give less than 50 % of their oxygen;‡ and 9-alkyl- and 9:10-dialkyl-anthracene photo-oxides do not liberate oxygen.§ Substituents in the aromatic rings can also affect the yield of oxygen, even in the presence of *meso*-aryl groups.

These results are summarized in table 9·1.

The photo-oxide of 9:10-dimethoxyanthracene is unusual as it can be sublimed without decomposition. On the other hand, the photo-oxide of 1:4-dimethoxy-9:10-diphenylanthracene is unusually unstable. It liberates 25 % of its combined oxygen in 10 days at 20°, 55 % in 30 days, and 78 % in 40 days.‖ The effect of other groups in the 1- and 4-positions of 9:10-diphenylanthracene has also been investigated. The favourable action of a 1-methoxy group, for example, is diminished by the introduction of a 4-chloro substituent.¶ 2:3-Dimethoxy-9:10-diphenylanthracene photo-oxide dissociates only slightly less readily than that of the 1:4 compound. Substituents in the 1-position do seem to have a greater influence than the same substituents in the 2-position. While the photo-oxide of 9:10-diphenylanthracene dissociates at 180°, those of the 1- and 2-methoxy derivatives dissociate at 150° and 160° respectively.**

---

* Dufraisse and Gérard, ibid. 1935, **201**, 428; 1936, **202**, 1859; *Bull. Soc. chim. Fr.* 1937, **4**, 2052; Dufraisse and Horclois, ibid. 1936, **3**, 1880.

† Dufraisse and Étienne, *C.R. Acad. Sci., Paris*, 1935, **201**, 280; Dufraisse and Le Bras, *Bull. Soc. chim. Fr.* 1937, **4**, 349; Duveen and Willemart, *J. Chem. Soc.* 1939, p. 116.

‡ Willemart, *C.R. Acad. Sci., Paris*, 1936, **203**, 1372; *Bull. Soc. chim. Fr.* 1937, **4**, 1447.

§ Willemart, *C.R. Acad. Sci., Paris*, 1937, **205**, 866; *Bull. Soc. chim. Fr.* 1938, **5**, 556.

‖ Dufraisse, Velluz and Velluz, *C.R. Acad. Sci., Paris*, 1939, **208**, 1822.

¶ Dufraisse, Velluz and Demuynck, ibid. 1942, **215**, 111.

** Dufraisse, Demuynck and Allais, ibid. 1942, **215**, 487.

TABLE 9·1. *Temperatures of decomposition and yields of oxygen liberated by decomposition of the photo-oxides of substituted anthracenes*[*]

Photo-oxide of	Temperature of decomposition	Yield of oxygen (%)
Anthracene	120°	0
9-Methylanthracene	80°	0
9-Ethylanthracene	—	0
9:10-Dimethylanthracene	—	0
9-Phenylanthracene	155°	12
9-Phenyl-10-methylanthracene	170°	23
9-Phenyl-10-carbomethoxyanthracene	170–190°	59
9-Phenyl-10-methoxyanthracene	170–175°	25
9-Phenyl-10-thienylanthracene	150°	95
9:10-Diphenylanthracene	180°	>95
9:10-Di(1′-Naphthyl)anthracene	180–200°	89
9:10-Di(2′-Naphthyl)anthracene	180–200°	94
9:10-Diphenyl-1-methylanthracene	170°	60
9:10-Diphenyl-2-methylanthracene	170–175°	94
9:10-Diphenyl-1-chloroanthracene	175–180°	75
9:10-Diphenyl-2-bromoanthracene	165°	91
9:10-Diphenyl-1-methoxyanthracene	150°	89
9:10-Diphenyl-2-methoxyanthracene	160–170°	93
9:10-Diphenyl-1:4-dimethoxyanthracene	80°	98
9:10-Dimethoxyanthracene	180°	Traces
9:10-Dinitroanthracene	95–100°	0

From a study of 9:10-diphenyl-1:4-dimethoxyanthracene photo-oxide, Audubert and Racz[†] concluded that the oxygen is released in the molecular state, and a general mechanism for the thermal decomposition has been postulated by Dufraisse, Pinazzi and Baget.[‡] It was suggested that there is a simultaneous rupture of the C—O linkages, and a momentary formation of a relatively stable structure of the hydrocarbon with a free electron pair at each *meso*-carbon atom.

The dissociation is certainly complex, and is accompanied by the emission of light. In the case of rubrene photo-oxide this emission of light was observed by Moureu, Dufraisse and Butler,[§] and, in the case of 1:4-dimethoxy-9:10-diphenylanthracene

---

[*] After Étienne in *Traité de chimie organique*, **17**, 1316, 1319.
[†] *Bull. Soc. chim. Fr.* 1943, **10**, 380.
[‡] *C.R. Acad. Sci., Paris*, 1943, **217**, 375.
[§] *Ibid.* 1926, **183**, 101.

photo-oxide, by Dufraisse, Velluz and Velluz.* The light is yellow-green in colour, and the energy change involved is of the order of 40 kcal./mole. It is also noteworthy that the dissociation is accompanied by an emanation which affects a photographic plate covered with black paper,† an observation which certainly merits further investigation.

REACTIONS OF PHOTO-OXIDES

Many of the photo-oxides undergo interesting transformations when treated with mineral acid. The photo-oxide of 9:10-diphenylanthracene, for example, adds a molecule of hydrogen chloride to give a chlorohydroperoxide (I).‡ A similar compound is no doubt formed as an intermediate when anthracene photo-oxide is treated with hydrogen chloride, but in this case water is eliminated and chloroanthrone (II) is obtained.§

Rubrene photo-oxide also reacts with hydrogen halides, but the halohydroperoxide first formed loses hydrogen halide and the hydroperoxide (III) is formed. This hydroperoxide gave dehydrorubrene (IV) on reduction with zinc and acetic acid.‖

Reference has already been made to the fact that several photo-oxides have been catalytically reduced to the corresponding diols. When rubrene photo-oxide is reduced with zinc and acetic acid, however, a rubrene mono-oxide is formed.¶ This

---

* Ibid. 1939, **209**, 516; *Bull. Soc. chim. Fr.* 1942, **9**, 171.
† *C.R. Acad. Sci., Paris*, 1939, **209**, 516.
‡ Pinazzi, ibid. 1947, **225**, 1012.
§ Dufraisse and Gérard, *Bull. Soc. chim. Fr.* 1937, **4**, 2052.
‖ Enderlin, *Ann. Chim.* 1938, **10**, 5; *C.R. Acad. Sci., Paris*, 1933, **197**, 691.
¶ Enderlin, *Ann. Chim.* 1938, **10**, 5; *C.R. Acad. Sci., Paris*, 1933, **197**, 691; Moureu, Dufraisse and Enderlin, ibid. 1929, **188**, 1528.

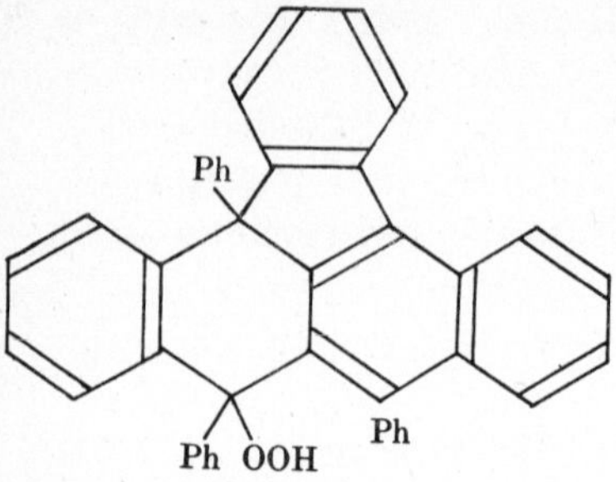

(III)

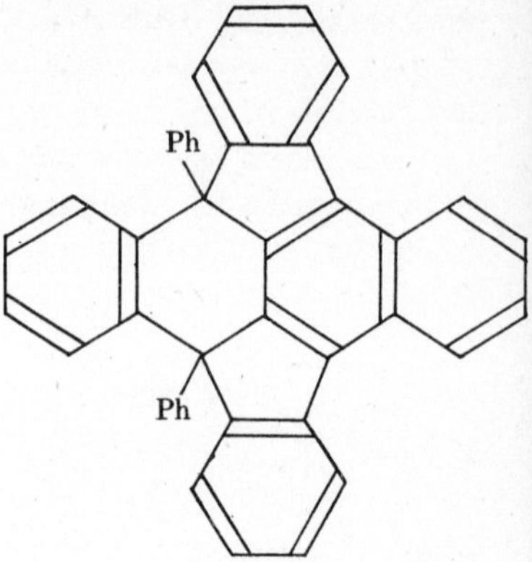

(IV)

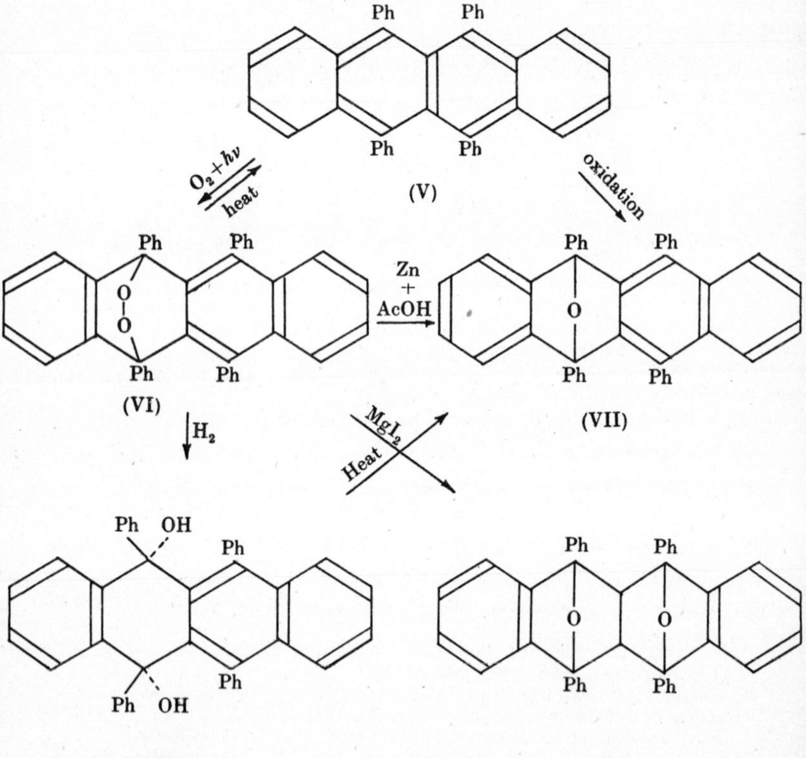

(V)

(VI)

(VII)

(VIII)

(IX)

mono-oxide has also been obtained by direct oxidation of rubrene (V) with nitric acid, or with potassium permanganate, and it has also been isolated following thermal dehydration of the *cis*-diol (VIII).* An *endo*-mono-oxide structure (VII) has been suggested for this compound. This structure seems to be in accord with all the available experimental evidence; but, if correct, the molecule must be very highly strained. It is impossible to have a 1:4-oxygen bridge in a *dihydro*-aromatic ring without very considerable distortion of the normal oxygen valency angle (*c.* 110°) and/or the C—O bond length.

Rubrene photo-oxide (VI) also undergoes another interesting transformation. When treated with an anhydrous ethereal solution of magnesium iodide, it is isomerized to rubrene *iso*-oxide (IX), for which a di-*endo*-oxide structure has been suggested.† However, this structure cannot be correct,‡ and a satisfactory alternative has not yet been suggested. Both rubrene mono-oxide and rubrene *iso*-oxide are stable compounds, and do not lose oxygen on heating. Indeed, the *iso*-oxide can be distilled *in vacuo*. For further details of these transformations see Bergmann and McLean.§

PHOTO-DIMERIZATION

When solutions of anthracene in benzene, xylene, acetic acid and other solvents are irradiated by sunlight or by ultra-violet light, a sparingly soluble photo-dimer called dianthracene (para-anthracene) is formed.‖ The reaction has been used to separate anthracene from crude fluorene, for fluorene is unaffected by such treatment.¶ As the reaction can be carried out in glass vessels, it follows that the wave-length of the activating light must be between 3100 Å., the transparency limit of glass, and 3750 Å.,

---

* Dufraisse and Enderlin, *Bull. Soc. chim. Fr.* 1932, **51**, 132.
† Dufraisse and Badoche, *C.R. Acad. Sci., Paris*, 1930, **191**, 104; Enderlin, *Ann. Chim.* 1938, **10**, 5.
‡ Gillet (*C.R. Acad. Sci., Paris*, 1949, **229**, 936) has produced spectrographic evidence that the *iso*-oxide cannot have the structure (IX). He has advanced an alternative structure which, however, seems equally unlikely.
§ *Chem. Rev.* 1941, **28**, 367.
‖ Fritzsche, *J. prakt. Chem.* 1867, **101**, 333.
¶ Capper and Marsh, *J. Chem. Soc.* 1926, p. 724.

the first absorption band of anthracene. This is the region of the four most characteristic absorption bands of anthracene.

Dianthracene almost certainly has the structure of a 9:10-9′:10′-dimer (I), and on heating it is dissociated to anthracene.

Several other anthracene derivatives have also been shown to form photo-dimers. 5-Methyl- and 4′-methyl-1:2-benzanthracenes also form photo-dimers, as does 1:2-benzanthracene itself. So does the highly cancer-producing hydrocarbon, methylcholanthrene.*

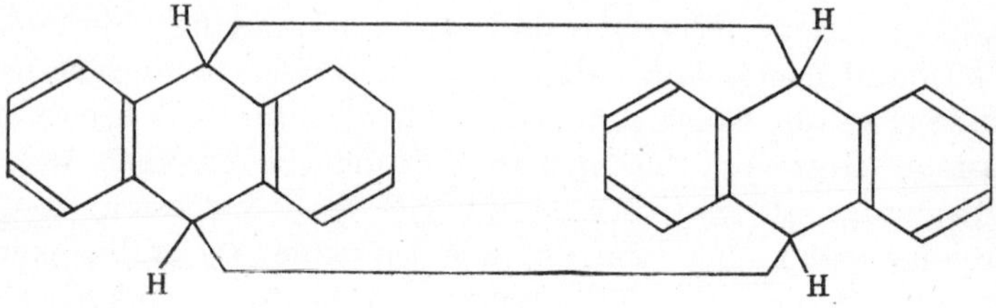

(I)

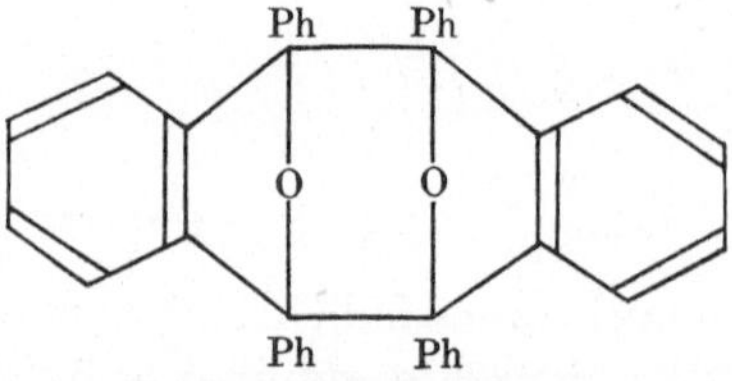

(II)

It is interesting to note that 1:3-diphenyl*iso*benzofuran also forms a photo-dimer on exposure to sunlight in the absence of air.† This compound evidently has the structure (II).

It seems that there is a close analogy between the photo-oxidation and photo-dimerization reactions, and, indeed, it was the tendency of anthracene to photo-dimerize in most solvents which so materially delayed the discovery of anthracene photo-oxide (which is best formed in carbon disulphide).

* Schönberg, Mustafa et al. J. Chem. Soc. 1948, p. 2126; 1949, p. 1039.
† Guyot and Catel, Bull. Soc. chim. Fr. 1906, 35, 1127.

On the other hand, relatively little work has yet been carried out to investigate the extent of the photo-dimerization reaction. Certain differences do seem to be apparent. For example, 9·10-diphenylanthracene, and rubrene, which are very readily photo-oxidized, do not form photo-dimers. There are at least two possible explanations: (i) the *meso*-phenyl groups tend to occupy a plane at right angles to that of the remainder of the molecule, and there may be considerable steric hindrance to photo-dimerization; (ii) the photo-dimers may form, but may be unusually labile, dissociating into the aromatic structures at room temperature. Some such explanation as this does seem to be necessary, for the kinetics of the two reactions appear to be remarkably similar.

THE MECHANISM OF PHOTO-OXIDATION

The analogy between the addition of oxygen to anthracene derivatives and the corresponding addition of dienophilic reagents such as maleic anhydride is certainly very close; but several important differences have been noted, and it seems unlikely that the *reacting* molecule can be identical in the two cases. It has often been observed, for example, that the presence of *meso*-phenyl groups *promotes* the addition of oxygen, but *retards* the addition of maleic anhydride.*

The major difference, however, is that although the activation energy for the addition of maleic anhydride can be supplied by heat alone, the addition of oxygen requires light. There is ample evidence that photo-oxidation cannot be brought about in the dark. Dufraisse, Badoche and Butler[†] found that rubrene does not react with oxygen in the dark when left in contact, in benzene solution, for as long as seven years. Even at high temperatures, in diphenylether as solvent, and with oxygen under pressure, rubrene does not react in the dark. Moreover, all attempts to find a catalyst (apart from light!) for this reaction have been entirely unsuccessful. Similar experiments which only served to confirm the photo-chemical character of the reaction have been carried out with anthracene and 9:10-diphenylanthracene.[‡]

* *C.R. Acad. Sci., Paris*, 1943, **216**, 60. † Ibid. 1943, **216**, 344.
‡ Dufraisse, Le Bras and Allais, ibid. 1943, **216**, 383.

Julian and his collaborators[*] claimed to have prepared a transannular photo-oxide of phenylanthranol in the dark, but this peroxide has now been shown to be a hydroperoxide of phenylanthrone.[†]

Another feature of both the photo-oxidation and photo-dimerization reactions is the important part played by the solvent. The relative rates of photo-oxidation of rubrene in various solvents have been determined, and considerable differences have been observed (table 9·2). The explanation of this solvent effect is still obscure, because the rate of oxidation cannot, as yet, be correlated with any other property of the solvent. It cannot, for example, be correlated with the dielectric constant, with the refractive index, or with the effect on the position of the absorption bands of rubrene.

TABLE 9·2. *Relative rates of photo-oxidation of rubrene in various solvents*[‡]

(Benzene = 1)

Solvent	Relative rate
Carbon disulphide	9
Chloroform	3
Ethyl iodide	1
Benzene	1
Acetone	1
Ether	0·5
Anisole	0·4
Pyridine	0·25
Nitrobenzene	0·1

The importance of the solvent is also illustrated by the fact that Dufraisse and Gérard[§] found that anthracene photo-oxide can be obtained in 70 % yield if the irradiation is carried out in carbon disulphide. In benzene, however, the photo-dimer is formed. In the same way, chloroform favours photo-oxidation and ether favours photo-dimerization. Again, Lauer and Oda[‖] found that the dimerization of anthracene proceeds at about the same rate in hexane, in *cyclo*hexane and in ethanol, but that in benzene and in acetic acid the reaction is faster.

[*] *J. Amer. Chem. Soc.* 1934, **56**, 2174; 1935, **57**, 1607, 2508; 1945, **67**, 1721, 1724.
[†] Dufraisse, Étienne and Rigaudy, *C.R. Acad. Sci., Paris*, 1948, **226**, 1773; *Bull. Soc. chim. Fr.* 1948, **15**, 804.
[‡] After Dufraisse and Badoche, *C.R. Acad. Sci., Paris*, 1935, **200**, 1103.
[§] *Bull. Soc. chim. Fr.* 1937, **4**, 2052.   [‖] *Ber. dtsch. chem. Ges.* 1936, **69**, 137.

The kinetics of the photo-dimerization of anthracene were investigated by Weigert.* On irradiation in benzene, toluene, or in xylene solution at various temperatures, anthracene was found to dimerize at a rate increasing with concentration up to an asymptotic value of about 2%. Weigert suggested that an anthracene molecule (A) absorbs light to give an activated molecule (A') which then collides with a normal molecule to form an activated molecule (A_{III}) of different type. He assumed that dianthracene (D) is formed by collision between two molecules of A_{III}. That is:

(i) $A + h\nu \rightarrow A'$,
(ii) $A' + A \rightarrow A + A_{III}$,
(iii) $A' \rightarrow A + h\nu_2$,
(iv) $A_{III} + A_{III} \rightarrow D$,
(v) $A_{III} \rightarrow A$.

The kinetics of the photo-oxidation reaction[†] indicate that the two processes are very similar. The following hypothetical mechanism has been suggested for the photo-oxidation of rubrene (R):

(i) $R + h\nu \rightarrow R^*$,
(ii) $R^* \rightarrow R + h\nu_2$,
(iii) $R^* + R \rightarrow R + R'$,
(iv) $R' \rightarrow R + h\nu_3$,
(v) $R^* + O_2 \rightarrow RO_2^*$,
(vi) $R' + O_2 \rightarrow RO_2$,
(vii) $RO_2^* \rightarrow R + O_2$,
(viii) $RO_2^* + R \rightarrow RO_2 + R$.

The activated molecule (R') corresponds to the activated molecule (A_{III}) which Weigert postulated in the photo-dimerization of anthracene.

* *Naturwissenschaften*, 1927, **15**, 124.
† Gaffron, *Biochem. Z.* 1933, **264**, 251; Bowen and Steadman, *J. Chem. Soc.* 1934, p. 1098; Koblitz and Schumacher, *Z. phys. Chem.* 1937, **35**B, 11; Bowen, *Quart. Rev. Chem. Soc., Lond.*, 1947, **1**, 1.

The simplest explanation would seem to be that A_{III} and R', and similar activated molecules formed by the irradiation of other substances, must be diradicals (as (I), (II)), and (i) that photo-dimerization occurs by the combination of two such diradicals, and (ii) that photo-oxidation occurs by the addition of oxygen to such a diradical.

(I)

(II)

The assumption that anthracene derivatives can be converted into diradicals (as (I), (II)), either by heating or by irradiation, has been made repeatedly.* No conclusive evidence has ever been presented, however, for observations such as those of Ingold and Marshall† are capable of other interpretations.

As a matter of fact, magnetic measurements have failed to reveal the presence of diradicals either in irradiated or in other solutions,‡ and even with the very reactive hydrocarbons such as pentacene, diradicals cannot be detected by magnetic susceptibility measurements.§ On the other hand, these experiments do not prove that a diradical cannot have a *transitory* existence; and, in any case, Born and Schönberg∥ have pointed out that the method of Müller could not be expected to detect free radicals at the concentrations involved.

It seems, therefore, that the most satisfactory interpretation of the reactions of photo-dimerization and photo-oxidation, at the present time, is that light energy *does* transform anthracene and anthracene derivatives into diradicals;¶ but further work on the mechanisms of these reactions is clearly required.

---

* For example, Clar, *Ber. dtsch. chem. Ges.* 1932, **65**, 503.
† *J. Chem. Soc.* 1926, p. 3080.
‡ Müller, *Z. Elektrochem.* 1934, **40**, 542.
§ Müller and Müller-Rodloff, *Liebigs Ann.* 1935, **517**, 134.
∥ *Nature, Lond.*, 1950, **166**, 307.
¶ Cf. Schönberg, *Trans. Faraday Soc.* 1936, **32**, 514.

CHAPTER 10

ABSORPTION AND FLUORESCENCE SPECTRA OF AROMATIC COMPOUNDS

BENZENE AND SUBSTITUTED BENZENES

The absorption and emission spectra of organic molecules owe their origin to changes in energy. A change from a lower to a higher energy state is accompanied by absorption of light, and a change from a higher to a lower energy state is accompanied by an emission of light. The frequency of the light emitted or absorbed is given by
$$\Delta E = h\nu,$$
where ΔE is the energy difference between the two levels, h is Planck's constant and ν is the frequency.

The energy of a molecule may be divided broadly into kinetic (rotational and vibration-rotational) energy and electronic (binding) energy. Changes in the rotational energy give absorption bands in the far infra-red region, and changes in the vibration-rotational energy in the near infra-red (see Chapter 1). The Raman spectrum also furnishes information on changes in kinetic energy; visible and near ultra-violet light is absorbed and 'scattered', part of the scattered light being at frequencies other than the incident light. These changes in frequency, called Raman shifts, are due to the energy absorbed in exciting atomic vibrations. Unfortunately, the infra-red and Raman spectra of aromatic compounds are complex and their interpretation is very often difficult at present; but rapid progress is being made in this field.

In the present state of knowledge, the visible and ultra-violet absorption spectra of aromatic compounds are possibly of the greatest value to organic chemists, for absorption in this region corresponds with electronic transitions. The simplest organic molecules have their main absorption bands in the far ultra-violet (1–2000 Å.), a region which has not yet been extensively investigated, as the experimental difficulties have only recently been largely overcome.* More complex substances, including the

* Walsh, *Ann. Rep. Chem. Soc.* 1947, **44**, 32.

aromatic compounds, have their main absorption bands in the near ultra-violet (2000–4000 Å.). Absorption in this region corresponds to energy increases of 70–140 kcal./mole, intermediate between that required for the photo-ionization of single covalent bonds and that required for vibrational or rotational transitions.*

Although the frequency, or wave-length, of the light absorbed by any molecule is determined by the energy of the transition involved, the intensity of absorption is apparently governed by the probability of that particular transition; that is, the maxima in the absorption curves correspond with the most probable transitions. With molecules such as the aromatic compounds, the absorption regions are broad, because there are a number of vibrational and rotational states all belonging to the same electronic transition. Complex aromatic compounds usually give complex absorption spectra, with many maxima.

Absorption curves usually take the form of a plot of the molecular extinction coefficient, ϵ (or $\log \epsilon$), against the frequency (or wave-length) of absorption, for ϵ is related to the intensity of absorption by the Beer-Lambert law:

$$\log (I_{\lambda_0}/I_\lambda) = \epsilon c l,$$

where I_{λ_0} is the intensity of incident light of wave-length λ, I_λ is the intensity of transmitted light, c is the concentration of absorbent in gram-moles per litre, and l is the thickness of the absorbing layer in centimetres. Both the theory relating ϵ to the molecular dimensions[†] and the classical theory of light absorption predict that ϵ can have a maximum value of 10^5–10^6, which is, in fact, the order of the highest values which are observed.

The main absorption bands of ethylene[‡] are in the region 1745 Å. to about 1600 Å. All compounds having an isolated ethylenic double bond show high-intensity absorption in this region, and unsaturated centres of this nature are called 'chromophores'. The effect of the presence of two or more chromophores in a molecule is normally additive if the centres of unsaturation

---

* Braude, *Ann. Rep. Chem. Soc.* 1945, **42**, 105.
† Braude, *J. Chem. Soc.* 1950, p. 379; *Nature, Lond.*, 1945, **155**, 753.
‡ Price and Tutte, *Proc. Roy. Soc.* A, 1940, **174**, 207.

are separated by two or more bonds; that is, if the double bonds are not conjugated.

The nature of the electronic transitions involved during absorption by ethylene and related compounds has been reviewed by Bowen.[*] As already indicated (Chapter 2), three electrons from each carbon atom provide three σ-type coplanar orbitals at an angle of 120° to one another. The p-orbitals of the two carbon atoms, having nodes in the plane of the σ-orbitals, interact with one another to form a π-orbital. This orbital is designated π_u, the subscript indicating that the orbital has opposite phases on each side of the node, and, according to Bowen, the longest-wave absorption corresponds to one electron passing from a π_u- to a π_g-orbital, that is, developing a new nodal plane across the C—C link.[†]

The absorption spectra of butadiene and of other conjugated polyenes can be interpreted in terms of rather similar electronic transitions, and as the energy change involved is *less* than that for ethylene the absorption maxima for polyenes occurs at longer wave-lengths. The greater the number of conjugated double bonds, the longer the wave-length of the absorbed light.

The absorption spectra of aromatic compounds differ from those of the conjugated polyenes, although certain similarities are evident. Benzene exhibits three main regions of absorption in the ultra-violet. The first is a band of intense absorption (maximum at 1790 Å.) in the far ultra-violet.[‡] The second is a region of high-intensity absorption around 2000 Å., and the third is a region of low-intensity absorption in the 2300–2600 Å. region. The latter evidently corresponds to that of hexatriene in the 2600 Å. region. The three main regions of benzene absorption may be referred to as the group I, II and III bands respectively, in agreement with Braude.[§] These subdivisions now seem to be generally accepted, but various authors have used other names for the regions, and some confusion has resulted. Table 10·1 summarizes some of the different nomenclatures which have been used.

[*] *Ann. Rep. Chem. Soc.* 1943, **40**, 12.
[†] The subscripts g and u are short for *gerade* (even) and *ungerade* (odd).
[‡] Price and Walsh, *Proc. Roy. Soc.* A, 1947, **191**, 22; Klevens and Platt, *J. Chem. Phys.* 1949, **17**, 470.
[§] *Ann. Rep. Chem. Soc.* 1945, **42**, 105.

TABLE 10·1. *Nomenclature of different regions of absorption in benzene*

Name of region				$\lambda_{max.}$ (in Å.)[†]
Braude[*]	Clar[†]	Doub and Vandenbelt[‡]	Klevens and Platt[§]	
Group I bands	β-bands	Second 'primary' bands	B_b	1790
Group II bands	*para* bands	'Primary' bands	L_a	1978 2034 2068
Group III bands	α-bands	'Secondary' bands	L_b	2290 2339 2378 2435 2487 2547 2607 2886

Considerable attention has been given to the effect of substituents on the absorption spectra of aromatic compounds, and Doub and Vandenbelt[‖] have made an extensive survey of the effect of substituents on the spectrum of benzene. It is found that there is often a pronounced effect on the *intensity* of the bands, but that the general shape of the absorption curve is usually unchanged. For example, the absorption spectra of toluene and of chlorobenzene are very similar to that of benzene, except that the absorption bands are shifted to longer wavelengths.[¶] A *bathochrome* is a substituent group which produces a shift of an absorption band towards longer wave-lengths, and a *hypsochrome* is one which produces a shift towards shorter wave-lengths. Under certain circumstances (e.g. methylazulenes) a given group can act either as a bathochrome or a hypsochrome, depending on the position of substitution and on the absorption band concerned. In general, however, sub-

---

[*] Braude, *Ann. Rep. Chem. Soc.* 1945, **42**, 123.
[†] Clar, *Aromatische Kohlenwasserstoffe*, 2nd ed., Berlin, Springer-Verlag, 1952.
[‡] Doub and Vandenbelt, *J. Amer. Chem. Soc.* 1947, **69**, 2714.
[§] Klevens and Platt, *J. Chem. Phys.* 1949, **17**, 470.
[‖] *J. Amer. Chem. Soc.* 1947, **69**, 2714; 1949, **71**, 2414.
[¶] Price and Walsh, *Proc. Roy. Soc.* A, 1947, **191**, 22.

stituents in carbocyclic six-membered rings produce bathochromic shifts, and hypsochromic effects in these compounds are rare.

The magnitude of the bathochromic shift varies considerably with the nature of the substituent. This is illustrated by the tabulation of the positions and intensities of the absorption bands of substituted benzenes, in aqueous solution, given in table 10·2.

TABLE 10·2. *Ultra-violet absorption characteristics of some mono-substituted benzenes in aqueous solution*[*]

Compound	Group II band		Group III band	
	$\lambda_{max.}$	$\epsilon_{max.}$	$\lambda_{max.}$	$\epsilon_{max.}$
Benzene	2,035	7,400	2,540	204
Aniline cation	2,030	7,500	2,540	160
Toluene	2,065	7,000	2,610	225
Chlorobenzene	2,095	7,400	2,635	190
Bromobenzene	2,100	7,900	2,610	192
Phenol	2,105	6,200	2,700	1,450
Anisole	2,170	6,400	2,690	1,480
Benzenesulphonamide	2,175	9,700	2,645	740
Benzonitrile	2,240	13,000	2,710	1,000
Benzoic acid anion	2,240	8,700	2,680	560
Benzoic acid	2,300	11,600	2,730	970
Aniline	2,300	8,600	2,800	1,430
Phenol anion	2,350	9,400	2,870	2,600
Acetophenone	2,455	9,800	—	—
Benzaldehyde	2,495	11,400	—	—
Nitrobenzene	2,685	7,800	—	—

Examination of the table shows that the group II ('primary') and group III ('secondary') bands are shifted to longer wavelengths by both electron-attracting and electron-donating substituents, and the two types of substituent may be arranged in order of the magnitude of the shift in wave-length which they produce:

Ortho-para-directing,

$$CH_3 < Cl < Br < OH < OCH_3 < NH_2 < O^-.$$

Meta-directing,

$$NH_3^+ < SO_2NH_2 < CO_2^- = CN < CO_2H < COCH_3 < CHO < NO_2.$$

[*] After Doub and Vandenbelt, *J. Amer. Chem. Soc.* 1947, **69**, 2716.

In both series the largest bathochromic shifts are produced by those groups having the greatest electronic interaction with the ring. It is noteworthy that the amine cation, which produces exclusive *meta* substitution, causes little or no bathochromic shift. On the other hand, the nitro group, which is also strongly *meta*-directing, produces a very large bathochromic shift. This difference indicates that the shift cannot be associated with the coulombic displacement, and Doub and Vandenbelt therefore associate the magnitude of the bathochromic shift largely with the extent of the mesomeric or resonance effect. This conclusion is supported by the work of Kiss, Molnár and Sandorfy,* who have pointed out that the magnitude of the shift in the absorption bands produced by —OH, —OCH$_3$ and by —CH$_3$ substituents may be correlated with the degree of conjugation of these groups as determined by other methods.

Doub and Vandenbelt also examined the spectra of large numbers of disubstituted compounds. In the *para*-disubstituted series it was found that when two groups of the same type are opposed, the bathochromic shift does not differ greatly from that of the most displaced mono-substituted compound. On the other hand, with compounds having two substituents of opposite type *para* to one another, the shift is considerably greater than that given by the most displaced mono-substituted compound, and the increase is greatest with those substituents causing the greatest shifts. In other words, the displacement is greatest for those groups having greatest interaction across the ring (*ortho-para-* against *meta*-directing) and least for those capable of least interaction across the ring (*ortho-para-* against *ortho-para-*, and *meta-* against *meta-*).

DIPHENYL AND THE POLYPHENYLS

The spectra of diphenyl (I) and the *para*-polyphenyls differ sharply from the spectrum of benzene, the group II region of absorption being progressively displaced towards longer wavelengths for every additional phenyl group. Pronounced conjugation between the rings, across the coannular bond, therefore

* *C.R. Acad. Sci., Paris,* 1948, **227**, 724.

seems to be indicated. In the customary valence-bond terminology, this may be interpreted in terms of the contribution of structures of type (II) and (III) to the resonance hybrid.

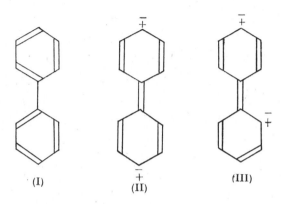

(I)   (II)   (III)

If this is the case, the coannular bond in diphenyl should be somewhat shorter than the normal value for a single bond, and this has been verified experimentally. X-ray diffraction measurements have shown that the length of this bond is 1·48 Å., as against the normal C—C value of 1·54 Å.* The *para*-polyphenyls are evidently similar, with extended conjugation through all the rings. The λ_{max}. of the group II region is shifted progressively to longer wave-lengths as the length of the molecule is increased (see fig. 10·1 and table 10·3).†

In the series of *meta*-polyphenyls, extended conjugation is not possible, and the wave-length of the group II region of absorption remains practically unchanged, at 2515 Å., for the entire series.‡

The pronounced conjugation between the rings which occurs in diphenyl, and the resulting double-bond character of the coannular bond, tends to fix, or restrict, the molecule to a coplanar configuration. Free rotation of one ring with respect

---

* Dhar, *Indian J. Phys.* 1932, **7**, 43.
† It is clearly incorrect to associate the diphenyl maximum at 2515Å. with the benzene maximum at c. 2550Å., as has been done by Wheland, *The Theory of Resonance*, p. 151, John Wiley, New York, 1944. The benzene maximum at c. 2000Å. is the comparable one, as is at once apparent from fig. 10·1.
‡ Gillam and Hey, *J. Chem. Soc.* 1939, p. 1170; Ferguson, *Chem. Rev.* 1948, **43**, 385.

TABLE 10·3. *Positions of absorption maxima in the spectra of para- and meta-polyphenyls**

n	$\lambda_{max.}$
Benzene	2068[a]
0	2515[b]
1	2800[b]
2	3000[b]
3	3100[b]
4	3175[b]

n	$\lambda_{max.}$
0	2515
1	2515
7–12	2530
13	2540
14	2550

(*a*) in hexane; (*b*) in chloroform.

to the other about this common bond is therefore inhibited. On the other hand, if suitably large *ortho* substituents are introduced into diphenyl, the resulting steric hindrance prevents the assumption of a coplanar configuration, and conjugation between the rings is thereby prevented, or inhibited. This is reflected in the absorption spectra. The spectrum of dimesityl (IV), for example, is unlike that of diphenyl (I) and resembles that of mesitylene. In this case the *ortho*-methyl groups (which have a van der Waals radius of 2·0 Å.) interfere with one another, and effectively prevent both free rotation and the assumption of a coplanar configuration.

* After Ferguson, *Chem. Rev.* 1948, **43**, 397.

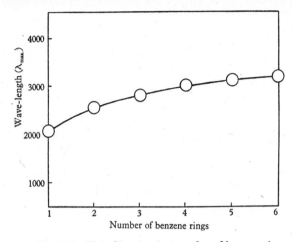

Fig. 10·1. Plot of $\lambda_{max.}$ against number of benzene rings for benzene and the *para*-polyphenyls.

An even simpler case of steric inhibition of resonance is exhibited by many phenyl-substituted anthracenes.* The spectra of 9-phenylanthracene, 9:10-diphenylanthracene, 1-phenylanthracene and of 1:4-diphenylanthracene are all very similar to

that of anthracene itself, indicating that the phenyl groups are sterically prevented from becoming coplanar with the anthracene ring system. Resonance phenomena of the type found in diphenyl

* Hirshberg, *Trans. Faraday Soc.* 1948, **44**, 285.

are therefore impossible, and the spectra of these derivatives are very largely the sum of the spectra of anthracene and benzene, the latter being negligible in the region of anthracene absorption. On the other hand, the spectrum of 2-phenylanthracene, in which there is no steric interference with coplanarity, shows a very different absorption curve. The absorption is less intense, and the maximum is shifted to longer wave-lengths. It is probable that resonance forms of type (V) contribute significantly to the resonance hybrid.

SPECTRAL RELATIONSHIPS IN AROMATIC HYDROCARBONS

Most of the more complex aromatic hydrocarbons give rise to absorption spectra having three well-defined regions of absorption in the ultra-violet and/or the visible regions. The first of these (group I bands) usually has $\epsilon_{max.}$ 4·5–5·2; the second (group II bands) 3·6–4·1; and the third (group III bands) is a region of low-intensity absorption, $\epsilon_{max.}$ 2·3–3·2. Investigations into the spectral relationships of the polycyclic aromatic hydrocarbons have been carried out by several authors, but particularly by Clar, who has reviewed his work in some detail.*

Clar has found that not only do the polycyclic hydrocarbons give rise to regions of absorption which are closely related to those of benzene, but that the shifts produced by linear or angular 'anellation' follow certain definite and simple rules. In the polyacenes, for example (that is, in the series of *linear* benzologues of benzene), all the absorption bands are progressively displaced towards longer wave-lengths as the number of rings increases. The first maximum in each group of absorption bands for benzene is at 1790, 2068 and 2607 Å., respectively. For naphthalene, the wave-lengths of the *corresponding* bands are 2210, 2850 and 3100 Å. As the length of the molecule increases, the group II bands are shifted to a greater extent than the group I and group III bands, with the result that in anthracene and the higher polyacenes the group III absorption is 'swamped' under that of group II. This is clear from the graphical representation of the absorption of the polyacenes in fig. 10·2.

* *Aromatische Kohlenwasserstoffe*, 2nd ed., pp. 25 et seq., Springer-Verlag, 1952.

The human eye is sensitive to the region from about 4000 to 7500 Å., and substances having important absorption bands in this region are therefore coloured, the visible colour being complementary to that absorbed, as given in table 10·4.

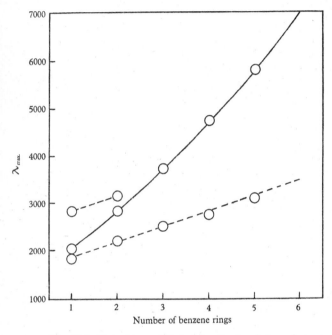

Fig. 10·2. Plot of $\lambda_{max.}$ of first absorption bands in each group against number of benzene rings in polyacenes.

It should be noted that this table is strictly only applicable to molecules having a single narrow absorption band; in practice, most aromatic substances have a number of bands of varying intensity, and it is therefore difficult to predict the visual colour *exactly*.

The first maximum in the group II region of the absorption spectrum of naphthacene is at 4735 Å., and this compound is orange. Similarly, the absorption bands of the higher benzologues, namely, pentacene and hexacene, are still further displaced, and these compounds are deep blue and deep green respectively. This is summarized in table 10·5.

TABLE 10·4. *Absorption wave-length and colour**

Wave-length (Å.)	Colour absorbed	Visible colour
4000–4350	Violet	Yellow-green
4350–4800	Blue	Yellow
4800–4900	Green-blue	Orange
4900–5000	Blue-green	Red
5000–5600	Green	Purple
5600–5800	Yellow-green	Violet
5800–5950	Yellow	Blue
5950–6050	Orange	Green-blue
6050–7500	Red	Blue-green

TABLE 10·5. *Absorption spectra of polyacenes*

Compound	Positions of first maxima			Colour of compound
	Group I	Group II	Group III	
Benzene	1790†	2068‡	2607‡	Colourless
Naphthalene	2210§	2850§	3100§	Colourless
Anthracene	2515§	3745§	—	Colourless
Naphthacene	2780‖	4735‖	—	Orange
Pentacene	3100‖	5800‖	—	Deep blue
Hexacene	—	—	—	Deep green

(I)   (II)   (III)

It is usually possible to discern three main regions of absorption in the spectra of 'angular' polycyclic compounds as well as in the polyacenes. Here again the bands shift to longer wave-length

---

* After Branch and Calvin, *The Theory of Organic Chemistry*, p. 156, Prentice-Hall, New York, 1941.
† From Price and Walsh, *Proc. Roy. Soc.* A, 1947, **191**, 22; all other figures from Clar, *Aromatische Kohlenwasserstoffe*.
‡ In hexane.   § In alcohol.   ‖ In benzene.

with increasing number of benzene rings. As already indicated, the first maxima in each group of absorption bands of benzene (I) are at 1790, 2068 and 2607 Å. For phenanthrene (II) the corresponding bands are at 2507, 2930 and 3450 Å. Similarly, for pentaphene (III) the corresponding bands are at 3170, 3590 and at 4230 Å. These and other results are given graphically in fig. 10·3, and it is apparent that angular anellation results in a much smaller shift in the group II bands than does linear anellation.

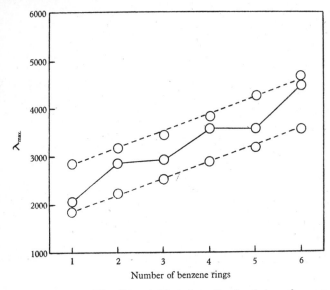

Fig. 10·3. Plot of $\lambda_{max.}$ of first absorption bands in each group against the number of benzene rings in the 'phenes'.

Clar has found that in both the linear and angular compounds the positions of the absorption bands may be calculated from equations of the type

$$\sqrt{\frac{R}{\nu}} = n,$$

where R is a constant, ν is the frequency, and n is an integer (called the 'order number') which can be determined by vector addition of the double bonds. The equation has been used to calculate the positions of the absorption bands of unknown hydrocarbons, and subsequent preparation of such hydrocarbons, and examination of their spectra, has invariably shown a very good agreement between the calculated and observed wave-lengths. The theoretical interpretation of the spectral relationships found by Clar is far from clear, but as empirical relationships, and for the identification of unknown hydrocarbons, these spectral relationships are of very considerable value.

SUBSTITUTED POLYCYCLIC AROMATIC HYDROCARBONS

The absorption spectra of substituted polycyclic compounds closely resemble those of the corresponding unsubstituted compounds, except that the absorption bands are usually shifted to somewhat longer wave-lengths. This is only to be expected by analogy with the substituted benzenes, but the substituted polycyclic compounds exhibit several other features of interest. For one thing, all the possible positions of substitution are not identical. Steric inhibition of resonance is often observed, and, furthermore, the various groups of absorption bands (I, II and III) are not always shifted to the same extent.

There are two main regions of absorption in the ultra-violet spectrum of anthracene, and eight fairly well-defined bands may be distinguished (fig. 10·4). In agreement with Jones* these may be labelled $A...H$, the bands A and C being points of inflexion. Bands $A-B$ constitute the group I absorption and correspond to the benzene maximum at 1790 Å. Bands $C-H$ constitute the group II absorption and correspond to the 2000 Å. region of absorption in benzene. It seems probable that the various maxima $(C-H)$ in the group II region are vibrational sublevels, and it is interesting to note that the frequency difference between the bands is practically constant and equal to about 1400 cm.$^{-1}$. Jones associates these vibrational sublevels with the $C-H$ bending vibration.

Substitution in the anthracene molecule usually gives rise to a bathochromic shift, but the two regions are not affected to the

* *Chem. Rev.* 1947, **41**, 353.

same extent. With *meso* substituents, the group II bands are shifted more than those of group I. The frequency difference between the various maxima in the group II region remains practically constant with substitution, and it is therefore possible to estimate the magnitude of the shift in this region of the spectrum by averaging the individual shifts on the frequency scale. Some results derived in this way have been given in table 10·6.

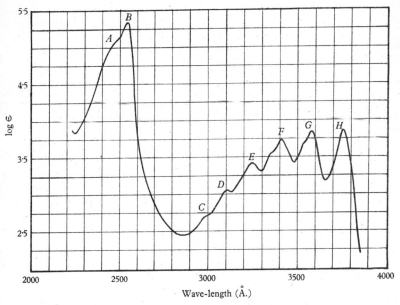

Fig. 10·4. Ultra-violet absorption spectrum of anthracene in ethanol. (After Jones, *Chem. Rev.* 1947, **41**, 353.)

It will be noted that, especially with the unsaturated substituents (—CN, —CHO) exhibiting pronounced conjugation effects, the group II bands are shifted much farther than the group I bands. It seems likely, therefore, that the group II bands are associated with transverse polarization, represented by classical ionic structures such as (I)–(III), and that the group I bands are associated with longitudinal polarization represented by structures such as (IV), (V).

TABLE 10·6. *Average frequency shift in each group of absorption bands produced by* meso *substitution in anthracene*

Substituent	Bathochromic shift	
	Group I (m.$^{-1}$)	Group II (m.$^{-1}$)
9-Methyl	5·4	7·6
9-Nitro	0·4	7·5
9-Cyano	6·5	18·65
9-Aldehyde	15·5	27·9
9:10-Dimethyl	12·6	15·1
9:10-Diphenyl	11·1	11·9

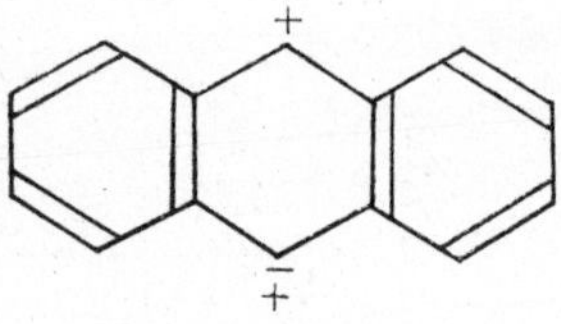

(I)

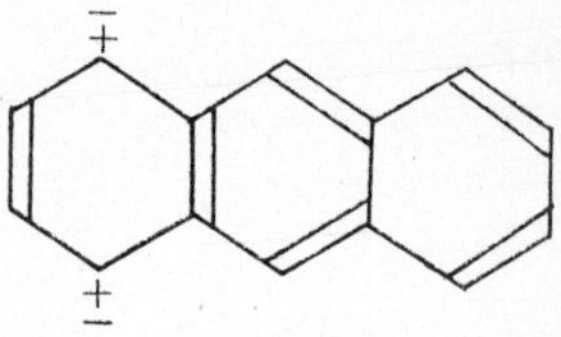

(II)

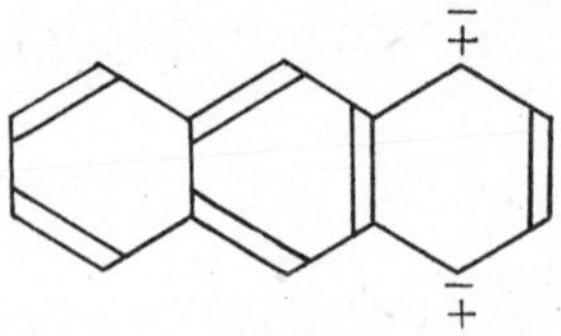

(III)

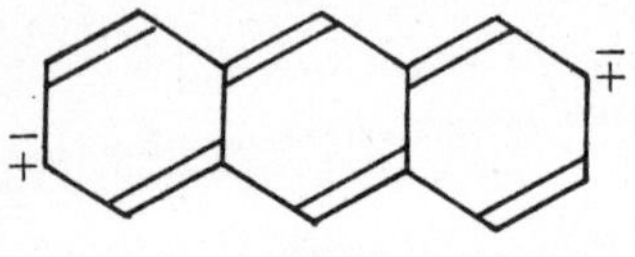

(IV)

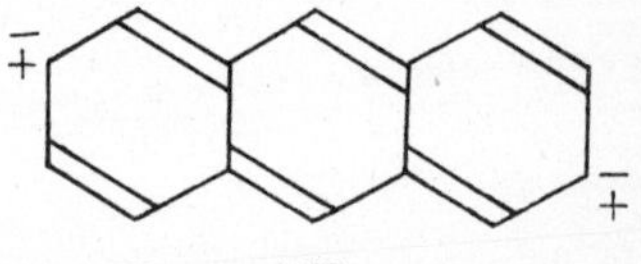

(V)

This being so, it might be expected that substituents in the 2-position would produce greater shifts in the group I region, and there is some experimental evidence to support this suggestion. For example, if one compares the spectra of 9:10-diphenylanthracene and 2:3:9:10-tetraphenylanthracene, it appears that the introduction of the two phenyl groups at positions 2 and 3 displaces the group II bands by only 8 m.$^{-1}$, but the group I bands by 40 m.$^{-1}$.

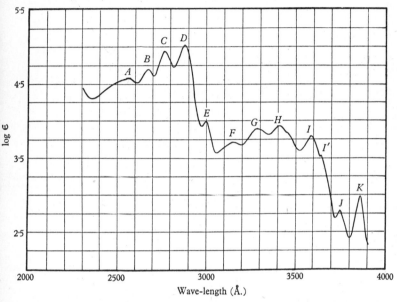

Fig. 10·5. Ultra-violet absorption spectrum of 1:2-benzanthracene in ethanol.

Similar results have been obtained with the spectra of *meso*-substituted 1:2-benzanthracenes.[*] There are twelve well-defined absorption bands in the spectrum of 1:2-benzanthracene, and these have been labelled $A...K$.[†] As with benzene, naphthalene and all the *angular* hydrocarbons, they fall into three groups. Group I includes the bands $ABCDE$. Group II includes the bands $FGHI$, and group III the bands $I'JK$. The average frequency difference between the maxima in group I is 1490 cm.$^{-1}$; that for group II is 1300 cm.$^{-1}$; but that for the group III maxima is only 700 cm.$^{-1}$. *Meso* substituents in benzanthracene were found to cause greatest displacement in the group II bands, the shift in group I and group III bands being somewhat smaller and approximately equivalent.

[*] Badger and Pearce, *J. Chem. Soc.* 1950, p. 3072.
[†] Jones, *Chem. Rev.* 1943, **32**, 1.

It seems reasonable to conclude that, as in benzene, the bathochromic shifts produced by substituents in anthracene and in benzanthracene are directly associated with the extent of conjugation of the substituent with the ring. However, the substituents which give rise to the greatest displacements in benzene sometimes produce relatively small displacements when present in the *meso* positions of anthracene and benzanthracene. For example, the nitro group, which produces a pronounced bathochromic shift in benzene, gives rise to only a relatively small displacement in the absorption bands of anthracene and benzanthracene (table 10·6). Furthermore, the spectrum of 9:10-diphenylanthracene resembles that of anthracene much more closely than would be expected by analogy with benzene and terphenyl. It is true that a bathochromic shift is produced (table 10·6), but this is only a fraction of the displacement which could be expected. In terphenyl, the conjugation may be represented by a number of ionic structures of type (VI), and the

(VI)

coannular bonds have considerable double-bond character, a fact which indicates that the molecule must be essentially in a coplanar configuration. As indicated in the previous section, however, the presence of suitable *ortho* substituents in diphenyls prevents the assumption of a coplanar configuration and extensive conjugation between the rings is prevented. This also appears to be the case with certain *meso*-substituted anthracenes. In 9:10-diphenylanthracene, steric interference occurs between the hydrogen atoms at the 1- and 8-positions of the anthracene ring system and the *ortho* positions of the *meso*-phenyl substituents. In this way the assumption of a coplanar configuration is prevented and conjugation is thereby markedly reduced.

Jones[*] has calculated that the plane of the phenyl rings cannot approach the plane of the anthracene ring system closer than

* *J. Amer. Chem. Soc.* 1945, **67**, 2127.

an angle of 57°. In the same way he has calculated that *meso*-methoxy groups are strained even when rotated 90° from the plane of the anthracene ring system, and other results of the same nature are given in table 10·7.

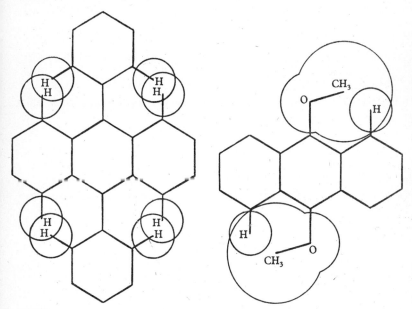

Fig. 10·6. 9:10-Diphenylanthracene.   Fig. 10·7. 9:10-Dimethoxyanthracene.

TABLE 10·7. *Minimum angle of approach between the plane of the substituent group and the plane of the anthracene ring system*

Compound	Angle
9-Methoxyanthracene	90°
9-Phenylanthracene	57°
9-Anthraldehyde	45°
9-Aminoanthracene	29°

This means that very many *meso*-substituted anthracenes, and very many other substituted polycyclic compounds, cannot show the degree of conjugation between the substituent and the ring system which might be expected from a consideration of the same substituents in benzene or in other molecules in which such steric inhibition of resonance does not occur.

Jones considered the steric interference solely in terms of the *rotation* of the substituent, but this ignores the fact that such groups may be displaced *as a whole* either above or below the plane of the ring system. Such displacements are now known to occur in many cases.*

It seems reasonable to conclude, therefore, that the chief reason for the difference in the relative magnitudes of the displacements given by a *series* of substituents in benzene and those given by the same series of substituents in certain polycyclic compounds is that the resonance interaction is sometimes sterically hindered in the latter.

The third factor of interest in connexion with the study of the absorption spectra of substituted polycyclic compounds concerns the effect of the same substituent at different positions in the molecule. In naphthalene, for example, a substituent can be present either in the 1- or in the 2-position, and the bathochromic shift appears to be greater if it is in the former. In anthracene, a substituent can be present either in the 1-, 2- or 9-positions, and the bathochromic shift is always greatest if it is in the 9-position.† This is evidently the reason for the fact that, in general, only *meso*-substituted anthracene derivatives are coloured. Only if the substituents are present in the *meso* positions is the displacement of the absorption sufficient to bring it into the visible region.

As there is little doubt that the magnitude of the displacement is associated with the extent of conjugation, it seems that a substituent in the 1-position of naphthalene must be conjugated with the ring system to a greater extent than the same substituent in the 2-position. This is perfectly reasonable, for, as explained in Chapter 6, seven resonance structures showing such conjugation may be written for 1-substituted naphthalenes, and only six for 2-substituted naphthalenes. It is to be expected, therefore, that the bond joining a substituent to the 1-position of naphthalene will have greater double-bond character than that joining

---

* Newman and Hussey, *J. Amer. Chem. Soc.* 1947, **69**, 978, 3023; Bell and Waring, *Chem. & Ind.* 1949, p. 321; Everard and Sutton, *J. Chem. Soc.* 1949, p. 2312.
† Pullman, *Bull. Soc. chim. Fr.* 1948, **15**, 533.

the same substituent to the 2-position. By analysis of the dipole moments of the halogenated naphthalenes, Ketelaar and Oosterhout* have shown that the C—Cl bond in 1-chloronaphthalene has 14·2% double-bond character, as against 13·0% for the isomeric 2-chloronaphthalene. Similarly, the C—X bond in 9-substituted anthracenes must have greater double-bond character than in either 1- or 2-substituted anthracenes. Theoretical studies indicate that this is probably the case, and to some extent the concept is supported by chemical evidence. For example, *meso*-methyl groups in anthracene and benzanthracene are very readily brominated, indicating extensive conjugation; but methyl groups substituted at other positions in the anthracene ring system do not seem to be affected under the same conditions.

Pullman† has suggested that the extent of conjugation of a given substituent, such as a methyl group, is determined by the free-valence number at the position of substitution. This conclusion has recently received some support from Daudel,‡ who has found that there is a linear relationship between the free-valence number and the bond order of the bond linking the substituent to the ring. It is to be expected, therefore, that in any given polycyclic compound the bathochromic shift produced by a methyl group should be related to the free-valence number at the position of substitution, and in a few compounds an approximate correlation has been observed.

SPECTRA OF NON-BENZENOID AROMATIC HYDROCARBONS

Among the non-benzenoid aromatic hydrocarbons, only azulene (I) and the alkyl-azulenes have been studied extensively. Azulene itself is a blue-violet compound, and its absorption spectrum extends from the ultra-violet region well into the visible.§ In the ultra-violet region the spectrum resembles that of naphthalene and other similar aromatic compounds, and the

---

* *Rec. trav. chim. Pays-Bas*, 1946, **65**, 448.
† *C.R. Acad. Sci., Paris*, 1946, **222**, 1396.
‡ Ibid. 1950, **230**, 99.
§ Cf. Clar, *J. Chem. Soc.* 1950, p. 1823.

introduction of methyl substituents invariably produces a bathochromic displacement of the absorption bands of this part of the spectrum. In the visible region, however, only two of the possible five mono-methyl-azulenes show the usual bathochromic effect, the three remaining isomers showing hypsochromic shifts, that is, a shift to *shorter* wave-lengths with substitution. Thus, substitution in the 1-, 3-, 5- and 7-positions gives a shift to longer wave-lengths (smaller frequencies), and substitution in the 2-, 4-, 6- and 8-positions has the opposite effect.* According to Brown and Lahey[†] the two sets of positions are those of (local) high and low π-electron densities respectively. Using the absorption band of longest wave-length, the wave-number shift for each position of the methyl group may be summarized as in (II).

For di- and poly- substitution, the shifts are approximately additive, and it is possible to predict both the position of the absorption band of longest wave-length and also the colour of the compound from the above data. Among the mono-methyl compounds, for example, the 1- and 5-methyl compounds are *blue*, in accordance with the shift in the absorption to longer wave-lengths. The 2-, 4- and 6-methyl derivatives are *violet*, in accordance with the shift in absorption to shorter wave-lengths.

The absorption spectra of the alkyl-azulenes in the ultra-violet region have been investigated by Plattner and Heilbronner.[‡] In this part of the spectrum, the introduction of a methyl group always produces a shift to longer wave-length, but the magnitude of the displacement varies with the position of substitution and is much greater when the five-membered ring is involved. The absorption bands are not all affected to the same extent. In the

---

* Plattner, *Helv. chim. acta*, 1941, **24**, 283E; Plattner and Heilbronner, ibid. 1947, **30**, 910.
† *Aust. J. Sci. Res.* A, 1950, **3**, 593.   ‡ *Helv. chim. acta*, 1948, **31**, 804.

spectra of 1-methyl- and 2-methyl-azulenes, the absorption bands in the region 3000–3800 Å. are displaced to a much greater extent than the bands in the region 2200–3000 Å.

TABLE 10·8. *Observed shifts for the longest wave-length absorption band for alkylated azulenes*

Compound	Wave number	Shift	Colour of compound
Azulene	14,350	—	Blue-violet
1- (or 3-) Methylazulene	13,550	−800	Blue
2-Methylazulene	14,790	+440	Violet
4- (or 8-) Methylazulene	14,710	+360	Violet
5- (or 7-) Methylazulene	13,940	−410	Blue
6-Methylazulene	14,690	+340	Violet
1:2-Dimethylazulene	13,950	−400	Blue
1:3-Dimethylazulene	12,990	−1360	Blue-green
1:4-Dimethylazulene	13,870	−480	Blue
4:8-Dimethylazulene	15,010	+660	Violet

ABSORPTION SPECTRA OF HETEROCYCLIC AROMATIC SYSTEMS

Among the compounds containing only six-membered rings, the heterocyclic compounds differ from the hydrocarbons only in the replacement of one or more CH groups by one or more nitrogen atoms. The absorption spectra of such heterocyclic compounds are very similar to those of the corresponding hydrocarbons, and although there is often some loss of fine structure, it is usually possible to distinguish the three main regions of absorption (groups I, II and III) which have previously been referred to with the hydrocarbons. Certain important differences can, however, be noted. These may be illustrated by comparison of the spectrum of benzene with that of pyridine and other nitrogen-containing analogues of benzene. With all these compounds the absorption *begins* at somewhat longer wave-lengths than with benzene, but the *maxima* in the group III region are usually shifted to shorter wave-lengths. The most noticeable result of the replacement of a CH group by a nitrogen atom, however, is the very pronounced increase in the *intensity* of the group III absorption bands. In pyridine the ϵ_{max} is about ten times greater than in benzene, and in pyrimidine the ϵ_{max} is still greater. In pyridazine, however, the ϵ_{max} is only about six times that for benzene.* In this region, therefore, the replacement of CH by N in the ring system intensifies absorption, shifts the position of the maxima to shorter wave-lengths, and the start of absorption to longer wave-lengths.

There is very little information on the group II region of absorption in these compounds, but it seems certain that there is no increase in the intensity of absorption in this region. Indeed, the published curve for pyridine seems to indicate that the absorption is shifted to shorter wave-lengths and the intensity somewhat reduced. The group I bands are also shifted to shorter wave-lengths. The λ_{max} is at c. 1700 Å., as against c. 1790 Å. for benzene.†

In the bicyclic compounds, the replacement of a CH group by N also results in a marked increase in the intensity of the group III absorption. The increase is greater in quinoline than in *iso*quinoline, and is considerably less than either in quinazoline. The effect of the hetero atom on the group II region of these spectra is only slight, and the positions of the maxima in the group I region of the heterocyclic analogues of naphthalene have not been determined.‡

In the tricyclic compounds, the effect of a nitrogen atom is again similar. The linear compounds, acridine, 1-azanthracene and phenazine, have only two regions of absorption (I and II), as has anthracene; but with the heterocyclic compounds, absorption begins at somewhat longer wave-lengths, and the maxima in the group II region are shifted to shorter wave-lengths. The intensity of absorption in this region increases progressively with the introduction of nitrogen atoms; but the increase is not as great as in the mono- and dicyclic compounds. The intensity of absorption in the group I region is somewhat reduced by the introduction of nitrogen, but there is very little effect on the position of the maximum.§

---

* Menczel, *Z. phys. Chem.* 1927, **125**, 161; Heyroth and Loofbourow, *J. Amer. Chem. Soc.* 1934, **56**, 1728; Evans and Wiselogle, ibid. 1945, **67**, 60.

† Price and Walsh, *Proc. Roy. Soc.* A, 1947, **191**, 22.

‡ Ewing and Steck, *J. Amer. Chem. Soc.* 1946, **68**, 2181; Elderfield, Williamson, Gensler and Kremer, *J. Org. Chem.* 1947, **12**, 405.

§ Rădulescu and Ostrogovich, *Ber. dtsch. chem. Ges.* 1931, **64**, 2233; Blout and Corley, *J. Amer. Chem. Soc.* 1947, **69**, 763; Craig and Short, *J. Chem. Soc.* 1945, p. 419.

Other data for the absorption spectra of nitrogen-containing six-membered heterocyclic systems also confirm the above conclusions; that is, the most important differences between the spectra of hydrocarbons and of their aza analogues are (i) that the absorption generally starts at somewhat longer wave-lengths, (ii) that the absorption bands of longest wave-length are much more intense than the corresponding bands from hydrocarbons, and (iii) that, in some cases, the maxima are shifted to shorter wave-lengths.*

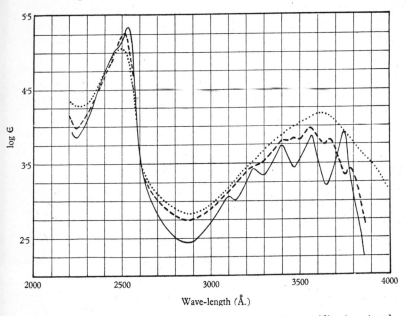

Fig. 10·8. Ultra-violet absorption spectra of anthracene (—), acridine (- - -) and phenazine (· · ·). (Mainly after Rădulescu and Ostrogovich, *Ber. dtsch. chem. Ges.* 1931, **64**, 2233; see also Badger, Pearce and Pettit, *J. Chem. Soc.* 1951, p. 3199.)

The simpler five-membered heterocyclic systems absorb in approximately the same region as the corresponding hydrocarbons, but it is very difficult to compare the spectra as the absorption bands are not well resolved. Thiophen has a low-intensity region of absorption at about 2600 Å., which corresponds to the benzene absorption in the region 2200–2600 Å. It

* Badger, Pearce and Pettit, *J. Chem. Soc.* 1951, p. 3199.

has a second region corresponding to the benzene 2000 Å. region, and an intense region of absorption at 1750–1800 Å., as has benzene. The spectra of pyrrole and furan are similar.*

FLUORESCENCE SPECTRA OF AROMATIC COMPOUNDS

To some extent fluorescence is the reverse of absorption. While absorption represents the excitation of π-electrons from the ground state to one or other of the vibrational sublevels of the excited state, fluorescence represents the return of π-electrons from the lowest vibrational level of the excited state to the various vibrational levels of the ground state. While the constant spacing (on the frequency scale) which is observed in the absorption bands corresponds to the vibrational sublevels of the *excited* state, the similar constant-frequency spacings which are observed in fluorescence spectra correspond to the vibrational sublevels of the *ground* state.† For benzene, therefore, the absorption and fluorescence processes may be represented by the diagrammatic sketch given in fig. 10·9. The fluorescence spectrum is seen to be approximately the 'mirror image' of the absorption spectrum.

The fluorescence spectra of polycyclic compounds are similarly related approximately to the absorption spectra as object is to mirror image, so that the same relationships as have been noted in the absorption spectra are also to be found in the fluorescence spectra. That is to say, (i) substitution generally shifts the fluorescence bands to longer wave-lengths, and (ii) the angular or linear fusion of further benzene rings also shifts the position of the bands to longer wave-lengths, the magnitude of the shift being greater for linear fusion.

Schoental and Scott,‡ for example, have investigated the fluorescence spectra of 1:2-benzanthracene and a number of its alkyl and other derivatives. In general, a methyl substituent was found to shift the bands to longer wave-lengths, the greatest shifts being observed with the 9- and 10-methyl derivatives, and

* Walsh, *Quart. Rev. Chem. Soc.*, Lond., 1948, **2**, 85; Price and Walsh, *Proc. Roy. Soc.* A, 1941, **179**, 201.
† Bowen, *Quart. Rev. Chem. Soc.*, Lond., 1947, **1**, 1.
‡ *J. Chem. Soc.* 1949, p. 1683.

the shift with 9:10-dimethyl-1:2-benzanthracene was found to be 1250 cm.$^{-1}$. Similarly, other substituents, such as —CO_2H, —CN and —NH_2, which are conjugated with the ring to a large extent, were also found to produce large shifts. Again, the

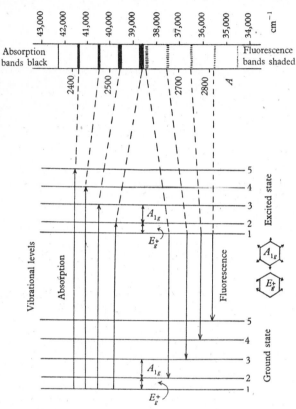

Fig. 10·9. Diagrammatic sketch showing the absorption and fluorescence bands for benzene, with the vibrational transitions. (From Bowen, *Quart. Rev. Chem. Soc., Lond.*, 1947, **1**, 4.)

fluorescence spectra of nitrogen-containing polycyclic compounds were found to resemble those of the corresponding hydrocarbons.

The fluorescence spectra of the polycyclic compounds usually consist of a main set of three or four bands, separated by an approximately constant frequency difference. In some cases a series of less intense bands can also be distinguished, but generally speaking the fluorescence spectra are relatively simple, and as already mentioned the bands are shifted to longer wave-lengths by both linear and angular anellation. Thus, for naphthalene, the main fluorescence bands are at 31,000, 30,000 and 28,900 cm.$^{-1}$; for anthracene, at 26,500, 25,050, 23,650 and 22,300 cm.$^{-1}$; for naphthacene, at 21,200 and 19,750 cm.$^{-1}$; and for pentacene at 17,300 cm.$^{-1}$.[*]

The *intensity* of the fluorescence exhibited by polycyclic compounds shows wide variations even among isomeric substances. For example, Schoental and Scott have drawn attention to the fact that the fluorescence intensities of three isomeric pentacyclic hydrocarbons, 1:2:7:8-dibenzanthracene, 1:2:5:6-dibenzanthracene and naphtho-(2':3'-1:2)-phenanthrene, are in the ratio 1:10:300. The presence of minute traces of impurities can sometimes affect the intensity of fluorescence, and it is well known that the presence of a small amount (1 in 10^4) of naphthacene in anthracene largely suppresses the anthracene fluorescence.[†]

On the other hand, the fluorescence exhibited by some compounds is so intense as to be distinguishable even in very complex mixtures, such as coal tar. Indeed, it was this fact, coupled with the knowledge that the positions of the fluorescence bands are characteristic for a given compound, which made possible the isolation of the very potent cancer-producing hydrocarbon, 3:4-benzopyrene, from coal tar.

The cancer-producing tars were all observed to have a characteristic fluorescence spectrum, and the isolation of the active substance from this complex material was achieved by examination of the fluorescence spectra of the various fractions obtained in the purification processes. The pure substance was eventually isolated[‡] and the fluorescence spectrum found to consist of three bands at 4000, 4180 and 4400 Å.

[*] *J. Chem. Soc.* 1949, p. 1683.
[†] Bowen, *J. Chem. Phys.* 1945, 13, 306.
[‡] Hieger, *Amer. J. Cancer*, 1937, 29, 705; Cook, Hewett and Hieger, *J. Chem. Soc.* 1933, p. 395.

CHAPTER 11

OPTICAL ACTIVITY IN AROMATIC COMPOUNDS

ASYMMETRY

Very many organic compounds are able to rotate the plane of polarized light. The phenomenon has been extensively investigated, and it has now been established that only those compounds having structures which are not superimposable upon their mirror images are optically active. Such substances are said to be *asymmetric*, and two compounds which are related to one another as object is to mirror image are called *enantiomorphs*. The chemical and physical properties of such enantiomorphs are identical, with the exception that one rotates plane-polarized light to the left, and the other rotates it to the right. Enantiomorphs may also react at different rates with other asymmetric molecules, and may differ in physiological properties.

The great majority of asymmetric compounds owe their activity to the presence of at least one individual asymmetrically substituted atom. However, the presence of an individual asymmetric atom is by no means essential, for the absence of *any* element of symmetry which permits a given structure to be superimposable upon its mirror image is sufficient. In recent years several classes of asymmetric compounds which do not possess an individual asymmetric atom have been discovered.*

It is not the purpose of the present book to discuss all types of optical activity. Only those cases of molecular asymmetry which are directly related to aromatic ring systems will be mentioned, and all examples of molecular asymmetry associated with an asymmetrically substituted tetrahedral carbon atom are excluded.

* For a review see Shriner, Adams and Marvel, in Gilman's *Organic Chemistry*, vol. 1, 2nd ed., John Wiley, New York, 1943.

STEREOCHEMISTRY OF THE DIPHENYLS

In 1922, Christie and Kenner* resolved 6:6'-dinitrodiphenic acid (I) into two optically active forms. There is no asymmetric carbon atom in this molecule, and the cause of the asymmetry was at first obscure.

Structure (I): 6,6'-dinitrodiphenic acid with NO_2 and O_2N groups in ortho positions and CO_2H and HO_2C groups.

Several other *ortho*-substituted diphenyls were soon resolved into optically active components in a similar fashion. Not *all* such compounds could be resolved, however, and after consideration of all the evidence available it became clear,† in 1926, that the asymmetry is associated with the fact that, in suitable *ortho*-substituted diphenyls, the two benzene rings are coaxial, but not coplanar, and that the isomers may therefore be represented as in structures (IIa) and (IIb).

Structures (IIa) and (IIb) shown as mirror images.

In diphenyl itself (III), the benzene rings are free to rotate about the pivot or coannular bond as axis. On the other hand, conjugation between the two rings is greatest when the two rings are coplanar, for resonance structures of type (IV) require a coplanar configuration.

---

* *J. Chem. Soc.* 1922, **121**, 614.
† Turner and Le Fèvre, *Chem. & Ind.* 1926, **45**, 831, 883; Bell and Kenyon, ibid. 1926, **45**, 864; Mills, ibid. 1926, **45**, 884, 905; Christie and Kenner also included the coaxial-non-coplanar theory in their discussion of the possible causes of asymmetry.

Examination of the ultra-violet absorption spectra of such compounds confirms that the conjugation between the rings is pronounced, so that structures of type (IV) must contribute significantly to the hybrid molecule. It follows that a considerable number of molecules must have a coplanar configuration at any given time, or, alternatively, that any given molecule must have a coplanar configuration for a large part of the time.*
X-ray crystallographic investigation confirms that diphenyl, in the crystalline state, does have a planar configuration.†

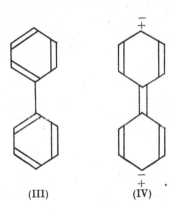

(III)   (IV)

Using the molecular dimensions of substituents obtained by X-ray crystallographic investigations, however, Mills‡ was able to show that certain *ortho* substituents in one ring of diphenyl can 'interfere' with *ortho* substituents in the other ring to such an extent that free rotation is prevented, and the molecule is no longer able to assume a coplanar configuration. In 6-chlorodiphenic acid, for example, the construction of scale diagrams (figs. 11·1 and 11·2) indicates that the rotation of one nucleus relative to the other must be limited to a restricted arc. There is insufficient room for the 2′-carboxy group to pass either the 2-carboxy group or the 6-chloro atom.

* Sutton, *Trans. Faraday Soc.* 1934, **30**, 791; Pauling and Sherman, *J. Chem. Phys.* 1933, **1**, 679.
† Dhar, *Indian J. Phys.* 1932, **7**, 43; Toussaint, *Acta Cryst.* 1948, **1**, 43; Saunder, *Proc. Roy. Soc.* A, 1946, **188**, 31; Van Niekerk and Saunder, *Acta Cryst.* 1948, **1**, 44.
‡ *Chem. & Ind.* 1926, **45**, 884, 905.

It is reasonable to suppose that the magnitude of the steric-hindrance effect is directly related to the size of the *ortho* substituents, and that not all *ortho*-substituted diphenyls will exhibit restricted rotation and the inability to assume a coplanar configuration. As indicated in Chapter 10, the ultra-violet absorption spectra often provide useful evidence on this point. If the *ortho* substituents are sufficiently bulky to prevent the coplanar configuration, conjugation between the rings is prevented. The ultra-violet absorption spectra of such compounds

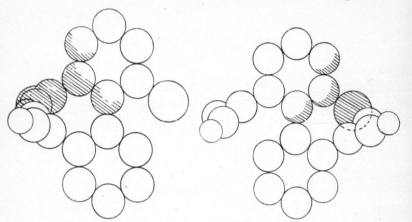

Fig. 11·1. Scale drawing showing the overlapping of the two carboxy groups in 6-chlorodiphenic acid. (From Mills, *Chem. & Ind.* 1926, **45**, 885.)

Fig. 11·2. Scale drawing showing the overlapping of the 2'-carboxy group and the 6-chloro atom in 6-chlorodiphenic acid. (From Mills, *Chem. & Ind.* 1926, **45**, 885.)

differ from that of diphenyl, but resemble that of a substituted benzene. This is the case with dimesityl, the absorption spectrum of which resembles that of mesitylene rather than diphenyl.

On the other hand, the presence of restricted rotation and non-coplanarity is not of itself a sufficient criterion for molecular asymmetry. The molecule must be substituted asymmetrically, or in such a way that the mirror image of the non-coplanar molecule cannot be superimposed on that of the original. Very many diphenyls have now been resolved. Most of them fall into one or other of the following classes (V*a*) to (XII*b*).

As already indicated, however, not all representatives of each group can be resolved. Among the tetrasubstituted diphenyls,

OPTICAL ACTIVITY IN AROMATIC COMPOUNDS

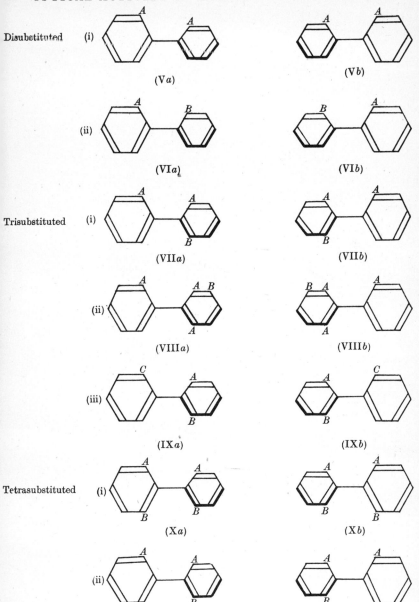

Disubstituted (i) (Va) (Vb)

(ii) (VIa) (VIb)

Trisubstituted (i) (VIIa) (VIIb)

(ii) (VIIIa) (VIIIb)

(iii) (IXa) (IXb)

Tetrasubstituted (i) (Xa) (Xb)

(ii) (XIa) (XIb)

Tetrasubstituted (iii)

(XIIa)   (XIIb)

fluoro and methoxy substituents appear to be too small to cause restricted rotation, and the compounds (XIII), (XIV), (XV) and (XVI) cannot be resolved:

(XIII)   (XIV)

(XV)   (XVI)

(XVII)   (XVIII)

If fluoro and methoxy substituents are associated with a very large *ortho* substituent such as a carboxy group, however, then resolution is possible, but the resulting optical isomerides are readily racemized. Examples of this class are (XVII) and (XVIII).

It must be assumed that the steric hindrance in these molecules is not very pronounced, and that rotation, although

OPTICAL ACTIVITY IN AROMATIC COMPOUNDS 417

restricted, is not entirely prevented. Racemization evidently occurs because the fluorine atom or methoxy group can occasionally 'slip past' the bulky carboxy group. A nitro group is even larger than a carboxy group, and the compound (XIX) does not racemize. Many other tetrasubstituted diphenyls, including the classical example investigated by Christie and Kenner, are also very stable, and do not racemize. In all such cases the substituents must be sufficiently large to prevent rotation about the axis of the pivot bond.

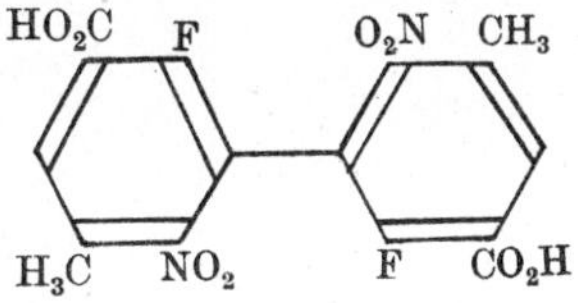

(XIX)

Among the trisubstituted diphenyls the results are similar. That is, resolution has only been achieved if the *ortho* substituents are sufficiently bulky. Adams and others* have investigated two series of tri-*ortho*-substituted diphenyls, which may be represented by the formulae (XX) and (XXI). In the first series, X represented OCH_3, CH_3, CO_2H or NO_2; in the second series Y represented Br, Cl or F. By studying the resolution of these compounds, and the relative rates of racemization of the optically active components, it was concluded that the relative interference values of the substituents fall in the order:

$$Br > CH_3 > Cl > NO_2 > CO_2H > OCH_3 > F.$$

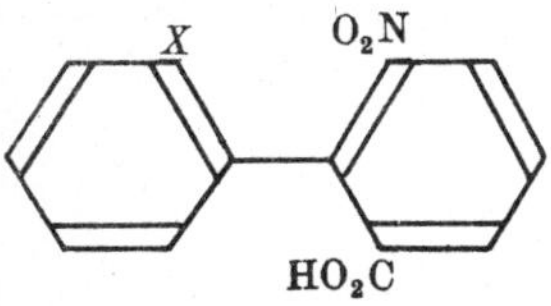

(XX) 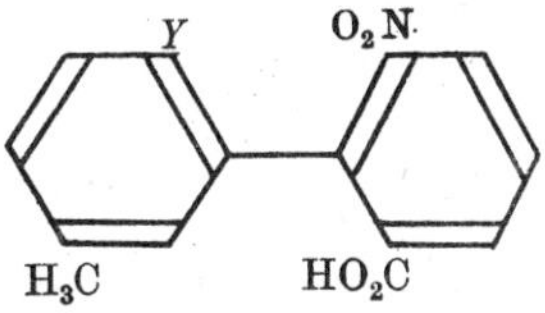   (XXI)

* Adams and Yuan, *Chem. Rev.* 1932, **12**, 261; Stoughton and Adams, *J. Amer. Chem. Soc.* 1932, **54**, 4426.

The possibility that 2:2′-disubstituted diphenyls might be resolved was first pointed out by Bell and Robinson.* It is clear that if a given substituent is sufficiently large it must be capable of showing interference with the *ortho* hydrogen atom in the other ring; in other words, the *ortho* hydrogen atoms may be considered as small substituents. Several resolutions of this type have been achieved. Lesslie and Turner† reported the optical resolution of (XXII) and (XXIII), and Searle and Adams‡ succeeded in resolving a derivative (XXIV) of 2:2′-diiodo-diphenyl. The iodine atom is, of course, very large, but it is interesting that the same investigators also observed mutarotation in a salt of the corresponding dibromo derivative (XXV), indicating a slight degree of restricted rotation.§

Shaw and Turner‖ pointed out that even a 2:3′-disubstituted diphenyl should show molecular asymmetry provided the *ortho* substituent is sufficiently large. The $-\text{As}^+\text{Me}_3\} X^-$ group is large enough to cause some restriction of rotation by interference with the hydrogen atoms in the other *ortho* positions, and Lesslie and Turner¶ obtained definite indications of resolution with 3′-bromodiphenylyl-2-trimethylarsonium iodide.

* *J. Chem. Soc.* 1927, p. 1695.   † *Ibid.* 1932, pp. 2021, 2394.
‡ *J. Amer. Chem. Soc.* 1933, **55**, 1649.
§ If the steric hindrance is only slight it may not be possible to isolate the optically active acid; but asymmetry may be demonstrated by a mutarotation of a salt with an optically active base, that is, by an asymmetric transformation. See Turner and Harris, *Quart. Rev. Chem. Soc., Lond.*, 1948, **1**, 299.
‖ *J. Chem. Soc.* 1933, p. 135.   ¶ *Ibid.* 1933, p. 1588.

A carboxy group is not large enough to interfere with a hydrogen atom in the *ortho* position of the adjacent ring, although two carboxy groups certainly interfere with one another when in the *cis* position (XXVI). This molecule can therefore assume a planar configuration in the *trans* but not in the *cis* position. In the normal rotational oscillations of the two phenyl groups of diphenic acid, there must be a continuous periodic formation of the symmetrical state (XXVII), and resolution is therefore impossible.

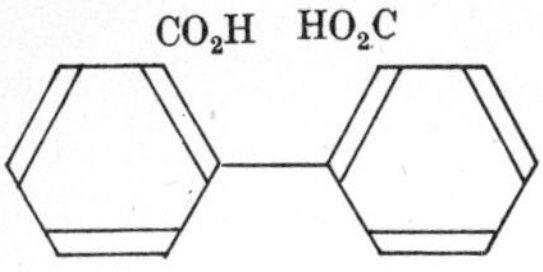

(XXVI) 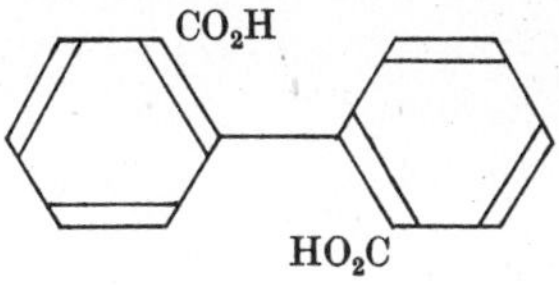   (XXVII)

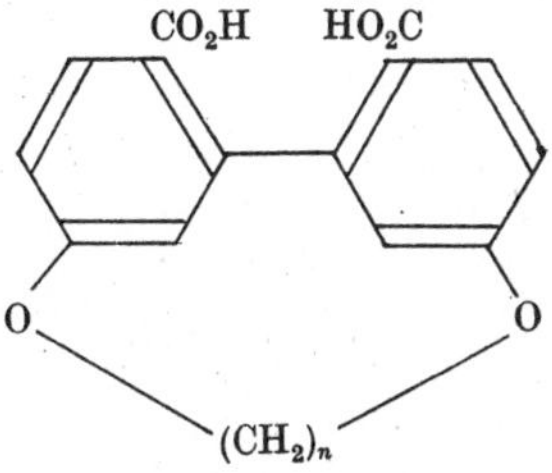

(XXVIII)

The introduction of additional *ortho* substituents leads to further restriction of rotation, as has already been described, and many *ortho*-substituted derivatives of diphenic acid have been successfully resolved. However, Adams and Kornblum* suggested that the desired restriction in rotation might also be achieved not by the introduction of further *ortho* substituents but by the attachment of a polymethylene bridge, as in (XXVIII).

Four possibilities can be recognized: (i) for very small values of n, the molecule may be so strained that the carboxy groups do not interfere; (ii) for slightly larger values of n, the molecule

* J. Amer. Chem. Soc. 1941, **63**, 188.

having the carboxy groups in the *cis* position, but 'staggered', should be resolvable; (iii) for medium values of n, a *trans* configuration should be possible, and this should be interconvertible with the *cis* form, but otherwise the rotation is restricted; (iv) for very large values of n, the diphenyl system should rotate freely inside the large ring, and such compounds should not be capable of resolution.

Adams and Kornblum successfully resolved two representatives of class (iii), namely, those having $n = 8$ and 10. The optically active compound having a polymethylene bridge of eight carbon atoms racemized more slowly than that having a bridge of ten carbon atoms.

(XXIX)   (XXX)

An interesting example of diphenyl stereoisomerism has been described by Robinson.* When the opium alkaloid thebaine (XXIX) is treated with phenylmagnesium bromide, a remarkable rearrangement takes place, and phenyldihydrothebaine (XXX) is formed. This new base is derived from thebaine by the addition of the elements of benzene (C_6H_5, H). The ultra-violet absorption spectrum does not show the characteristics of diphenyl, for restricted rotation occurs, and, in fact, the molecule possesses two sources of asymmetry. One of these is a non-coplanar diphenyl configuration, and the other is an asymmetric carbon atom, marked with an asterisk. This is the first example of a compound possessing both kinds of asymmetry.

* *Proc. Roy. Soc.* A, 1948, **192**, xiv.

OPTICAL ACTIVITY IN AROMATIC COMPOUNDS

Molecular asymmetry of the diphenyl type is not confined to derivatives of diphenyl itself. Similar stereoisomerism occurs in suitably substituted polyphenyls, in phenylquinones and in substituted dipyridyls. For example, Woodruff and Adams* have resolved 2:4:2':4'-tetracarboxy-6:6'-diphenyl-3:3'-dipyridyl (XXXI).

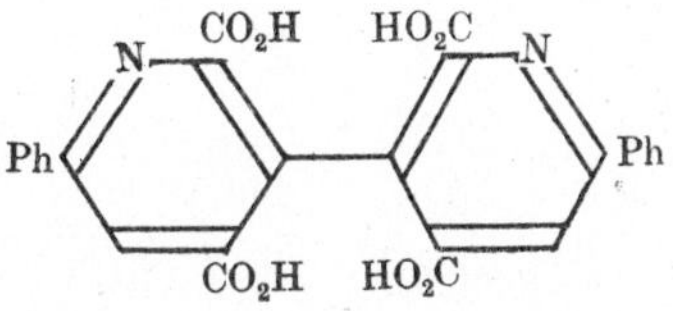

(XXXI)

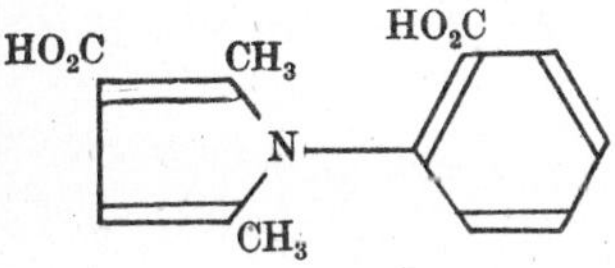

(XXXII) 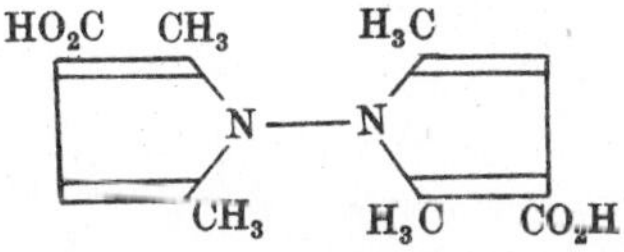   (XXXIII)

Even more interesting, however, were the successful resolutions of a N-phenylpyrrole and a dipyrryl. Bock and Adams† resolved N-2'-carboxyphenyl-2:5-dimethyl-3-carboxypyrrole (XXXII) into stable optically active forms; and Chang and Adams‡ obtained the optical isomerides of $N:N'$-2:5:2':5'-tetramethyl-3:3'-dicarboxydipyrryl (XXXIII), and these were also found to be remarkably resistant to racemization.

In other investigations, Bock and Adams§ found that the ability of phenylpyrroles to undergo resolution is determined by the same sort of factor as has been found for similar diphenyls, and there can be no doubt that these heterocyclic compounds are entirely analogous to the diphenyls themselves.

Several interesting examples of molecular asymmetry have also been observed among dinaphthyls and dianthryls. 8:8'-Dicarboxy-1:1'-dinaphthyl (XXXIV), which may conveniently be regarded as a disubstituted diphenyl with two *ortho* sub-

---

* J. Amer. Chem. Soc. 1932, **54**, 1977.　† Ibid. 1931, **53**, 374.
‡ Ibid. 1931, **53**, 2353.　§ Ibid. 1931, **53**, 3519.

stituents of the type —C—CO_2H, was resolved in 1931 by several workers.* This resolution indicates that the substituents are sufficiently large to interfere with the *ortho* hydrogen atoms; but the optically active compound rapidly racemized. In this connexion it is interesting that the isomeric substance 2:2′-dicarboxy-1:1′-dinaphthyl (XXXV) can also be resolved, and the isomers are very stable to racemization.†

(XXXIV)   (XXXV)

In anthracene derivatives, non-coplanarity is most easily obtained. Even in 9-phenylanthracene, the 2′-hydrogen atom interferes with the hydrogen atoms at positions 1 and 8 of the anthracene ring system, and free rotation is prevented. The ultra-violet absorption spectra of anthracene and of 9-phenyl-, 9:10-diphenyl- and of 1:4-diphenyl-anthracenes are all very similar, indicating that there is little conjugation between the phenyl rings and the anthracene ring system,‡ and there is no doubt that compounds of this nature exhibit hindered rotation. The spectrum of 2-phenylanthracene, on the other hand, indicates that there is considerable conjugation, for there is no steric interference with rotation, and the molecule can assume a coplanar configuration (Chapter 10). In the dianthryls, the steric interference is considerable, and some important work has been carried out in this field. Bell and Waring,§ for example,

* Stanley, *J. Amer. Chem. Soc.* 1931, **53**, 3104; Corbellini, *R.C. Accad. Lincei*, 1931, **13**, 702; Meisenheimer and Beisswenger, *Ber. dtsch. chem. Ges.* 1932, **65**, 32.
† Kuhn and Albrecht, *Liebigs Ann.* 1928, **465**, 282.
‡ Hirshberg, *Trans. Faraday Soc.* 1948, **44**, 285; Jones, *J. Amer. Chem. Soc.* 1941, **63**, 1658.
§ *J. Chem. Soc.* 1949, p. 1579.

have prepared 1:1'-dianthryl-2:2'-dicarboxylic acid (XXXVI), and have resolved it into its optically active forms. The very considerable steric interference was demonstrated by conversion of an optically active form into active 2:2'-diamino-1:1'-dianthryl (XXXVII), and this was again converted into the original acid by two different routes, via (XXXVIII) and (XXXIX), without change in either the magnitude or sign of the rotation.

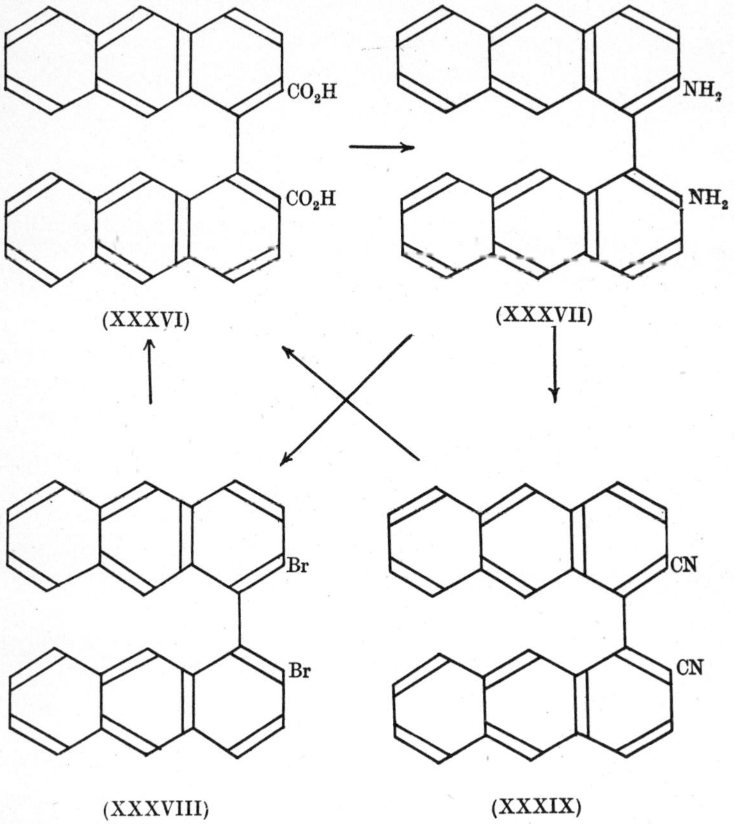

(XXXVI)   (XXXVII)

(XXXVIII)   (XXXIX)

The same authors prepared (+)- and (−)-1:1'-dianthryl from the corresponding optically active 2:2'-diamino derivatives. This optically active hydrocarbon was found to be racemized in hot solvents.

THE INFLUENCE OF THE SIZE AND NATURE OF SUBSTITUENTS

Stanley and Adams[*] realized that if the optical activity in the diphenyl series is to be associated with restricted rotation due to the presence of large *ortho* substituents, then it should be possible to predict which compounds are capable of optical resolution, and which do not exhibit sufficient steric interference for resolution to be possible. From X-ray crystallographic data on a number of different compounds they first constructed a list of bond distances, as in table 11·1.

TABLE 11·1. *Distances (in Å.) from the nucleus of the carbon atom of the benzene ring to the centre of the substituent atom or group*[†]

Bond	Length (Å.)	Bond	Length (Å.)
C—H	0·94	C—CH$_3$	1·73
C—F	1·39	C—Cl	1·89
C—OH	1·54	C—NO$_2$	1·92
C—CO$_2$H	1·56	C—Br	2·11
C—NH$_2$	1·56	C—I	2·20

The vertical distance between the two *ortho* carbon atoms when the rings are coplanar was estimated to be 2·90 Å.,[‡] and the difference between this value and the sum of the two 'internuclear distances', as obtained from table 11·1, was called the 'interference value'. In other words, the interference value (I.V.) is obtained from the equation

$$\text{I.V.} = (d_A + d_B) - 2·90,$$

where d_A and d_B are the internuclear distances from the *ortho* carbon atoms to the centres of the substituents A and B. If the interference value is positive, the compound should be capable of resolution. In the case of 6:6'-dinitrodiphenic acid, the interference is between (i) NO$_2$ and CO$_2$H, (ii) NO$_2$ and NO$_2$ and (iii) CO$_2$H and CO$_2$H. In the first case the interference value is

$$(1·56 + 1·92) - 2·90 = +0·58;$$

[*] J. Amer. Chem. Soc. 1930, 52, 1200.
[†] After Stanley and Adams, ibid.
[‡] Dhar, *Indian J. Phys.* 1932, 7, 43.

in the second, $(1\cdot92+1\cdot92)-2\cdot90 = +0\cdot94$;
and in the third, $(1\cdot56+1\cdot56)-2\cdot90 = +0\cdot22$.

In other words, there is always considerable steric interference in dinitrodiphenic acid, and successful resolution could have been predicted. In table 11·2, various other interference values for certain pairs of atoms or groups are collected.

TABLE 11·2. *'Interference values' of certain pairs of atoms or groups*

Groups at 2:2′ position	Interference value (in Å.)
H, H	−1·02
H, CO_2H	−0·40
H, Cl	−0·07
H, NO_2	−0·04
F, NH_2	+0·05
F, CO_2H	+0·05
Cl, CO_2H	+0·55
NO_2, CO_2H	+0·58

If the interference value is negative, no resolution is to be expected, for there is no steric interference to rotation. If the interference value has a small positive value, resolution, but easy racemization, can be predicted. If the interference value has a large positive value, resolution should not only be possible, but the resulting optical isomerides should be quite stable. These conclusions are illustrated by the following three compounds. In the first (I) the interference value is negative, and this compound cannot be resolved. In the second (II) the positive interference is small, and this compound can be resolved, but the optical isomerides are easily racemized. In the third (III) the interference value is large and the optical isomerides of this compound cannot be racemized.

By experiments of this nature the theory of restricted rotation by virtue of the size of the *ortho* substituents has been confirmed many times. However, the stability of the optical isomerides to racemization does not depend *solely* on the magnitude of the *ortho* substituents, but is also affected by substituents at other positions. For example, Kuhn and Albrecht[*] found that optically

[*] *Liebigs Ann.* 1927, **458**, 221.

active 2:4'-dinitro-2':6-dicarboxydiphenyl (IV) is considerably less stable to racemization than 2:4:4'-trinitro-2':6-dicarboxydiphenyl (V).

The only difference between these compounds is the additional 4-nitro group in the latter, and it was not expected that a substituent in this position would influence the rate of racemization. Many other examples of such an effect have since been observed, and several extensive investigations of the problem have been carried out. For example, Adams and his co-workers* have studied the rates of racemization of the optical isomerides of compounds of types (VI), (VII) and (VIII).

2-Nitro-6-carboxy-2'-methoxydiphenyl itself racemizes readily at room temperature, and various other substituents were found to have a profound effect on the rate of racemization. Of the substituents studied the effect was found to be in the order $H < OCH_3 < CH_3 < Cl < Br < NO_2$, regardless of whether the substituent was in the 3'-, 4'- or 5'-position. However, the stability

* Chien and Adams, *J. Amer. Chem. Soc.* 1934, **56**, 1787, 2112; Hanford and Adams, ibid. 1935, **57**, 1592; Yuan and Adams, ibid. 1932, **54**, 2966, 4434.

of the 3'-substituted compounds was much greater than that of the corresponding 5'-substituted compounds, and, in general, the 4'-substituted derivatives were found to be less stable (table 11·3). The effect of most of these substituents in the 4-position has also been reported.*

TABLE 11·3. *Half-life periods, in minutes, of substituted 2-nitro-6-carboxy-2'-methoxydiphenyls*

Position of substituent	Nitro-	Bromo-	Chloro-	Methyl-	Methoxy-
3'	1905	827	711	331	98
5'	35	32	31	11·5	10·8
4'	115	25	12	2·6	3·6
4	4·3	9	11·6	5·1	—

In some cases the presence of substituents in other than the 2:2':6:6' positions governs whether or not a certain diphenyl can be resolved. For example, 2:2'-dibromo-4:4'-diaminodiphenyl (IX) cannot be separated into its optical isomerides, but the related compound, 2:2'-dibromo-4:4'-dicarboxydiphenyl (X), has been resolved.†

The way in which substituents other than in the 2:2':6:6' positions can affect the ease of resolution and the stability of the resulting optical isomerides is still very obscure. Several

* Adams and Snyder, ibid. 1938, **60**, 1411.
† Searle and Adams, ibid. 1934, **56**, 2112.

suggestions have been made, and the effect may depend on several factors. Adams suggests: (i) that an additional substituent may affect the valency angle at which an *ortho* substituent is attached; (ii) that it may lengthen or shorten the internuclear distance; (iii) that it may modify the 1:1' bond distance; (iv) that it may result in bending of the coannular bond in such a way that the rings are no longer coaxial; and (v) that it may modify the oscillation of the two rings in such a way as to diminish the chances of rotation.

OTHER EXAMPLES OF RESTRICTED ROTATION

In 1928, Mills and Elliott[*] drew attention to the fact that if the molecular asymmetry of the diphenyls is due to the obstruction of rotation of the phenyl groups by *ortho* substituents, then it should be possible to obtain evidence of similar restriction of rotation in other classes of compounds. In particular, they suggested that compounds of type (I) might show restricted rotation about the axis of the C—N bond by virtue of the interference of the groups R and R_1 with the large *peri*-nitro group. The construction of models indicated that, provided the substituents R and R_1 are large, this restriction would be considerable, and that, in suitable cases, resolution should be possible.

N-benzenesulphonyl-8-nitro-1-naphthylglycine (II) was selected, and it was successfully resolved into two optically active forms by means of the brucine salt. Both enantiomorphs rapidly racemized. That the asymmetry is due to restricted rotation about the axis of the C—N bond, and interference of the large substituents on the nitrogen with the *peri*-nitro group, is shown by the fact that the parent compound, N-benzenesulphonyl-1-naphthylglycine (III), in which there is no such restriction, cannot be resolved.

Many other examples of the same, or similar, molecular asymmetry have since been observed. Mills[†] observed that the 1-ethylquinolinium iodide derivative (IV) can be resolved,

---

[*] *J. Chem. Soc.* 1928, p. 1291.
[†] *Trans. Faraday Soc.* 1930, 26, 431; Mills and Breckenridge, *J. Chem. Soc.* 1932, p. 2209.

OPTICAL ACTIVITY IN AROMATIC COMPOUNDS

indicating interference between the N-ethyl group and the substituents on the amino-nitrogen.

Neither the parent tertiary compound (V) nor the N-methylquinolinium iodide derivative could be resolved. Furthermore, Mills found that substituted 1-naphthylamine-8-sulphonic acids can be resolved, and that the stability of the resulting optically active forms depends on the nature of the substituents. In particular, the acetyl-methyl derivative (VI) possessed so high a degree of optical stability that similarly substituted benzene derivatives were investigated.

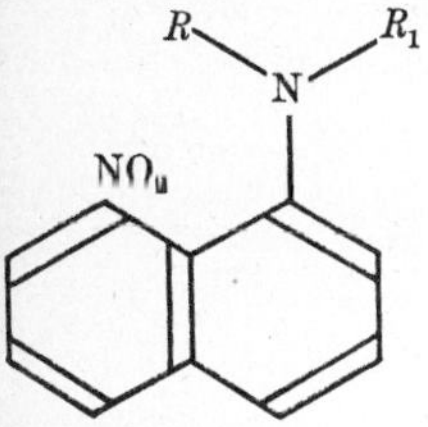

(I) 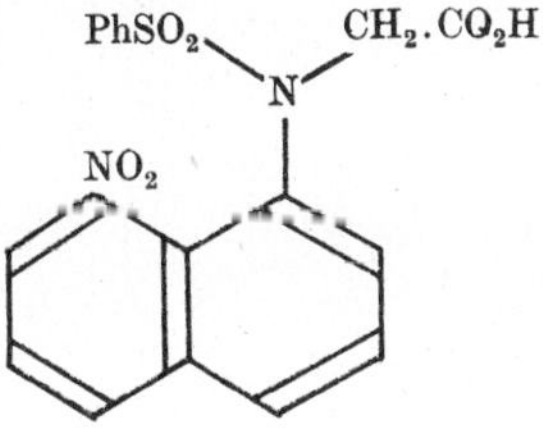   (II) 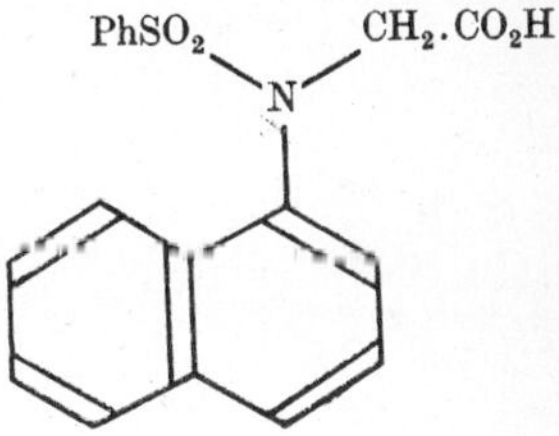   (III)

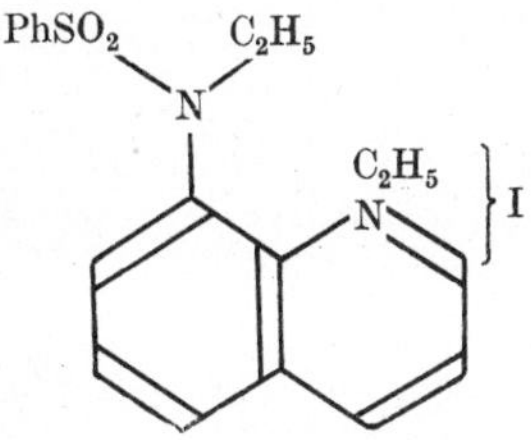

(IV) 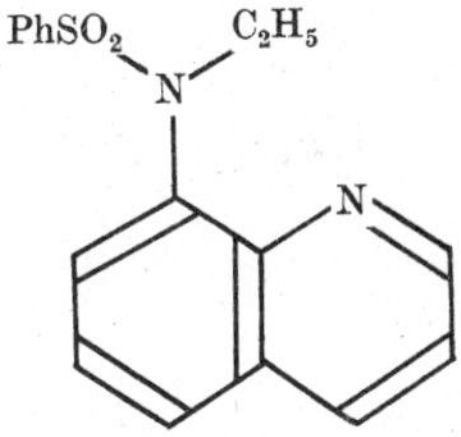   (V)

Mills and Kelham* prepared N-acetyl-N-methyl-*para*-toluidine-3-sulphonic acid (VII), and found that optically active salts showing mutarotation could be obtained. The extent of interference is shown by the accompanying scale drawing (fig. 11·3).

A carboxy group is smaller than a sulphonic acid group, and as no steric interference of this type occurs, the resolution of

* *J. Chem. Soc.* 1937, p. 274.

such compounds would not be anticipated. This was confirmed by the observation that the brucine salt of *N*-acetyl-*N*-methyl-anthranilic acid showed no sign of mutarotation.

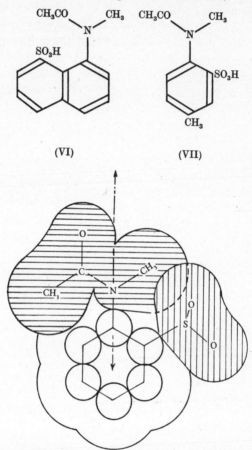

Fig. 11·3. Scale drawing showing the extent of steric hindrance in *N*-acetyl-*N*-methyl-*para*-toluidine-3-sulphonic acid. (From Mills and Kelham, *J. Chem. Soc.* 1937, p. 275.)

Several other compounds exhibiting restricted rotation of the same type have been investigated by Adams and his co-workers.*

* Adams and Dankert, *J. Amer. Chem. Soc.* 1940, **62**, 2191; Adams and Stewart, ibid. 1941, **63**, 2859; Adams and Albert, ibid. 1942, **64**, 1475; Adams and Sundholm, ibid. 1948, **70**, 2667.

OPTICAL ACTIVITY IN AROMATIC COMPOUNDS

Several N-methyl- (or N-ethyl-) N-succinyl derivatives of *ortho*-substituted amines have been successfully resolved, and the rates of racemization studied, usually in boiling butanol. The half-life periods can be used as an index of optical stability. For example, the compounds (VIII)–(XVI) were resolved and were found to have the half-life periods stated.

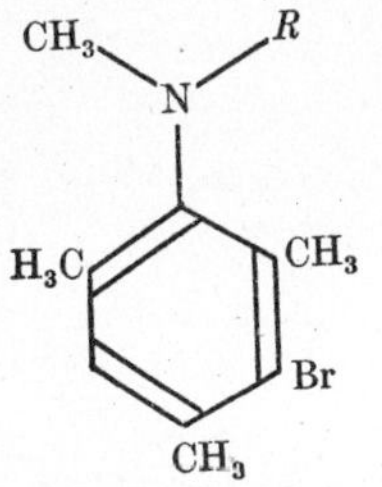

(VIII) 9 hours (n-butanol)

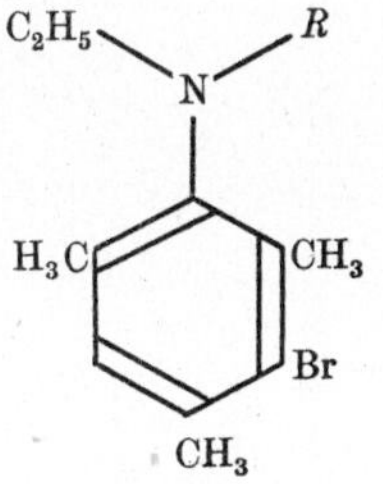

(IX) 28 hours (n-butanol)

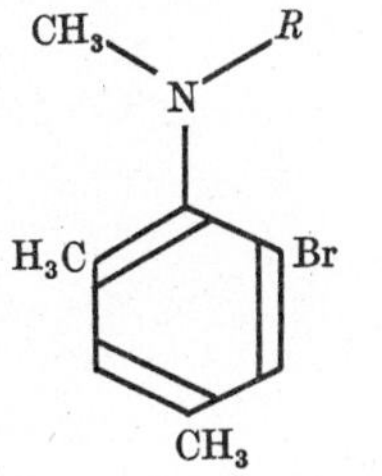

(X) 3·1 hours (n-butanol)

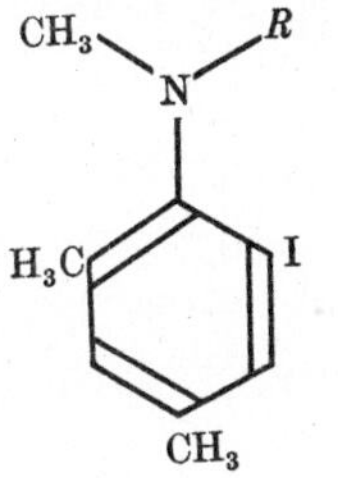

(XI) 20·5 hours (n-butanol)

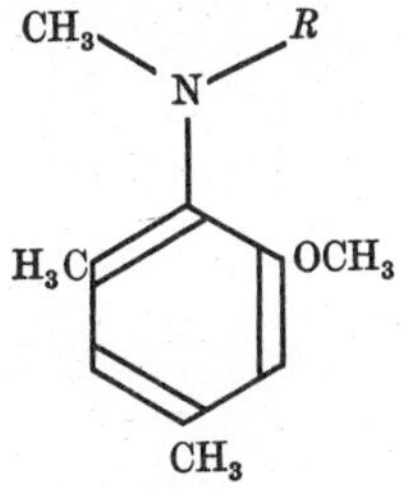

(XII) 2·7 hours (methyl acetate)

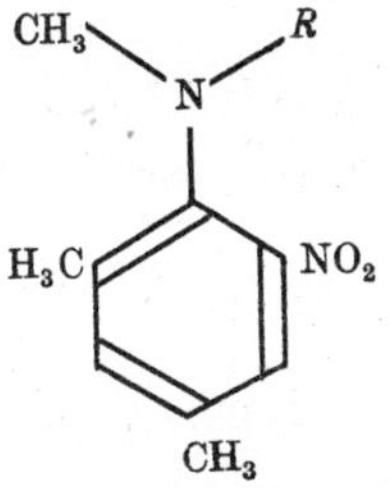

(XIII) 0·6 hour (methyl acetate)

(XIV) 1·1 hours (n-butanol)

(XV) 5·7 hours (n-butanol)

(XVI) 4·1 hours (n-butanol)

As the optical activity of compounds of this type can be destroyed by symmetrical substitution in the ring, there can be little doubt that the molecular asymmetry is due to restricted rotation about the axis of the C—N bond, rather than to asymmetry of the nitrogen atom. However, the compounds are interesting in that the relative effectiveness of various *ortho* substituents is not always the same as the relative effectiveness of the same groups in the diphenyl series. Bromine is only about one-third as effective as methyl, and a nitro group is only about one-fifth as effective as a methoxy group, although in the diphenyl series a nitro group is more effective than a methoxy group, and bromine is more effective than methyl. Adams has pointed out that a decrease in the expected stability of the active amines is observed with substituents of an electronegative character, which markedly reduce the basicity of arylamines. It seems that the increased double-bond character of the C—N bond in these amines of decreased basicity facilitates racemization. That is, the tendency to form a double bond aids in forcing the substituents on the amino-nitrogen into a coplanar configuration with the ring.

In an attempt to evaluate the effect of the *nature* of a substituent, as opposed to its *size*, the stability of the bromo derivative (XVII) was compared with that of the dibromo derivative (XVIII).

The optically active form of the latter had a half-life in boiling butanol of 1·1 hours as compared with 3·1 hours for the former; and similar results were also obtained with the corresponding chloro compound.* It was also found that the analogous

* Adams and Gordon, *J. Amer. Chem. Soc.* 1950, 72, 2454.

derivative of 1-amino-2-methylnaphthalene is more stable than 1-amino-2-methyl-4-chloronaphthalene. All these results are consistent with the view that substituents which decrease the basicity of the amine, and increase the double-bond character of the C—N bond, reduce the optical stability of these compounds even when they are in *para* positions.

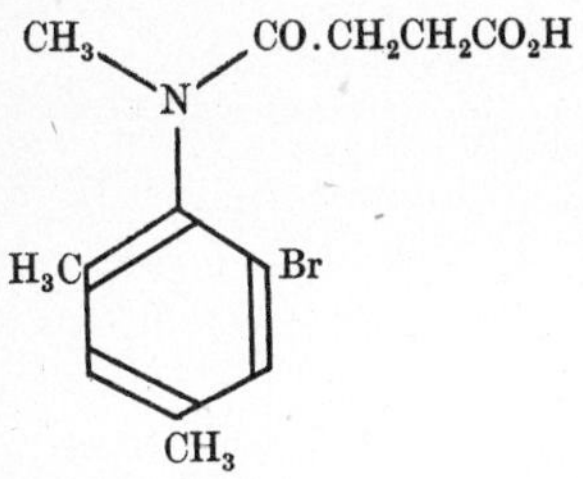

(XVII) 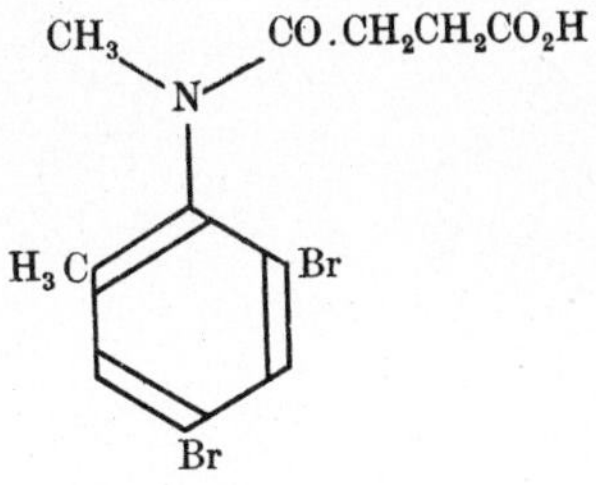   (XVIII)

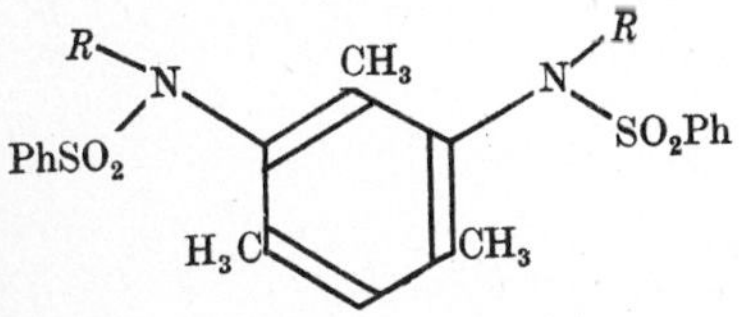

(XIX) 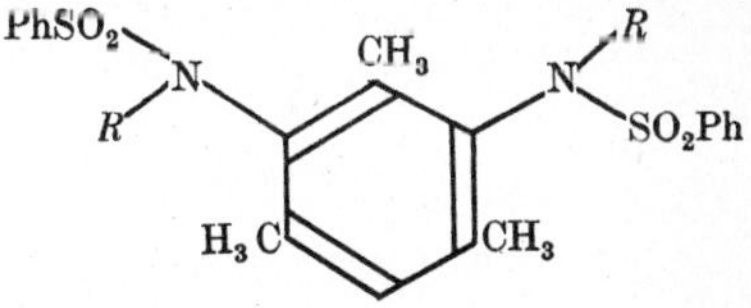   (XX)

Adams and Tjepkema* have also studied a series of substituted diamines having *two* centres of restricted rotation. When dibenzsulphonyldiaminomesitylene was alkylated, for example, *two* alkyl derivatives evidently having the structures (XIX) and (XX) were obtained. Pairs of isomers were obtained with compounds of this type in which the alkyl group (R) was methyl, ethyl-, n-butyl, n-dodecyl and benzyl; and pairs of isomers were also obtained when R was methyl and the benzenesulphonyl group was replaced by substituted benzenesulphonyl groups.

The pair of isomers obtained from dibenzenesulphonyldiaminomesitylene by treatment with ethyl bromoacetate is especially noteworthy, for hydrolysis gave the two acids (XXI) and (XXII). The higher-melting isomer must have a *trans*-racemic configuration (XXI), for it was successfully resolved into

* Ibid. 1948, **70**, 4204.

two optically active forms. The lower-melting isomer, on the other hand, could not be resolved, and must therefore have a *cis-meso* configuration (XXII).

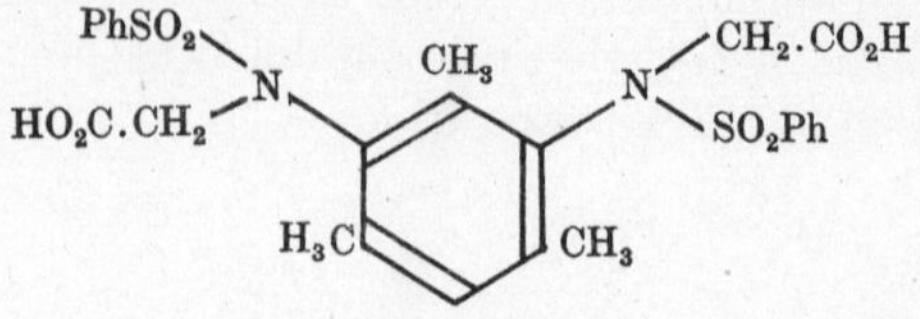

(XXI)

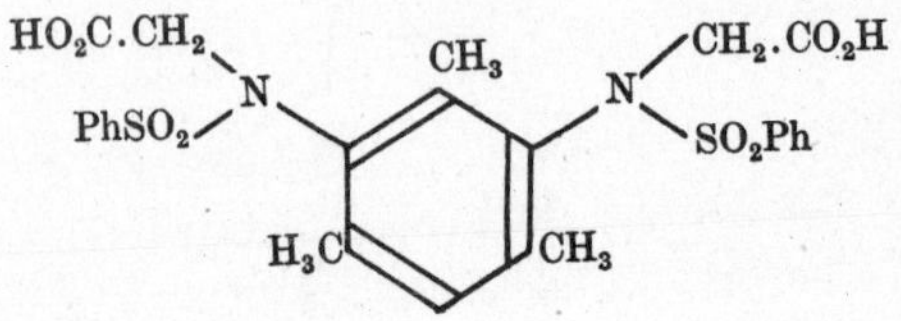

(XXII)

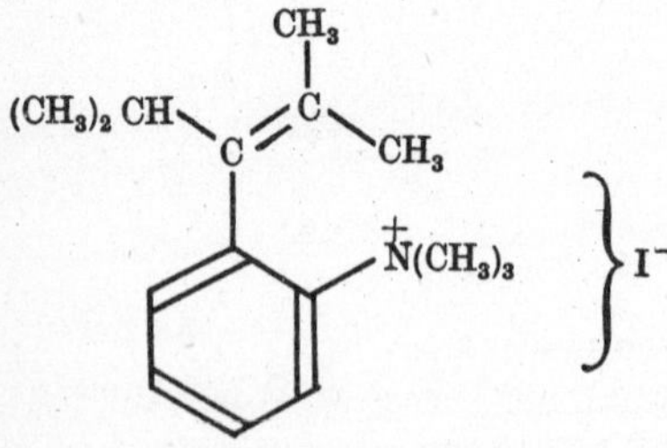

(XXIII) 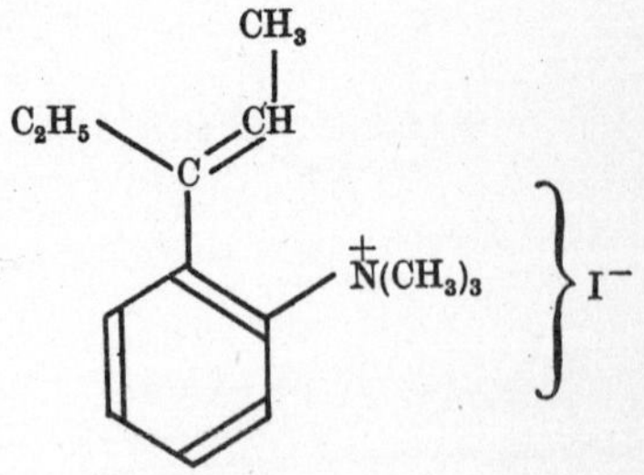   (XXIV)

Restricted rotation probably occurs in many other classes of organic compounds. It has been demonstrated with certainty in a number of aryl olefins and also in certain oximes.

As an example of the former type it may be mentioned that Mills and Dazeley* successfully resolved the compound (XXIII), in which rotation about the axis of the $C_{arom.}$—$C_{aliph.}$ bond is restricted by the presence of a substituted vinyl group and an *ortho*-N^+Me_3 group. In contrast to the compounds mentioned earlier, in which a substituted amino group is the rotating entity,

* *J. Chem. Soc.* 1939, p. 460.

OPTICAL ACTIVITY IN AROMATIC COMPOUNDS

the optical isomerides of this compound showed considerable stability. Heating in aqueous solution at 100° for 8 hours resulted in a loss of only 3% of the optical activity, so the steric interference must be considerable. With smaller substituents on the

(XXV)
Not racemized
(boiling n-butanol)

(XXVI)
200 minutes
(boiling n-butanol)

(XXVII)
173 minutes
(44° in n-butanol)

(XXVIII)
9 minutes
(20° in n-butanol)

vinyl group, however, the interference is less marked, and the related compound (XXIV) could not be resolved.

This work was extended by Adams and his co-workers,* and the rates of racemization of a series of related compounds were studied, with a view to the semi-quantitative determination of

* Adams and Miller, *J. Amer. Chem. Soc.* 1940, **62**, 53; Adams, Anderson and Miller, ibid. 1941, **63**, 1589; Adams and Binder, ibid. 1941, **63**, 2773; Adams and Gross, ibid. 1942, **64**, 1786.

the relative interference of various substituents. The compounds (XXV)–(XXX) were resolved, and the optical isomerides were found to have the half-life periods indicated.

(XXIX)
70 hours
(boiling n-butanol)

(XXX)
70 minutes
(boiling n-butanol)

(XXXI)

(XXXII)

(XXXIII)

The results show that *all* the substituents on the vinyl group are important. Of the various substituents, a chloro group is more effective than a methyl group, and a methoxy group is least effective.

OPTICAL ACTIVITY IN AROMATIC COMPOUNDS

In the related series of oximes, only one of the two *geometrical* isomers would be expected to exhibit restricted rotation. Thus the α-form of 2-hydroxy-3-carboxy-1-naphthylmethylketoxime (XXXI) could not be resolved, but salts of the β-form (XXXII) exhibited mutarotation in solution. The derivative (XXXIII), corresponding to the β-form, also formed a salt which showed mutarotation, and by carrying out the decomposition of the salt at a low temperature, an active compound was obtained.*

OPTICAL ACTIVITY OF THE 4:5-PHENANTHRENE TYPE

Unsuccessful attempts to prepare 4:5-dimethylphenanthrene (I) and other similar compounds by standard reactions led to the belief that it is impossible to introduce two methyl groups in the 4:5 positions of a phenanthrene ring system.† In 1940, however, Newman‡ successfully prepared the analogous compound 6:7-dimethylchrysene (II) (or 4:5-dimethylchrysene in the American system of numbering).

If a scale model of 4:5-dimethylphenanthrene is drawn in which the aromatic rings are coplanar, and the methyl groups occupy the same volume as in toluene, it is at once apparent that the two methyl substituents must offer considerable steric interference to one another. Newman discussed three possibilities: (i) that the methyl groups lie bent away from each other but in

---

* Meisenheimer, Theilacker and Beisswenger, *Liebigs Ann.* 1932, **495**, 249.
† Haworth and Sheldrick, *J. Chem. Soc.* 1934, p. 1950; Fieser and Seligman, *J. Amer. Chem. Soc.* 1938, **60**, 170; 1939, **61**, 136; Cook and Kennaway, *Amer. J. Cancer*, 1937, **33**, 55.
‡ *J. Amer. Chem. Soc.* 1940, **62**, 2295.

the same plane as the aromatic rings; (ii) that the aromatic rings themselves are distorted; and (iii) that the methyl groups are bent out of the plane of the aromatic rings. The latter possibility seemed the most likely, and it opened up a new field of research; for if the methyl groups *are* bent out of the plane of the phenanthrene ring system, then optical activity should be possible. In other words, two forms ((III) and (IV)), which are mirror images, but not superimposable, should be obtainable.

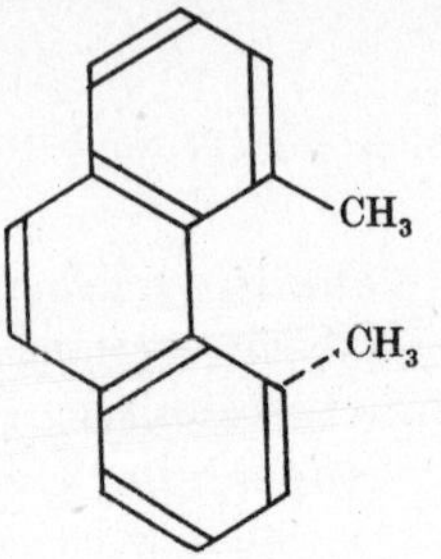

(III) 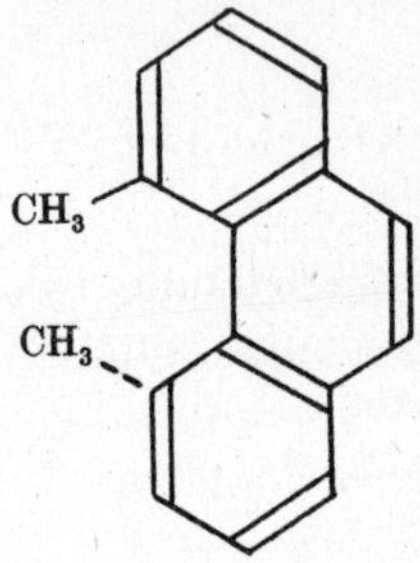 (IV)

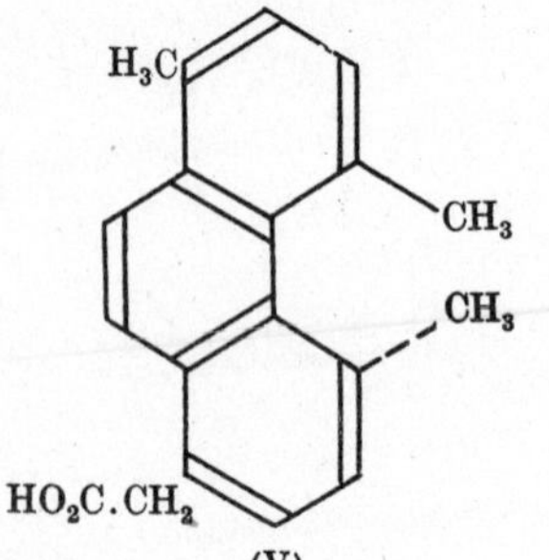

(V) 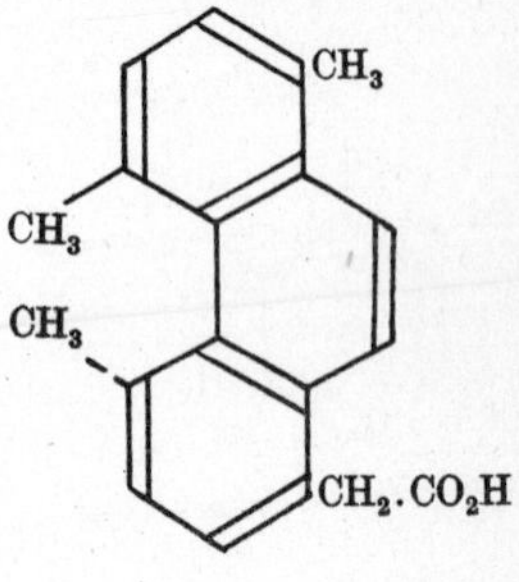 (VI)

The successful resolution of a compound of the 4:5-dimethylphenanthrene type was first achieved by Newman and Hussey,* who obtained the optically active forms (V), (VI) of 4:5:8-trimethyl-1-phenanthrylacetic acid. This success rules out the possibility that the methyl groups are simply bent away from one another; and examination of the ultra-violet absorption spectra of such compounds makes it unlikely that the aromatic

* *J. Amer. Chem. Soc.* 1947, **69**, 3023.

OPTICAL ACTIVITY IN AROMATIC COMPOUNDS

rings are distorted to any great extent. There seems little doubt, therefore, that the optical activity is to be associated with molecular asymmetry as illustrated by structures (V) and (VI).

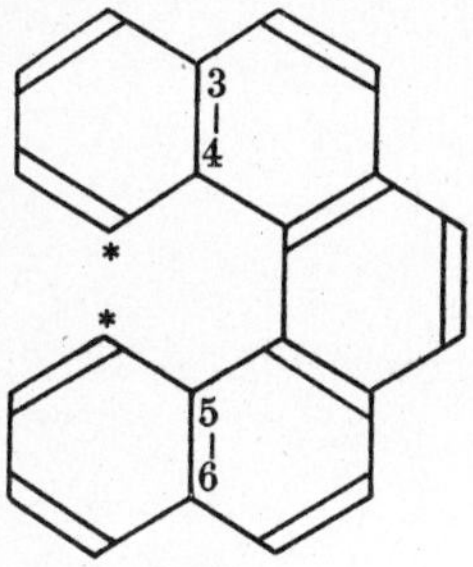

(VII) 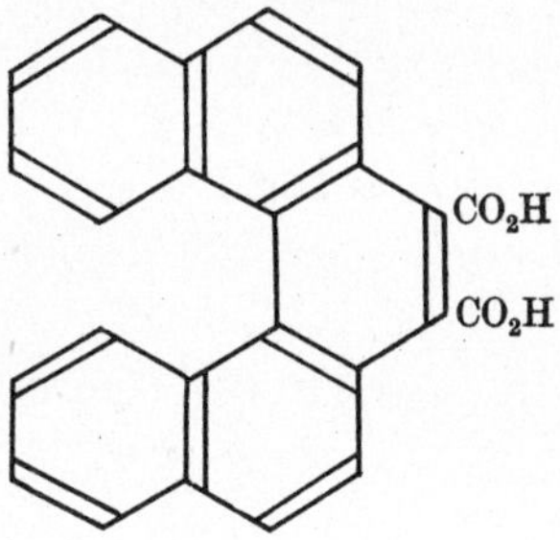   (VIII)

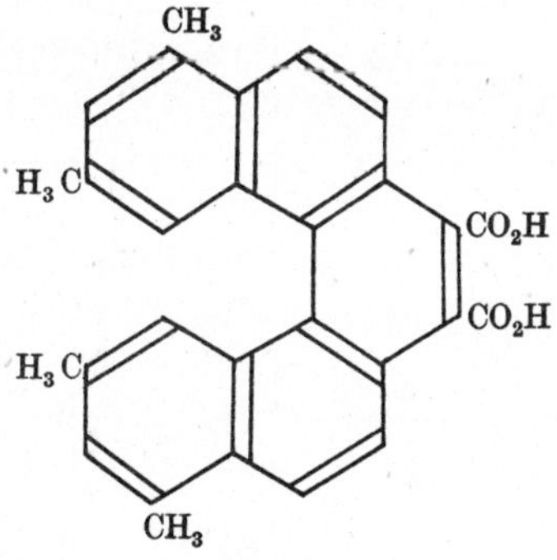

(IX)

The parent compound, 4:5-dimethylphenanthrene (I), has also been prepared, so that there is no longer any reason to regard the 4:5 positions in phenanthrene as 'impossible'.* Similar overcrowding of two methyl groups occurs in 2:2′-dimethyl-*meso*-benzodianthrone,† in 2:2′-dimethyl-*meso*-naphthodianthrone‡ and in some other compounds.§

The case of 3:4:5:6-dibenzophenanthrene (VII) is essentially

---

* Newman and Whitehouse, ibid. 1949, **71**, 3664; see also Badger, Campbell, Cook, Raphael and Scott, *J. Chem. Soc.* 1950, p. 2326.
† Scholl and Tänzer, *Liebigs Ann.* 1923, **433**, 163.
‡ German Patent 458,710.
§ Newman, *J. Amer. Chem. Soc.* 1940, **62**, 1683.

the same as that of 4:5-dimethylphenanthrene. Cook* first pointed out that this ring system must be distorted, because in the undistorted molecule there is no room for the hydrogen atoms at the positions shown with asterisks.

(X)   (XI)

Experimental evidence that this distortion does occur was provided by the observation of the mutarotation of the morphine salt of 3:4:5:6-dibenzophenanthrene-9:10-dicarboxylic acid (VIII) The related tetramethyl derivative (IX) (in which the analogy to 4:5-dimethylphenanthrene is even closer) shows greater steric interference, and has been resolved; but the optically active forms rapidly lost activity on standing in solution.[†] The distortion in 3:4:5:6-dibenzophenanthrene has also been confirmed by X-ray crystallography.[‡]

A third and analogous type of molecular asymmetry is exhibited by 5-methyl-3:4-benzophenanthrene. Here the methyl group must be displaced out of the plane of the remainder of the molecule by the proximity of the 'benz'-ring, for Newman and Wheatley[§] have successfully resolved 5-methyl-3:4-benzophenanthrene-8-acetic acid into its optically active components (X) and (XI).

* *Ann. Rep. Chem. Soc.* 1942, **39**, 173.
† Bell and Waring, *Chem. & Ind.* 1949, p. 321; *J. Chem. Soc.* 1949, p. 2689.
‡ McIntosh, Robertson and Vand, *Nature, Lond.*, 1952, **169**, 322.
§ *J. Amer. Chem. Soc.* 1948, **70**, 1913.

INDEX OF SUBJECTS

Absorption spectra, origin of, 383
absorption wave-length, and colour, 394
aceanthrenoaceanthrene, 56
acetylene,
   C—C bond length in, 12
   heat of formation of, 40
acid dissociation constant, 76
acridine, 63, 77, 78, 244
   absorption spectrum of, 406–7
acridone, 66
alizarin, 243, 309
alkyl substituents, 212
allomucic acid, 9
allyl radical, 38
amination, 309–11
aminopyridines, 243, 310
aminoquinolines, 310
   diazo coupling of, 300
amino*iso*quinoline, 310
aniline, contributing structures, 200
   diazo coupling of, 295, 297
anthracene, 7, 37, 43, 44–5, 52–3, 63, 95, 98, 107, 141, 158, 162, 166–7, 172, 174, 182, 350, 354, 358, 363, 382
   absorption spectrum of, 392, 394, 396–7, 407
   Diels-Alder reaction with, 349–50
   fluorescence spectrum of, 410
   reduction of, 7, 109
   sulphonation of, 306
anthracene dibromide, 106–7, 112
anthracene photo-oxide, 367, 378, 380
meso-anthradianthrene, 345
anthraquinone, 89, 90, 91, 290
   sulphonation of, 306
9-anthrol, 47, 50
anthrone, 47, 50, 65
anti-bonding orbital, 24, 33
aromatic character, 2
aromatic compound, definition of, 43
atomic refractivity, 209
1-azanthracene, absorption spectrum of, 406–7
azonaphthalene, 224
azulene, 43, 56, 57
   absorption spectrum of, 403–4
azulenes, synthesis of, 143–7

bathochrome, 386
bathochromic shifts, 387–8
Beer-Lambert law, 384
1:2-benzacridine, 77

2:3-benzacridine, 77
3:4-benzacridine, 77
1:2-benzazulene, 147
1:2-benzanthracene, 141, 159–60, 175, 282, 286, 350, 352, 408
   absorption spectrum of, 399
1:2-benzanthraquinone, 89, 90
benzene, absorption spectrum of, 385, 392, 394, 395, 409
   application of wave mechanics to, 31
   bond localization energy in, 179, 182
   C—C bond length in, 12
   canonical set for, 36
   catalytic oxidation of, 9
   classical formulas for, 3–7
   discovery of, 1
   fluorescence spectrum of, 409
   heterocyclic analogues of, 5–6
   metabolic oxidation of, 9
   molecular diagram for, 172
   para-localization energy in, 363
   reduction of, 7
   resonance energy of, 39, 41, 43
   vibration forms, 15
benzofuran, 73
1:2-benzonaphthacene, 326
1:12-benzoperylene, 163, 345
2:3-benzophenazine, 63
benzophenone, 50
3:4-benzopyrene, 158, 277
   fluorescence spectrum of, 410
6:7-benzoquinoline, 77–8
5:6-benzoquinoline, 77, 117
7:8-benzoquinoline, 77
benzoquinone, 87, 89, 90, 354, 355
biphenylene, 53
bond energies, 40
bond fixation, 116
bond lengths, in conjugated compounds, 12
bond-length variations, 161
   in anthracene, 162
   in 1:12-benzoperylene, 163
   in coronene, 163
   in 1:2:5:6-dibenzanthracene, 164
   in furan, 162
   in naphthalene, 162
   in ovalene, 163, 177
   in pyrene, 163
   in pyrrole, 162
   in thiophen, 162

bond localization energy, 178
  in benzene, 179
  and bond order, 180–2
  in naphthalene, 179–80
  in phenanthrene, 180
bonding orbital, 23, 33
bond order (molecular orbital method), 173
  and bond length, 175
  and redox potential, 181–2
bond order (Penney type), 169–70
bond order (valence-bond method), 172
bond orders (molecular orbital method),
  in anthracene, 174
  in 1:2-benzanthracene, 175
  in benzene, 174
  in chrysene, 175
  in naphthalene, 174
  in phenanthrene, 174
  in pyrene, 175
bond orders (valence-bond method), in anthracene, 172
  in benzene, 172
  in naphthalene, 172
  in phenanthrene, 172
bromination, 237–8
  of bromobenzene, 263
  of naphthalene, 260–2
  of pyridine, 264
bromine cation, 238, 258
9-bromoanthracene, 107, 112
bromobenzene, 7, 263
9-bromophenanthrene, 51, 103
butadiene,
  absorption spectrum of, 385
  electronic structure of, 30, 35
cyclobutadiene, 53
tert.-butylbenzene, sulphonation of, 305

cadalene, 57
canonical set, 36
cantharidin, 341
carbazole, 73, 91
carbon-carbon bonds, 17
carbonium ions, 238, 268–9, 270, 272
chelation, 130
chloroazanthracene, 117
chloroanthrone, 375
6-chlorodiphenic acid, 414
chlorophyll, 74
2-chloropyridine, 65
chrysene, 95, 175
cinnoline, 77, 78
Claisen rearrangement, 126
clathrate complexes, 86
coal tar, 1, 410
conjugation,
  and free-valence number, 226
  with ring system, 223

complex formation, 79
coronene, 163
cyano group, introduction of, 311
cyclization, 117, 289–94
  influence of catalyst, 122
  steric factors in, 121

decalin, 7
delocalization energy, 37
9:10-diacetoxyanthracene, 369
diamond, C—C bond length in, 12
dianthracene, 377–8
1:1′-dianthryl, 423
9:9′-dianthryl, 369
dianthryls, substituted, 423
1:1-diarylethylenes, 360
diazoacetic ester, 138
  and alkyl benzenes, 142
  and anthracene, 141
  and 1:2-benzanthracene, 141
  and benzene, 139
  and indanes, 143, 145
  and naphthalene, 139
  and phenanthrene, 140
  and pyrene, 141
diazo coupling, 123, 294–301
1:2:5:6-dibenzanthracene, 164, 350, 352
  fluorescence spectrum of, 410
  synthesis of, 326
1:2:7:8-dibenzanthracene, fluorescence spectrum of, 410
1:2:5:6-dibenzanthraquinone, 89
dibenzocyclobutadiene, 53
dibenzofuran, 73
1:2:4:5-dibenzopentalene, 56
3:4:5:6-dibenzophenanthrene, 439–40
2:3:6:7-dibenzophenazine, 63
dibenzothiophen, 73
9:10-dibromoanthracene, 103, 112
9:10-dibromo-9:10-dihydroanthracene, 45, 106
1:5-dichloroanthracene, 107
1:8-dichloroanthracene, 108
9:10-dichloroanthracene, 102
Diels-Alder reaction, 334–63
  aromatic dienes in, 334–54
  and free-valence number, 363
  mechanism of, 360–3
  and para-localization energy, 362
  with pentacyclic compounds, 350
  reversible nature of, 341
  stereochemistry of, 358–60
9:10-diethyl-1:2-benzanthracene, 352
digonal hybridization, 25
dihydroanthracene, 7, 45, 91
dihydrobenzene, 45, 88, 110
dihydronaphthalene, 45, 88
dihydrotoluene, 110
dihydro-meta-xylene, 110

INDEX OF SUBJECTS 443

2:6-dihydroxybenzoic acid, 214
dimesityl, absorption spectrum of, 390
dimethylaniilines,
  molecular refractivities of, 219
  strength as bases, 220
9:10-dimethylanthracene, 351, 352, 369
  bromination of, 225
  reduction of, 109
9:10-dimethyl-1:2-benzanthracene, 351, 409
2:2′-dimethyl-*meso*-benzodianthrone, 439
2:3-dimethylbutadiene, 335
6:7-dimethylchrysene, 437
2:2′-dimethyl-*meso*-naphthodianthrone, 439
4:5-dimethylphenanthrene, 437
2:6-dimethyl-γ-pyridone, 309
9:10-dimethoxyanthracene, 369, 373, 401
1:4-dimethoxy-9:10-diphenylanthracene, 373–4
1:4-dimethoxynaphthalene, 228
1:5-dimethoxynaphthalone, 228
dinaphthylethylene, 225
dinaphthyls, substituted, 421–2
9:10-dinitro-9:10-dihydroanthracene, 108
6:6′-dinitrodiphenic acid, 412, 424
2:6-dinitrophenol, 243
dioxadiene, 71
diphenyl, 412–13
  absorption spectrum of, 388
1:4-diphenylanthracene, absorption spectrum of, 391
9:10-diphenylanthracene, 107, 351, 352, 369, 373, 375, 401
  absorption spectrum of, 391
9:10-diphenyl-1-azanthracene, 369
9:10-diphenyl-2-azanthracene, 369
1:3-diphenyl*iso*benzofuran, 341, 342, 368, 378
10:10′-diphenyl-9:9′-dianthryl, 370
1:1-diphenylethylene, 335
meso-diphenylhelianthrene, 370–1
diphenyl-*meso*-naphthodianthrene, 372
6:13-diphenylpentacene, 352
diphenyls, stereochemistry of substituted, 412–23
dipole moments, 184–90, 198
  of *ortho*-substituted compounds, 216–17
dipyrryl, 421
double-bond character, 36, 165
  in anthracene, 166
  in benzene, 165
  and bond length, 167–9
  in graphite, 165
  in naphthalene, 166
  in phenanthrene, 166

Elbs reaction, 323–30
  elimination of methyl groups during, 327
  mechanism of, 328–30
electromeric effect, 208
electrophilic substitution, 233–43
  addition-elimination theory of, 236
electron diffraction, 11
electron displacements and bond fixation, 131
electron pairs, 18
electronegativity, 186
electronic configuration, 22
elimination reactions, 112
  effect of solvent on, 113
ethyl benzoate, phenylation of, 314
ethyl salicylate, 215
ethylene, absorption spectrum of, 384
C—C bond length in, 12
  electronic structure of, 27
enantiomorph, 411
oxaltation, 210
extinction coefficient, 384

fluorene, 55
  purification of, 377
fluorescence spectra, 408–10
formal bond, 35
free-valence number (molecular orbital method), 176–7
free-valence number (valence-bond method), 172
free-valence numbers (molecular orbital method), in anthracene, 242, 363
  in benzene, 177, 363
  in butadiene, 347
  in chrysene, 242
  and the Diels-Alder reaction, 363
  in diphenyl, 346
  in naphthalene, 242, 363
  in perylene, 347
  in phenanthrene, 347
  in pyrene, 242
  in pyridine, 231
free-valence numbers (valence-bond method), in anthracene, 172
  in benzene, 172
  and conjugation, 403
  in naphthalene, 172
  in phenanthrene, 172
Friedel-Crafts reaction, 238, 264–94
  abnormal orientation in, 273, 285–6
  acylations, 278–89
  alkylations, 268–78
  isomerizations during, 269–72, 275–7, 287
  mechanism of, 264–7
  mechanism of acylation, 283
Fries rearrangement, 275

INDEX OF SUBJECTS

Fries rule, 52
fulvalene, 54
fulvene, 54
furan, 43, 67–71, 162, 359
  electron displacements, 230
  reactions of, 100
  reduction of, 100

Gomberg reaction, 313–16
guaiazulene, 57

haemin, 74
half-life periods, 427, 431, 435
halogenation, 257–64
  of phenols, 123
halogen substituents, 210–12
heat of hydrogenation, 41–2
helianthrene, 371
heptacene, 95, 363
heptalene, 56
cycloheptatriene, 54
heterocyclic bases, strength of, 77
hexabromocyclohexane, 8
hexacene, 45–6, 63, 95, 363
hexacene photo-oxide, 367
hexachlorocyclohexane, 8, 97
hexadeuterobenzene, 14
hexamethylbenzene, bond lengths in, 13
hybrid, 35
hybrid bond, 12
hybridization, 25
hydroxyacetylnaphthalene, 130
hydroxyacridine, 66
9-hydroxyanthracene, 47, 65
5-hydroxynaphthacene, 65
2-hydroxypyridine, 65
3-hydroxypyridine, 65
4-hydroxypyridine, 65
hydroxyquinolines, 65
hydroxyisoquinolines, 65
hydroxylation, 307–9
hyperconjugation, 213
hypsochrome, 386

imidazole, 69, 72, 101
indane, 134, 143
indanequinone, 356
indene, 55
indole, 43, 78
indolizine, 75
inductive effect, 184–94
inductomeric effects, 208
infra-red spectroscopy, 14, 383
interference value, 424
intramolecular acylation, 289–94
isoxazole, 72

Jacobsen reaction, 277–8

lithium, addition to anthracene, 108
para-localization energy, 49, 362–3
long-wave spectroscopy, 13

mesomeric effect, 194–204
2-methoxypyridine, 65
ortho-methylanisole, reduction of, 111
9-methylanthracene, 47, 351
5-methylazulene, 144
6-methylazulene, 143
methylazulenes, absorption spectra of, 404–5
5-methyl-3:4-benzophenanthrene, 440
methylpentacene, 48
N-methyl-2-pyridone, 65
N-methylpyrrole, 343
methylsuccinic anhydride, 284
Mills-Nixon effect, 133–8, 142–4
  chelation and, 135
  Claisen rearrangement and, 135
  diazo coupling and, 135
  ozonization and, 156
molecular complexes, 79
molecular orbital, 22
molecular refractivity, 209, 218–19
muconic acid, 9

naphthacene, 43, 45, 63, 95, 350–1, 363
  absorption spectrum of, 393, 394
  fluorescence spectrum of, 410
naphthacene photo-oxide, 367
naphthacenequinone, 88
meso-naphtha-dianthrene, 346
naphthalene, 43, 44–5, 63, 95, 139, 153, 162, 171, 172, 174, 179, 180, 286, 363
  absorption spectrum of, 392
  bond localization energies in, 179–80, 182
  bromination of, 260–3
  canonical set for, 37
  Diels-Alder reaction with, 347
  fluorescence spectrum of, 409
  reduction of, 7
  sulphonation of, 305
naphthalene tetrabromide, 102
naphthalene tetrachloride, 102
naphthaquinolinequinone, 92
naphthaquinone, 87, 89, 90
naphthatriazolequinone, 92
1-naphthol, 46
1-α-naphthyl-1-cyclopentene, 339
nitration, 235–7, 252–6
9-nitroanthracene, 113
nitrobenzene,
  contributing structures, 200
  phenylation of, 314
nitronaphthylamines, 131
nitronium ion, 238, 253

INDEX OF SUBJECTS 445

ortho-nitroacetanilide, 215
3-nitrocatechol, 215
ortho-nitrophenol, 215, 243, 309
2-nitroresorcinol, 215
nucleophilic substitution, 243-9

octahydroanthracene, 7, 276
octahydrophenanthrene, 276
cyclooctatetraene, 55
orbital, 17
orbital wave function, 17
order number, 395
'ortho' effect, 214-23
ortho-para ratio, 330-3
ortho substituents,
  and base strength, 222
  and dipole moments, 216-17
osmium tetroxide, 9
  addition to anthracene, 158
  addition to 1:2-benzanthracene, 158
  addition to 3:4-benzopyrene, 158
  addition to dinaphthylethylenes, 225
  addition to phenanthrene, 157
  addition to pyrene, 158
  relative rates of addition, 160
ovalene, 163, 176, 346
oxalic acid, 9
oxalyl chloride, 279-81
para-oxazine, 71
oxazole, 72
oxidation-reduction potentials, 87
ozone, addition to benzanthrone, 149
  addition to benzene, 10, 148
  addition to 4:7-dimethylindane, 156
  addition to 5:6-dimethylindane, 156
  addition to 1:4-dimethylnaphthalene, 153
  addition to 2:3-dimethylnaphthalene, 153
  addition to 2:3-dimethylpyridine, 152
  addition to 5:8-dimethylquinoline, 156
  addition to 6:7-dimethylquinoline, 156
  addition to fluoranthene, 149
  addition to naphthalene, 148
  addition to pyrene, 148
  addition to quinoline, 155
  addition to 2:3:4-trimethylpyridine, 152
  addition to ortho-xylene, 10, 150
ozonides, 147

π-bond, 27
  in benzene, 32
π-electron densities, 34
  in acridine, 232
  in furan, 231
  in pyridine, 231

  in pyrrole, 231
  in quinoline, 232
  in isoquinoline, 232
  in thiophen, 231
π-electrons, 28
para-anthracene, 377-8
Pauli exclusion principle, 18
pentacene, 43, 45, 63, 95, 355, 363, 394
  fluorescence spectrum of, 410
pentacene photo-oxide, 367
pentacenequinone, 88
pentalene, 56
pentaphene, absorption spectrum of, 395
cyclopentadiene, 54, 359
cyclopentenylphenanthrene, 339
perbenzoic acid, 224
percentage ionization, 79
perfluorocyclohexane, 98
perhydroanthracene, 7
perylene, 95
phenanthrene, 43, 44, 50, 95, 103-6, 140, 158, 167, 172, 174, 180, 371
  absorption spectrum of, 395
  bond localization energies in, 179-80, 182
  sulphonation of, 306
phenanthrene dibromide, 51, 96, 103-6
phenanthrene dichloride, 103
phenanthrahydroquinone, 96
phenanthraquinone, 88, 89, 90, 96
phenanthridine, 63, 77, 78, 244, 310
phenanthroline, 77, 78
phenazhydrin, 63, 83
phenazine, 63, 77, 78
  absorption spectrum of, 406-7
2-phenazinol, 66
phenol, 46, 50, 201
  contributing structures, 200
phenols, diazo coupling of, 295, 297-9
1-phenylanthracene, absorption spectrum of, 391
9-phenylanthracene, absorption spectrum of, 391
phenylpyrrole, 421
phenylsuccinic anhydride, 284-5
phloroglucinol, 46
phthalazine, 77, 78
photo-dimerization, 377-9
  influence of solvent on, 380
photo-oxidation, influence of solvent on, 380
  mechanism of, 379-82
photo-oxide, of rubrene, 364
photo-oxides,
  of heterocyclic compounds, 367
  of polyacenes, 367
  of substituted compounds, 369
  structure of, 364-7
  thermal stability of, 372-5

bis-photo-oxides, 369–70
phthalic anhydride, 282
picoline, 59
picrates, 79
pK_a, 76
polyacenes, absorption spectra of, 393–4
polyphenyls, absorption spectra of, 389–91
porphyrin, 74
probability distribution function, 17
2-*iso*propenylanthracene, 349
Pschorr synthesis, 316–23
  of benzanthracene, 319
  of benzophenanthrene, 319
  of dibenzanthracene, 319
  of dibenzophenanthrene, 319
  of fluorenes, 317
  of fluorenones, 317
  of phenanthrene, 316–17, 320–2
pyran, 71
pyrazine, 62–3, 77, 78
  dipole moment of, 229
pyrazole, 72, 79, 101
pyrene, 44, 51, 52, 95, 141, 158, 163, 175
  reduction of, 110
  substitution of, 51
pyridazine, 62, 77, 78
  dipole moment of, 229
pyridine, 43, 59–62, 77
  absorption spectrum of, 405
  bromination of, 264
  dipole moment of, 229
  electron displacements, 230
  electronic structure of, 61–2
  phenylation of, 314
  reduction of, 98
α-pyridone, 243, 309
pyrimidazole, 75
pyrimidine, 62, 77, 78
  dipole moment of, 229
pyrrole, 43, 67–70, 78, 162
  electron displacements, 230
  reduction of, 101

quinazoline, 77, 78
quinoline, 43, 59, 63, 77, 78, 155–6, 244
  reduction of, 99
quinolinequinone, 91
*iso*quinoline, 59, 63, 77, 244
*iso*quinolinequinone, 91
quinones as dienophiles, 354–7
quinoxaline, 63, 77, 78

radical substitution, 249–52
Raman spectroscopy, 14, 383
redox potentials, 87
  'corrected', 93
resonance, 35
resonance energies, 37
  of aromatic compounds, 43
  of polycyclic compounds, 95
resonance hybrid, 35
resorcinol, 46
rubrene, 364
rubrene *iso*-oxide, 377
rubrene mono-oxide, 375
rubrene photo-oxide, 364
  isomerization of, 377
  reduction of, 375

σ-bonds, 27
σ-constants, 204–8
salicylic acid, 214
salicylaldehyde, 215
self-polarizability, 227
Skraup reaction, 117
sodium,
  addition to anthracene, 108
  in liquid ammonia, 110
specific exaltation, 210
specific refractivity, 209
spectral relationships, 392–6
spin, 18
steric effects, 121
stipitatic acid, 58
strengths of acids, 190, 193, 194, 195, 212
strengths of amines, 193
strengths of anilines, 196, 221–2
strengths of phenols, 197, 219–20
substituted acetic acids, 190
substituted anthracenes, steric hindrance in, 400–2
substituted anthracenes, absorption spectra of, 398
substituted 1:2-benzanthracenes, absorption spectra of, 399
substituent constants, 204–8
succinic anhydride, 282
sulphonation, 301–7
  of anthracene, 306
  of anthraquinone, 306
  of naphthalene, 305
  of phenanthrene, 306
  steric hindrance in, 305
  of toluene, 304

meso-tartaric acid, 9
tetrabromotetrahydronaphthalene, 102
tetrachlorotetrahydronaphthalene, 102
tetrahedral hybridization, 25
tetrahydroanthracene, 7
tetrahydronaphthalene, 7, 134
tetralone, 290
tetramethylnaphthalene, Diels-Alder reaction with, 348
9:9′:10:10′-tetraphenyl-1:1′-dianthryl, 370

INDEX OF SUBJECTS

5:7:12:14-tetrazapentacene, 63
tetrazine, 62
thebaine, 420
thiapyran, 71
triazine, 62
para-thiazine, 71
thiazole, 72
thionaphthen, 73
thionaphthenquinone, 91, 92
thienothiophen, 73
thiophanthrenquinone, 91
thiophen, 43, 67-71, 162
    absorption spectrum of, 407
    chlorination of, 99
    electron displacements, 230
    reduction of, 99
thiophthen, 73
toluene, alkylation of, 273-4

trigonal hybridization, 25
triphenylene, 95
5:9:10-trimethyl-1:2-benzanthracene, 351
triptycene, 355
triptycenequinone, 356
tropolone, 58, 147
Tschitschibabin reaction, 309-11

vetivazulene, 145
2-vinylfuran, 339
1-vinylnaphthalene, 335-8, 361
2-vinylnaphthalene, 338
9-vinylphenanthrene, 339

X-ray diffraction, 11
ortho-xylene, 10, 150-2

INDEX OF AUTHORS

Abrahams, 84, 162
Adams, 98, 272, 283, 314, 411, 417, 418, 419, 421, 424, 426, 427, 430, 432, 433, 435
Adkins, 51, 98, 101
Ahmad, 284
Albert, 66, 76, 77, 79, 192, 196, 224, 430
Albrecht, 422, 425
Alder, 334, 340, 343, 349, 360
Allais, 373, 379
Allen, 141, 343, 352
Ames, 91, 92
Anderson, 59, 103, 234, 435
Andrews, 87
Angeli, 68
Anzilotti, 255
Archer, 255, 272
Ardagh, 67
Armit, 6
Armstrong, 5
Arnold, 137, 143, 146
Arntzen, 105
Aspinall, 56
Audsley, 211
Audubert, 374
Augood, 249
Auwers, 210
Avenarius, 329
Ayling, 213
Azorlosa, 329

Bachmann, 108, 121, 122, 293, 313, 318, 324, 336, 351, 352, 354, 359
Baddar, 84
Baddeley, 274, 275, 285, 286, 287
Badger, 64, 66, 83, 109, 116, 141, 149, 158, 160, 180, 181, 183, 224, 225, 227, 281, 363, 365, 369, 399, 407, 439
Badoche, 377, 379, 380
Baer, 360
Baeyer, 5, 8, 309
Baget, 374
Baker, 43, 53, 56, 130, 135, 211, 213, 215
Baldock, 346
Balle, 316, 318
Balsohn, 268
Baltzer, 59
Bamberger, 5, 314
Barber, 98
Barker, 137
Barnes, 14
Barnett, 107, 108, 112, 113, 114, 115, 225, 341, 354

Bartels-Keith, 147
Bartlett, 296, 354, 355, 356
Battegay, 306
Baumgarten, 302
Baur, 269
Beach, 162, 167, 199
Becker, 51, 148, 149
Beisswenger, 422, 437
Bell, 76, 117, 124, 352, 402, 412, 418, 422, 440
Benesi, 87
Benfey, 35
Benford, 254, 257
Bennett, 81, 82, 197, 254
Bergmann, 107, 133, 342, 349, 360, 361, 362, 365, 377
Bergstrom, 243, 244, 310
Berliner, 214, 283, 284, 290
Bernhauer, 271
Bersch, 340
Berthier, 58, 138, 157, 174, 346
Bhatia, 284
Bickel, 60
Bickford, 347
Bigelow, 98
Binder, 435
Birch, 99, 110
Birtles, 216, 217
Bleier, 318
Bloem, 263
Blood, 56
Blout, 83, 406
Bock, 421
Bodendorf, 268
Boedtker, 274
Boer, 155
Bogert, 238
Boggust, 119
Böhme, 268
Bondhus, 214
Bonner, 14
Boord, 141
Borcherdt, 274
Borisoff, 98
Born, 382
Borrows, 75
Borsche, 297
Bowden, 271
Bowen, 381, 385, 408, 409, 410
Bowman, 67
Boyer, 141
Brackmann, 83
Bradfield, 259, 304

INDEX OF AUTHORS

Bradley, 243
Bradsher, 291
Branch, 87, 89, 90, 93, 190, 215, 394
Brand, 56, 254, 302
Brandsma, 260
Brandt, 306
Brass, 85
Braude, 253, 303, 384, 385, 386
Breckenridge, 428
Bremner, 42
Breslow, 296
Brewer, 329
Briegleb, 83
Briner, 57
Brockmann, 371
Brockway, 13, 167, 199, 367
Brönsted, 76
Brooks, 197
Brown, 49, 53, 58, 85, 179, 180, 213, 221, 222, 234, 362, 363, 404
Brownell, 220
Brühl, 209
Bryson, 133
Buchner, 101, 139, 143
Bunnett, 312
Burger, 51, 98
Burkhardt, 204, 302
Burkitt, 176, 363
Burnett, 243
Burrows, 98
Burt, 141
Burtner, 269
Burton, 289
Butler, 374, 379
Butz, 334
Buu-Hoï, 243, 252, 346
Buzeman, 163, 175

Cadre, 81
Cady, 98
Cahn, 221, 222
Caland, 304
Calcott, 288
Calloway, 265
Calvin, 87, 89, 90, 93, 131, 190, 394
Cameron, 110
Campbell, 55, 110, 131, 149, 151, 439
Capper, 377
Carlson, 236
Caro, 309
Carr, 53, 141
Carruthers, 130
Carter, 94
Cass, 278
Catel, 378
Cattadori, 69
Chalvet, 243, 252
Chang, 421
Chatterjee, 258

Chédin, 254
Chemerda, 352
Chien, 426
Chopin, 147
Chow, 101
Christie, 412
Ciamician, 68, 69, 101
Claisen, 126, 127
Clapp, 250, 342
Clar, 45, 47, 56, 87, 95, 141, 325, 326, 327, 329, 344, 345, 346, 349, 350, 353, 354, 355, 367, 382, 386, 392, 394, 403
Claus, 5
Clauson-Kaas, 100
Clement, 235
Clemo, 83, 118
Cocker, 119
Cohen, 336, 338, 354, 355, 356
Cole, 10, 150
Compagnon, 342
Conant, 92, 101, 300
Connerade, 108, 236
Cook, 9, 58, 80, 107, 110, 112, 113, 114, 115, 121, 140, 141, 149, 158, 225, 286, 317, 319, 326, 327, 328, 365, 369, 410, 437, 439, 440
Coonradt, 99, 100
Coop, 198, 199
Corbellini, 422
Corell, 51, 148, 149
Corley, 83, 406
Cormack, 312
Cornubert, 210
Corson, 271
Cottrell, 40
Coulson, 17, 24, 25, 34, 38, 39, 110, 137, 157, 173, 174, 176, 225, 227, 231, 232, 329, 346, 363
Courtin, 125
Cowdrey, 239
Cox, 102, 104
Crafts, 264
Craig, 137, 406
Crawford, 213
Criegee, 157
Cromwell, 87
Crouch, 152
Crowell, 269

Dankert, 430
Datta, 258
Dauben, 329
Daudel, 170, 171, 172, 201, 227, 232, 243, 252, 346, 403
Daudt, 250
Davidson, 238
Davies, 259
Dayton, 348
Dazeley, 434

INDEX OF AUTHORS

Dean, 364, 372
de Bruyn, 312
de Heer, 181
De Jong, 216, 217
de la Mare, 211, 212, 258, 331, 333
Demuynck, 373
Den Hertog, 60
Dennstedt, 69
Deno, 336
Derbyshire, 238, 258, 259, 307
Derfer, 141
DeTar, 250, 314, 320
Dewar, 3, 40, 42, 58, 77, 217, 248, 300, 331
Deyrup, 266
Dhar, 389, 413, 424
Dickson, 140
Diels, 340, 343, 349
Dietz, 325, 326, 329
Dimroth, 295
Dinger, 317
Dippy, 191, 195, 211
Dobbin, 59
Doering, 58, 147
Dollear, 347
Dolliver, 137
Donaldson, 163
Doub, 386, 387
Downs, 9
Drake, 140
Dresel, 302
Drysdale, 235
Dufraisse, 100, 341, 342, 364, 365, 367, 368, 369, 370, 372, 373, 374, 375, 377, 379, 380
Durland, 51
Duveen, 373
Dyatkina, 189

Ecary, 368
Eddy, 85
Edgerton, 121, 293
Egoroff, 269
Eisenlohr, 209
Elbs, 258, 323
Elderfield, 406
Elliott, 428
Enderlin, 365, 375, 377
Erickson, 243
Ernst, 68, 99
Eschinazi, 361
Étienne, 369, 373, 374, 380
Euler, 253
Euwes, 307
Evans, 86, 181, 204, 213, 302, 406
Everard, 228, 229, 402
Everhart, 102
Ewing, 66, 406
Eyring, 253

Fairbrother, 266
Fakstorp, 100
Fankuchen, 56
Fanta, 85
Farenhorst, 252
Farlow, 98
Faust, 102
Feldmann, 139
Fenton, 308
Ferguson, 389, 390
Fernelius, 244
Ferris, 141
Fierens, 265
Fierz-David, 304, 306
Fieser, 52, 87, 91, 92, 96, 104, 121, 122, 123, 124, 125, 126, 128, 135, 148, 149, 237, 250, 288, 295, 306, 323, 324, 325, 326, 327, 329, 437
Finkelstein, 257
Firla, 265
Fischer, 68, 317, 347
Fittig, 51, 103
Flett, 131
Ford, 204
France, 313, 314
Francis, 258
Freymann, 130
Friedel, 264
Friedman, 300
Fries, 52, 117, 124, 125
Friess, 146
Fritsch, 101
Fritzsche, 377
Frobenius, 314
Frühling, 210
Fukuhara, 98
Fuller, 280
Fürst, 144, 145, 147
Fuson, 274

Gaffron, 381
Gajewski, 300
Gallaway, 137
Gates, 343
Gattermann, 299
Gelissen, 314
Gensler, 406
Gérard, 365, 373, 375, 380
Gergely, 181
Gibb, 58, 141, 281
Gibson, 83
Giguère, 366
Gillam, 389
Gillespie, 104, 238, 253, 255
Gillet, 365, 377
Gilman, 52, 68, 87, 100, 104, 123, 124, 237, 269, 295
Girard, 364, 372
Glasstone, 197

INDEX OF AUTHORS

Goalby, 86
Goddard, 254
Gold, 256
Goldacre, 77
Goldberg, 8
Goldschmidt, 295
Goldstein, 280
Golumbic, 137
Gomberg, 265, 313
Goodall, 83
Goodman, 244, 310
Goodway, 354
Gordon, 213, 432
Goss, 211
Götzky, 277
Goulden, 109
Graebe, 51, 103, 317
Graham, 110, 253
Grainger, 114, 115
Granara, 243
Grandmougin, 103
Greenlee, 141
Greenwood, 174
Gresham, 137
Grieve, 313, 314
Groggins, 283
Gross, 435
Grosse, 98
Groves, 197, 198, 202, 211, 212
Guile, 271
Gustavson, 269
Guyot, 303, 378

Haak, 110
Haaijman, 13, 151
Haber, 308
Halberstadt, 256
Hall, 196
Halse, 274
Hamilton, 98
Hammett, 204, 205, 266
Hampson, 137, 216, 217
Hancox, 138
Hanford, 426
Hantzsch, 266, 314
Harries, 10, 148, 153
Harris, 418
Hartmann, 295
Hartough, 67, 99, 100
Hartshorn, 100
Hathway, 131
Hauser, 296
Haworth, 58, 121, 250, 293, 314, 437
Hawran, 329
Hediger, 139
Heidelberger, 55, 329
Heilbron, 250, 313, 314
Heilbronner, 43, 57, 404
Heilmann, 130

Heise, 270, 271, 272
Heitler, 23
Hemming, 213
Hennig, 56
Hennion, 255, 272
Henrich, 132
Hermans, 314
Hershberg, 288, 324
Herzog, 348, 349
Hewett, 286, 410
Hey, 249, 250, 313, 314, 315, 320, 322, 389
Heymann, 121
Heyroth, 406
Hieger, 410
Higgins, 354
Hildebrand, 87
Hill, 100
Hinsberg, 63
Hinshelwood, 302
Hippchen, 146
Hirshberg, 133, 391, 422
Hobson, 58
Höchtlen, 86
Hoffman, 313
Hofmann, 86
Højendahl, 196
Holland, 75
Holleman, 234, 304
Hopkins, 211
Horclois, 367, 373
Horn, 147
Houpillart, 365, 369
Huang, 121
Hückel, 28, 229
Hudson, 335, 337, 360
Hughes, 213, 239, 241, 253, 254, 256, 258
Hugill, 198, 199
Huisgen, 118
Hunsberger, 130
Hunter, 117, 124, 215
Hurd, 329
Huse, 83
Hussey, 402, 438
Huston, 271
Hüter, 146

Ingham, 216, 217
Ingold, 4, 14, 15, 70, 196, 208, 211, 213, 235, 239, 241, 253, 254, 256, 257, 258, 292, 382
Ipatieff, 270, 271, 272

Jacques, 171
Jacobsen, 274, 278
Jaffé, 9
Jaroslavzev, 258
Jean, 171
Jefferies, 58

INDEX OF AUTHORS

John, 325, 329, 355, 367
Johnson, 55, 100, 118, 122, 147, 289, 340, 341
Jones, 109, 228, 269, 304, 396, 397, 399, 400, 422
Judson, 197
Julian, 380

Kadesch, 216, 217
Kampschmidt, 148
Kanyaev, 258
Kästner, 289
Kaufman, 56
Keefer, 87
Kehrmann, 265
Kekulé, 1, 3, 4, 116, 269
Kelbe, 269
Kelham, 429, 430
Kennaway, 276, 437
Kennelley, 91
Kenner, 412
Kenyon, 412
Kermack, 6
Ketelaar, 151, 178, 179, 180, 223, 403
Khalil, 284
Kharasch, 104, 237
Kiefer, 302
Kilpatrick, 197
King, 98, 102, 104
Kischner, 141
Kiss, 388
Kistiakowsky, 41, 137
Kizhner, 279
Klages, 258
Klatt, 85
Klemenc, 256
Klevens, 385, 386
Klingler, 319
Kloetzel, 121, 334, 348, 349, 351, 352, 354
Klopp, 100
Knorr, 101
Knox, 58, 147
Koblitz, 381
Kofod, 216, 217
Kohler, 279
Kokeguti, 128
Konowaloff, 269, 270, 271, 272
Kooyman, 151, 152, 153, 178, 252
Kornblum, 419
Kossiakoff, 136
Kowalski, 265
Krasova, 305
Kraus, 110
Kremer, 406
Krollpfeiffer, 284
Krueger, 121
Kuhn, 422, 425
Kuick, 98

Ladenburg, 4, 5, 98
Lahey, 404
Lappin, 71
Lapworth, 302, 331
Lauer, 307, 380
Laurent, 102
Lawrence, 354
Lazier, 98
Le Bras, 369, 373, 379
Leeds, 102
Le Fèvre, 184, 187, 305, 412
Leffler, 243, 310
Legg, 118
Lennard-Jones, 23, 28, 38
Lesslie, 418
Leuchs, 294
Levine, 10, 150
Levy, 284
Lewis, 185, 191, 211, 224, 356
Liecke, 258
Linstead, 56, 267, 286
Lippincott, 56
Lips, 263
Lochte, 152
Loebl, 249, 308, 309
Loeffler, 83
Lombardi, 350
Long, 10, 147, 259
Longuet-Higgins, 71, 137, 157, 174, 176, 227, 231, 346, 363
Lonsdale, 13
Loofbourow, 406
Lord, 56
Lothrop, 53, 125, 128, 135
Loudon, 58, 140, 317
Lowen, 256
Lowry, 76
Lu, 53
Luther, 8
Lux, 278
Lynn, 160

McAllan, 99
Maccoll, 26, 55
McElhill, 207
McGechen, 269
McIlwain, 83
McIntosh, 440
McKay, 312
McKenna, 272
McLean, 365, 377
McLeish, 131
Makin, 315
Mallet, 318
Mamalis, 314
Mander-Jones, 128, 129
Manz, 293
Marchand, 157
Markwald, 116, 117

INDEX OF AUTHORS

Mariella, 141
Mark, 56
Marsh, 377
Marshall, 382
Martin, 91, 92, 201, 215, 243, 252, 346, 365, 369
Martinsen, 254, 301
Marvel, 98, 411
Masterman, 239
Masters, 305
Mathews, 118
Mathieson, 162
Mathieu, 369
Matthews, 113, 114, 225
Mavin, 121
May, 137
Mayer, 316, 318
Mayo, 104, 220, 237
Meisenheimer, 108, 236, 246, 422, 437
Melander, 255
Melchior, 131
Menczel, 406
Meyer, 4, 47, 103, 271, 296, 346
Michael, 236
Michaelis, 297
Mikhail, 84
Millen, 104, 238, 254
Miller, 243, 318, 435
Mills, 132, 133, 143, 412, 413, 414, 428, 429, 430, 434
Modest, 339
Moffitt, 174
Molnár, 388
Moore, 83
Morgan, 310, 329
Mosettig, 51, 121
Moureu, 100, 364, 372, 374, 375
Moyer, 102
Moyle, 277
Mühlmann, 371
Muhr, 284
Müller, 121, 144, 257, 382
Müller-Rodloff, 382
Mulliken, 22, 28, 87, 213
Murray, 256
Mustafa, 378
Mustafin, 82

Nagasawa, 72
Nagel, 283
Nathan, 191, 192, 213
Naujoks, 340
Nechvatal, 249
Nelson, 234
Newman, 149, 327, 402, 437, 438, 439, 440
Nisizawa, 129
Nixon, 133, 143
Noller, 283
Norris, 265, 273

Norton, 334
Novello, 148, 149
Nunn, 147

Ochiai, 72, 128, 129
Oda, 380
Ogata, 246
Okano, 246
Oosterhout, 223, 403
Oparina, 244, 310
Orchin, 137
Ormrod, 86
Orndorff, 102
Osbond, 320, 322
Ostermayer, 51, 103
Ostrogovich, 406, 407

Paden, 297
Pajeau, 265
Palin, 86
Palmer, 199
Papendieck, 101
Parker, 314
Parkes, 135
Passino, 255
Patterson, 97
Paul, 339, 340
Pauling, 5, 12, 13, 17, 18, 20, 23, 25, 26, 35, 36, 39, 40, 43, 70, 71, 137, 162, 167, 171, 186, 197, 199, 227, 366, 413
Pearce, 66, 109, 227, 399, 407
Pearsall, 85
Peeling, 253
Pence, 108
Penney, 17, 31, 56, 137, 169
Perkin, 106
Perrottet, 57
Peters, 329
Peterson, 141, 300
Pettit, 64, 66, 227, 407
Pfeiffer, 82, 83, 269
Phillips, 77, 305
Pickett, 53
Piggott, 292
Pinazzi, 374, 375
Pinck, 255
Pines, 270, 271, 272
Pink, 56
Pinnow, 301
Plancher, 69
Plant, 275
Platt, 385, 386
Plattner, 144, 145, 146, 147, 404
Poole, 254
Powell, 83, 86
Praill, 289
Price, 51, 96, 104, 105, 106, 238, 253, 268, 273, 384, 385, 386, 394, 406, 408
Prier, 146

INDEX OF AUTHORS

Priou, 341, 365, 369
Prosen, 55
Pschorr, 316, 317
Pullman, 58, 138, 157, 171, 174, 201, 227, 231, 346, 402, 403
Putokhin, 101

Qua, 281

Rabe, 101
Racz, 374
Rădulescu, 406, 407
Rahn, 236
Ramsey, 97
Raphael, 58, 149, 439
Rapson, 147
Raudnitz, 255
Rayner, 86
Reed, 160, 253, 256, 271
Reinders, 295
Reinsch, 293
Remick, 230
Renshaw, 300
Reppe, 55
Ri, 253
Rice, 40
Richter, 137
Riechers, 244, 310
Riedel, 59
Rieke, 213
Rigaudy, 380
Rinkes, 68
Robert, 369
Roberts, 207, 235
Robertson, 11, 13, 29, 30, 84, 162, 163, 164, 199, 201, 211, 212, 259, 331, 440
Robinson, 6, 80, 121, 141, 243, 249, 250, 331, 335, 337, 360, 418, 420
Roe, 369
Roeske, 265
Rogers, 141
Roitt, 149
Rolfes, 299
Rondestvedt, 137
Roniger, 146
Rossini, 55
Rubenkoenig, 244, 310
Rubidge, 281
Rubinstein, 273
Ruggli, 125, 317
Ryan, 355, 356
Rytina, 334

Saame, 102
Saboor, 283
Sachs, 248, 311
Sagmanli, 320
Sakellarios, 236
Salcher, 295

Sampey, 102, 104
Sanders, 265
Sandorfy, 243, 252, 388
Saunder, 413
Saunders, 294
Sauvage, 370, 372
Schäfer, 284
Scheffer, 260
Scheifele, 250, 314
Schenck, 100, 368
Schilling, 52, 117, 125
Schlenker, 280
Schmerling, 270, 271, 272
Schmid, 145
Schmidt, 317
Schneider, 229
Schoental, 9, 158, 408
Scholl, 346, 439
Schöller, 256
Schomaker, 13, 53, 70, 71, 162, 366
Schönberg, 378, 382
Schottenhammer, 143
Schroeter, 121, 276, 277
Schrotter, 269
Schumaker, 381
Scott, 87, 149, 239, 359, 408, 439
Sculati, 271
Searle, 418, 427
Seide, 309
Seidler, 83
Seligman, 324, 437
Senkowski, 305
Shaw, 418
Sheldon, 98
Sheldrick, 121, 437
Shepherd, 83
Sherman, 17, 39, 70, 413
Sherndal, 57
Shilov, 258
Shive, 152
Shoesmith, 269
Short, 66, 285, 406
Shorter, 195
Shreve, 244, 310
Shriner, 411
Sidgwick, 136
Signaigo, 101
Silber, 69
Simons, 255, 272
Sinclair, 162
Singleton, 204
Sixma, 241, 251, 260, 261, 262, 263
Slade, 97
Smith, 55, 98, 132, 143, 257, 258, 277, 278, 297
Smyth, 198
Snyder, 427
Somerville, 58, 144
Soper, 259

INDEX OF AUTHORS

Sowa, 272
Spedding, 137
Spielmann, 146
Spindler, 253
Springall, 136, 137
Sprinkle, 196
St Pfau, 145
Stanley, 422, 424
Staub, 317
Staudinger, 280
Staveley, 86
Steadman, 381
Steck, 66, 406
Stein, 249, 308, 309, 360
Stewart, 430
Stoughton, 417
Streeck, 51, 148, 149
Stromberg, 285
Struve, 122
Stubbs, 195, 302
Sudborough, 80, 81, 83
Sugden, 197, 198, 202, 211, 212
Summerbell, 71
Sumuleanu, 317
Sundholm, 430
Suter, 301, 302
Sutton, 40, 137, 197, 198, 199, 202, 216, 217, 228, 229, 367, 402, 413
Suyver, 260, 262, 263
Sweeney, 140
Swift, 347
Syrkin, 189
Szmuszkovicz, 339, 360, 361

Taher, 213
Tänzer, 439
Tappen, 316
Tarbell, 127
Taub, 57
Tawney, 143
Taylor, 147
Tengler, 85
Theilacker, 437
Thiele, 6
Thomas, 42, 85, 152, 255, 265
Thompson, 55, 137
Thomson, 6, 83, 218, 219, 221
Tinker, 288
Tjepkema, 433
Tochtermann, 296
Töhl, 277
Tomisek, 273
Toussaint, 272, 413
Treibs, 147
Trikojus, 128, 129
Tschitschibabin, 243, 244, 309, 310
Turner, 69, 412, 418
Tutte, 384

Ubbelohde, 56, 199, 201
Ullmann, 317, 318

Vand, 440
Vandenbelt, 386, 387
van Campen, 340, 341
van der Linden, 97
Van Dijk, 151, 153
van Dranen, 179, 180
van Geuns, 312
Van Niekerk, 413
Van Vleck, 17
van Volkenburgh, 141
Vass, 257, 258
Vaughan, 137
Velluz, 364, 369, 373, 375
Verkade, 216, 217
Vernon, 258
Vicary, 302
Vincent, 55
Vingiello, 291
Vogel, 209
Vollmann, 51, 148, 149
von Auwers, 297
von Braun, 293, 294
von Bruchhausen, 340
von Pechmann, 59, 314
von Richter, 312
Voris, 53
Vorozhtzov, 305
Vroelant, 170, 243, 252

Wadsworth, 302
Wagner, 302
Wagner-Jauregg, 146, 335
Wain, 81, 82
Walden, 253
Wali, 284
Walker, 1, 314
Wallenstein, 329
Walls, 310
Walsh, 141, 169, 383, 385, 386, 394, 406, 408
Walter, 52, 117, 125
Wannowius, 157
Warhurst, 204
Waring, 402, 422, 440
Warren, 109, 336, 338
Waser, 53, 55
Wassermann, 360, 361
Wasson, 271
Waters, 87, 138, 149, 183, 238, 258, 259, 307, 315, 332, 333
Watson, 191, 192, 213
Wechsler, 314
Weigert, 381
Weinmayr, 288
Weiss, 10, 83, 84, 148, 153, 249, 308, 309
Weitzenböck, 319

INDEX OF AUTHORS

Weizmann, 107
Weller, 216, 217
Wentzel, 265
Wepster, 216, 217
Wertyporoch, 265
Weston, 301
Wheatley, 440
Wheland, 5, 12, 17, 36, 40, 71, 138, 169, 171, 220, 227, 239, 247, 251, 389
White, 104, 162, 163, 164, 237
Whitehouse, 149, 439
Wibaut, 13, 60, 72, 110, 148, 151, 152, 153, 155, 156, 251, 260, 261, 262, 263
Wieland, 43, 57, 236
Wiles, 285
Willemart, 373
Williams, 41, 249, 254, 256, 302
Williamson, 406
Willstätter, 55
Wilson, 259, 280
Winteler, 147
Wiselogle, 406

Wistar, 296
Wiswall, 198
Witt, 304
Witter, 101
Wohl, 243, 265, 309
Wojcik, 98
Woodruff, 421
Woodward, 274, 360
Wright, 47, 100
Wyss, 101, 145

Yabroff, 215
Young, 68, 126, 128
Yuan, 417, 426
Yvan, 243, 252

Zahler, 243
Zahn, 103
Zakharov, 303
Zataepina, 244, 310
Zelinsky, 98
Zincke, 124